Lecture Notes in Mechanical Engineering

Lecture Notes in Mechanical Engineering (LNME) publishes the latest developments in Mechanical Engineering—quickly, informally and with high quality. Original research reported in proceedings and post-proceedings represents the core of LNME. Volumes published in LNME embrace all aspects, subfields and new challenges of mechanical engineering. Topics in the series include:

- Engineering Design
- Machinery and Machine Elements
- Mechanical Structures and Stress Analysis
- Automotive Engineering
- Engine Technology
- Aerospace Technology and Astronautics
- Nanotechnology and Microengineering
- Control, Robotics, Mechatronics
- MEMS
- Theoretical and Applied Mechanics
- Dynamical Systems, Control
- Fluid Mechanics
- Engineering Thermodynamics, Heat and Mass Transfer
- Manufacturing
- Precision Engineering, Instrumentation, Measurement
- Materials Engineering
- Tribology and Surface Technology

To submit a proposal or request further information, please contact the Springer Editor of your location:

China: Dr. Mengchu Huang at mengchu.huang@springer.com
India: Priya Vyas at priya.vyas@springer.com
Rest of Asia, Australia, New Zealand: Swati Meherishi at swati.meherishi@springer.com
All other countries: Dr. Leontina Di Cecco at Leontina.dicecco@springer.com

To submit a proposal for a monograph, please check our Springer Tracts in Mechanical Engineering at http://www.springer.com/series/11693 or contact Leontina.dicecco@springer.com

Indexed by SCOPUS. All books published in the series are submitted for consideration in Web of Science.

More information about this series at http://www.springer.com/series/11236

Uwe Reisgen · Dietmar Drummer ·
Holger Marschall
Editors

Enhanced Material, Parts Optimization and Process Intensification

Proceedings of the First International Joint Conference on Enhanced Material and Part Optimization and Process Intensification, EMPOrIA 2020, May 19–20, 2020, Aachen, Germany

Springer

Editors
Uwe Reisgen
Institut für Schweißtechnik und Fügetechnik
Aachen, Nordrhein-Westfalen, Germany

Holger Marschall
Mathematical Modeling and Analysis
Darmstadt, Hessen, Germany

Dietmar Drummer
Lehrstuhl für Kunststofftechnik
der Universität Erlangen-Nürnberg
Erlangen, Bayern, Germany

ISSN 2195-4356 ISSN 2195-4364 (electronic)
Lecture Notes in Mechanical Engineering
ISBN 978-3-030-70331-8 ISBN 978-3-030-70332-5 (eBook)
https://doi.org/10.1007/978-3-030-70332-5

This Springer imprint is published by the registered company Springer Nature Switzerland AG
The registered company address is: Gewerbestrasse 11, 6330 Cham, Switzerland

Preface

This volume of Lecture Notes in Mechanical Engineering contains selected papers of the planned International Joint Conference on Enhanced Material and Part Optimization and Process Intensification (EMPOrIA2020), which would have taken place in Aachen/Germany on May 19–20, 2020, but had to be canceled due to the Corona Pandemic. The conference was organized by SFB 1120 Aachen (Precision Melt Engineering), SFB 814 Erlangen (Additive Manufacturing) and CCE Darmstadt (Center for Computational Engineering).

Many of the results published in this volume were funded by the Deutsche Forschungsgemeinschaft e.V. (DFG, German Research Foundation). We would like to take this opportunity to express our gratitude for this support.

Highest precision in the manufacturing process is preferably simple process chains with a low number of process steps–that is one of the main demands made by manufacturing companies that seek to maintain and expand stable and sustainable production in high-wage countries such as Germany at competitive costs. The Collaborative Research Center 1120 (SFB 1120) addresses these research topics with the aim to give a comprehensive description of melt-based manufacturing technologies such as casting, injection molding, welding, cutting, additive manufacturing and melt-based coating. For these processes, in which the material is at least temporarily in a liquid phase, a multi-scale description of the involved physical and material-based processes will be developed, so as to increase part precision by at least one order of magnitude.

Wherever innovations and personalized products are desired: With almost unlimited freedom of design, additive manufacturing technologies open up new perspectives to achieve constructive solutions. These types of manufacturing techniques barely set any limits to the spirit of innovation. Additive manufacturing techniques follow the trend toward individual customized products and will allow for serial production in the future. To take advantage of the potentials included in additive manufacturing techniques, the Collaborative Research Center 814 (SFB 814) does fundamental research on this technology, so it can be used for the production of multifunctional components. The most important thing is to analyze the process chain from beginning to end. This not only includes design and process simulation but also especially characteristics, creation and modification of suitable materials and their reactions in the fabrication process, up to the final component.

Computational Engineering (CE) is an integral part of the research profile of TU Darmstadt. CE is a modern and multidisciplinary science for computer-based modeling, simulation, analysis and optimization of complex engineering applications and processes and phenomena in nature. The Centre for Computational Engineering (CCE) concentrates all activities in CE at TU Darmstadt. It is organized orthogonally and in addition to existing departments covering the relevant topical disciplines in science and engineering. By defining relevant research areas, the CCE coordinates all research activities in Computational Engineering at TU Darmstadt. CCE offers a stable platform for multidisciplinary cooperation of its members and associated investigators, which initiates joint research projects directly, also on a large scale. Broadness and relevance of CE are well documented by many research projects at TU Darmstadt where CE is involved.

The EMPOrIA2020 is a perfect platform for efficient knowledge transfer in the field of materials processing and their applications. The EMPOrIA2020 is devoted to scientific presentations on the latest research results.

The EMPOrIA2020 focuses on the latest developments as well as future trends in the field of materials processing. The conference topics address anyone who is interested in the potential of manufacturing in theory and application. It is the aim of the EMPOrIA2020 to bring together international experts from research and industry in order to match scientific advances and economic needs for mutual benefit.

EMPOrIA2020 received 37 contributions. After a thorough review process, the program committee accepted 25 papers. Thank you very much to the authors for their contribution.

We would like to thank members of the program committee for their efforts and expertise in contributing to reviewing, without which it would impossible to maintain the high standards of reviewed papers.

This volume consists of seven parts: Welding with five papers, brazing with two papers, coating with three papers, additive manufacturing with five papers, casting with three papers, molding with five papers and cutting with two papers.

We appreciate the partnership with Springer for their essential support during the preparation of EMPOrIA2020.

Thank you very much for EMPOrIA Team. Their involvement and hard work were crucial to the success of the EMPOrIA2020 conference.

December 2020

Uwe Reisgen
Dietmar Drummer
Holger Marschall

Organization

Steering Committee

General Chair

Reisgen, Uwe	RWTH Aachen University, "Welding and Joining Institute," Germany

Co-chairs

Drummer, Dietmar	Friedrich-Alexander-Universität Erlangen-Nürnberg, "Institute of Polymer Technology," Germany
Marschall, Holger	Technical University Darmstadt, "Mathematical Modeling and Analysis," Germany

Members

Akyel, Fatma	RWTH Aachen University, "Welding and Joining Institute," Germany
Apel, Markus	ACCESS e.V., Germany
Bobzin, Kerstin	RWTH Aachen University, "Chair for Laser Technology," Germany
Elgeti, Stefanie	TU Wien, "Institute of Lightweight Design and Structural Biomechanics," Austria
Gillner, Arnold	RWTH Aachen University, "Chair of Laser Technology," Germany
Hopmann, Christian	RWTH Aachen University, "Institute for Plastics Processing," Germany

Mayer, Joachim	RWTH Aachen University, "Community Laboratory for Electron Microscopy," Germany
Olschok, Simon	RWTH Aachen University, "Welding and Joining Institute," Germany

Program Committee

Apel, Markus	ACCESS e.V., Germany
Behr, Marek	RWTH Aachen University, "Chair for Computational Analysis of Technical Systems," Germany
Bobzin, Kerstin	RWTH Aachen University, "Surface Engineering Institute," Germany
Bück, Andreas	Friedrich-Alexander-Universität Erlangen-Nürnberg, "Chair of Particle Technology," Germany
Bührig-Polaczek, Andreas	RWTH Aachen University, "Chair for Foundry Science and Foundry Institute," Germany
Drummer, Dietmar	Friedrich-Alexander-Universität Erlangen-Nürnberg, "Institute of Polymer Technology," Germany
Elgeti, Stefanie	TU Wien, "Institute of Lightweight Design and Structural Biomechanics," Austria
Gillner, Arnold	RWTH Aachen University, "Chair of Laser Technology," Germany
Häfner, Constantin	RWTH Aachen University, "Chair of Laser Technology," Germany
Hopmann, Christian	RWTH Aachen University, "Institute for Plastics Processing," Germany
Körner, Carolin	Friedrich-Alexander-Universität Erlangen-Nürnberg, "Chair of Materials Science and Engineering for Metals," Germany
Marschall, Holger	Technical University Darmstadt, "Mathematical Modeling and Analysis," Germany
Mayer, Joachim	RWTH Aachen University, "Community Laboratory for Electron Microscopy," Germany
Mergheim, Julia	Friedrich-Alexander-Universität Erlangen-Nürnberg, "Institute of Applied Mechanics," Germany

Merklein, Marion	Friedrich-Alexander-Universität Erlangen-Nürnberg, "Chair of Manufacturing Technology," Germany
Olschok, Simon	RWTH Aachen University, "Welding and Joining Institute," Germany
Reisgen, Uwe	RWTH Aachen University, "Welding and Joining Institute," Germany
Schäfer, Michael	Technical University Darmstadt "Numerical Methods in Mechanical Engineering," Germany
Schmidt, Michael	Friedrich-Alexander-Universität Erlangen-Nürnberg, "Chair of Photonic Technologies," Germany
Steinmann, Paul	Friedrich-Alexander-Universität Erlangen-Nürnberg, "Institute of Applied Mechanics," Germany
Stingl, Michael	Friedrich-Alexander-Universität Erlangen-Nürnberg, "Department of Mathematics," Germany
Weeger, Oliver	Technical University Darmstadt, "Numerical Methods in Mechanical Engineering," Germany
Zhang, Hongbin	Technical University Darmstadt, "Department of Materials and Geosciences," Germany

EMPOrIA Team

Akyel, Fatma	RWTH Aachen University, "Welding and Joining Institute," Germany
Lewandowsky, Marie-Luise	RWTH Aachen University, "Welding and Joining Institute," Germany
Olschok, Simon	RWTH Aachen University, "Welding and Joining Institute," Germany
Reisgen, Uwe	RWTH Aachen University, "Welding and Joining Institute," Germany

Contents

Coating

Additive Manufacturing

Casting

Welding

Simulation of Phase Transformation and Residual Stress of Low Alloy Steel in Laser Beam Welding

Fatma Akyel(✉), Uwe Reisgen, Simon Olschok, and Karthik Murthy

Welding and Joining Institute, RWTH Aachen University, Pontstr. 49, 52062 Aachen, Germany
akyel@isf.rwth-aachen.de

Abstract. The inhomogeneous temperature distribution in welding processes leads to high temperature gradients between the weld seam and the base material. A heterogeneous phase transformation takes place between the areas where the austenitic transformation temperature is exceeded and those where it remains below this temperature. This leads to a residual stress state which results in distortion when the yield strength is exceeded. In order to understand the thermal history of a welded specimen and the phase transformation that has taken place, numerical simulation is used.

This work focuses on the temperature field simulation and the resulting phase transformation in laser beam welding. Two heat sources are combined to simulate the weld pool. Typical models from the literature are used to represent the phase transformation. Altogether a model is developed which can be used as a basis for the calculation of residual stress formation due to thermal load and phase transformation.

Keywords: FEM simulation · Residual stress · Conical heat source · Goldak heat source · Phase transformation

1 Introduction

The use of fusion welding in manufacturing industry is wide spread. However, the local heat input in the welding process leads to an inhomogeneous temperature distribution resulting in varying temperature gradients within the whole sample. While the temperature surpasses the liquidus temperature in the fusion zone (FZ), resulting in molten material and even evaporation in this area, the temperature in the heat affected zone (HAZ) is below the solidus temperature. Areas that surpassed the austenitization temperature decompose in the cooling step into different phases depending upon the chemical composition and the cooling rate.

The phase transformation and the thermal shrinkage in the cooling process superpose to residual stress or, in the case the yield point is surpassed, result in distortion of the welded specimen.

A new approach to reduce the residual stresses is through targeted alloy composition. During the transformation from austenite to martensite the lattice elongates. This

U. Reisgen et al. (Eds.): EMPOrIA 2020, LNME, pp. 3–13, 2021.
https://doi.org/10.1007/978-3-030-70332-5_1

volume expansion ensures that compressive stresses are generated around the transformation zone. With a low-alloy material such as S235JR, the martensite transformation takes place at about 400 °C. At this temperature compressive stresses build up until the transformation is complete. However, as the component cools further, the tensile stresses, generated by shrinkage, dominate the compressive stresses. By using the alloy composition, it is possible to shift the martensite start temperature (M_s) to lower temperature. Thus, compressive stresses are induced in the tensile stress dominated weld seam, until room temperature is reached. Because the transformation takes place at reduced temperatures, this phenomenon is called the Low Transformation Temperature (LTT) effect. In order to understand this effect, a simulation model is generated, which shows the phase transformation in the base material and the weld seam. For this purpose, a model must first be created, which depicts the thermal history and the phase transformations in a specimen.

In literature, many approaches are present using FEM simulation to describe the formation of residual stress within a welded specimen [1–5]. However, most simulations are used for the arc welding processes. In this work, the heat distribution of carbon-manganese steel is simulated combining a typical heat source for arc welding and beam welding, in order to depict the distribution in a laser beam welded specimen. Furthermore, the microstructure transformation is simulated by building a model, which describes the phase transformations based on the heat input.

2 State of the Art

During production, each component is subject to certain environmental influences, for example static or dynamic stress. Apart from these environmental influences, additional conditions influencing internal stresses exist in the component. These are called residual stresses. Residual stresses act without external loads and can at most be in the range of the local material yield strength. If the yield strength is exceeded, plastic deformation of the component occurs. [6, 7].

The local heat input in the welding process causes metallurgical processes in ferritic-pearlitic steels, for instance in the weld pool and the heat-affected zone, [8]. As the austenite start temperature A_{c1} is reached, the base microstructure begins to transform into austenite. As soon as the austenite finish temperature A_{c3} is reached or exceeded, the microstructure consists only of austenite. Due to the inhomogeneous temperature distribution, the phase transformation and stress distribution is also inhomogeneous. These processes are restricted by the surrounding cool base material, which leads to local incompatibility [9].

Apart from experimental investigations, numerical simulation is used to numerically visualize phenomena that otherwise take place too quickly in the welding process itself. A significant advantage is the mathematical prediction of the residual stresses and distortion caused by the welding process.

According to [10], welding simulation is divided into 3 areas. In the *process simulation* the fusion zone geometry and the temperature field are simulated. Thermal boundary conditions are defined and thermal material characteristics are included. With the information about the temperature cycles the *material simulation* can be performed. Here, the processes that take place in the material are simulated. The phase volume fraction in the melt zone throughout the heat-affected zone up to the base material is depicted. With the results of the material characteristics and the thermal history, the *structure simulation* is used to map the residual stresses and distortion in the component.

In order to be able to display the metallurgical processes and the stress states in the workpiece, it is important to determine the transient temperature field as accurately as possible. For this purpose, a so-called equivalent heat source is used to represent the heat coupling. In this context, the heat source model according to Goldak is used in particular [11]. The model describes a volumetric double ellipsoid Gaussian distribution of the heat flux density and is often used for arc welding processes. For the description of the deep welding effect in the beam welding processes, a volumetric Gaussian heat source is also used, which describes the shape of a cone [12].

Various numerical models are used to simulate the phase transformation. For diffusion controlled transformations (ferrite, perlite and bainite) the microstructure components are described by the equation according to Johnson-Mehl-Avrami-Kolomogorov (short JMAK or Avrami equation) [13–17]. Furthermore, the equation according to Koistinen and Maburger is used to describe the diffusionless martensitic phase transformation as a function of the martensite starting temperature [18]. These and other models are used to represent the boundary conditions for the numerical analysis of phase transformations.

3 Welding Trials

The material used in this work is a low alloy carbon-manganese steel sheet (S235JR) with the dimensions 100 × 50 × 5 mm. No filler wire was used and a bead on plate was carried out (autogenous welding). For the welding tests, a Trumpf TruDisk 16002 disk laser was used as beam generator, which has a maximum beam power of 16 kW and a minimum fibre diameter of 200 μm.

The chemical composition of the base material was investigated by an OES analysis, Table 1.

Table 1. Alloying elements of the base material (S235JR) in m%

	Fe	C	Si	Mn	Cr	Ni	Mo	P	S
S235JR	98.2	0.08	0.056	1.01	0.413	0.041	0.014	0.034	0.003

To record the temperature distribution in the specimen, type K thermocouples were tacked to the top of the component, Fig. 1. The distance between each thermocouple was 5 mm. The temperature of both sides of the weld was recorded and an average built.

Fig. 1. Temperature measurement in the welding process with type K thermocouples

4 Numerical Simulation Model

A numerical model is a replica of the physical model. In order to realise a numerical model an exact definition of material properties and boundary conditions is necessary. Since the welding was performed lengthwise over the middle of the sample, only half of the sample needs to be simulated because of the symmetry. Symmetry does not only apply to geometry. Since the weld is made in the middle of the sample, the heat conduction to the sides is also the same. The temperature-dependent mechanical properties as well as thermal properties are subject to the same temperature gradients. For this reason it is sufficient to simulate half of the sample and still get a result for the whole sample. This is a common procedure to reduce the simulation calculation time. Thus, the simulated dimension of the model is 25 × 100 × 5 mm with a symmetry line along the path of the heat source.

An additional method to reduce the computing time is to adjust the mesh density. In this context, the mesh density in the welding zone is defined very finely, since the temperature gradients are particularly high here, and a mesh that is too coarse can lead to errors in the calculation, Fig. 2. The mesh density reduces gradually with further distance to the weld zone.

Fig. 2. Meshed model with high mesh density in the welding zone and a total geometry of 25 × 100 × 5 mm

The 3D finite element mesh was generated in ABAQUS CAE using the eight-node linear heat transfer brick element (DC3D8) totalling over 32.428 nodes and 27.930 elements. The smallest element size measures 0.5 × 0.5 × 0.5 in the weld zone. For the volumetric heat flux the subroutine DFLUX is used. The subroutine used for the phase transformation is UMAT with solution dependent variables (SDV).

4.1 Heat Source Model

The primary objective of the temperature field simulation is to determine the geometrical characteristics of the melt pool, as well as the temperature cycles in the melt pool and in its immediate vicinity (HAZ). The heat source is used to apply the thermal load in the weld zone. In this study a combination of a spherical and a conical heat source model is used to describe a transient heat source model, Fig. 3.

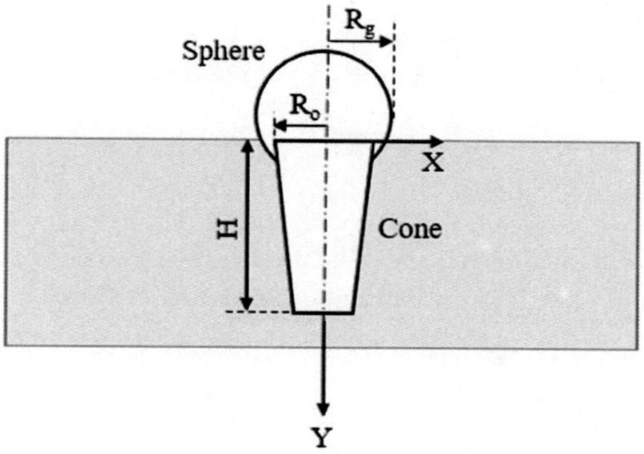

Fig. 3. Combined spherical and conical heat source model

The spherical heat source is suitable for modelling the heat input into the width of the weld seam. However, the geometrical description is not sufficient to describe the heat input into the depth of the workpiece. For this reason, the cone heat source is added.

$$Q_{Sperical} = (1 - \Psi) \underbrace{\frac{6\sqrt{3}\,Q}{R_g^3\,\pi\,\sqrt{\pi}}}_{Energy\ input} \quad exp\left[-3\left(\frac{x^2}{R_g^2} + \frac{y^2}{R_g^2} + \frac{[z - v\,\Delta t]^2}{R_g^2}\right)\right] \tag{1}$$

Equation 1 describes a Goldak heat source with geometric variables defined to be equal in size, thus representing a sphere with the radius R_g. This serves to simplify the equation and to reduce the variables in the numerical model. The beam power is described by the Parameter Q where by $Q = UIk_{eff}$; defined by the acceleration voltage U, the beam current I and the efficiency coefficient k_{eff}. The Parameter Ψ describes the portion of the conical heat source and is ≤ 1. This means for example with a $\Psi = 0.7$ the conical part is effective at 70% and the spherical part at 30%. The movement of the heat source is described by the expression $u = x - vt$. Here t is the welding time and v

the welding speed.

$$Q_{Conical} = \underbrace{\frac{3\,\Psi\,Q}{\pi\,H\,R_0^2}}_{Energy\ input} \exp\left[-3\left(\frac{x^2}{R_0^2}+\frac{(z-v\,\Delta t)^2}{R_0^2}\right)\right]\underbrace{\left[1+0.01\left(\frac{H-y}{H}\right)\right]}_{Inclination\ of\ flanks} step\left(\frac{H-y}{1mm}\right) \quad (2)$$

$$step = \begin{cases} 0 \text{ if } H-y < 0.0 \\ 1 \text{ if } H-y > 0.0 \\ 0.5 \text{ if } H-y = 0.0 \end{cases}$$

Equation 2 is used for the conical heat source. The geometrical contour of the cone is described by the radius R_0 and the cone height H. In addition, a term is provided in which the inclination of the cone flanks is described. The energy density distribution in the direction of the weld depth is regulated by the dimensionless *step*-function. The step-function is a simulation specification so that the heat input in y-direction never exceeds the cone height H. If y > H, the step-function is set as 0. If y < H, the step-function is set as 1. If y = H, the function is set to 0.5.

To represent the total heat source, the conical and spherical heat sources are added up, Eq. 3.

$$Q_{LB} = Q_{Spherical} + Q_{Conical} \quad (3)$$

In order to carry out the thermal analysis, it is necessary to define the thermo-physical properties of the weld plate and initial temperature distribution.

4.2 Phase Transformation Model

The kinetics of a phase transformation in the heating and cooling cycle are described by the JMAK-equation, Eq. 4.

$$V_A = \sum\nolimits_i V_i^0 (1 - \exp(-b_i t^{n_i})) \quad (4)$$

Where *VA* describes the austenite fraction, Vi^0 constitutes an initial fraction of ferrite (i ≡ F), pearlite (i ≡ P) and bainite (i ≡ B), while the constants b_i and n_i are material parameters and t the time.

The base material (S235JR) consists in its initial state of around 94% of ferrite and 6% of pearlite phase fraction. The kinetics of the phase transformation in the heating cycle from pearlite to austenite (P → A) and ferrite to austenite (F → A) have to be considered. Both transformations take place by nucleation and growth processes. The mathematical modelling of a P → A transformation was introduced by [19] and it can be adopted to the base material. With this, **Eq.** 5 describes the austenite volume fraction from pearlite dissolution $V\gamma^P$ during the continuous heating.

$$V_\gamma^P = V_{P_0}\left\{1 - \exp\left(\int_{A_{c_1}}^{T} \frac{4\pi}{3\dot{T}^4}\dot{N}G^3\Delta T^3 dT\right)\right\} \quad (5)$$

Where V_{P_0} is the initial pearlite fraction. The nucleation rate $\dot{N}$ and the growth rate G were expressed as a function of the activation energies Q_N and Q_G. **Equation** 6 describes the activation energies as follows:

$$\dot{N} = f_N \exp\left(-\frac{Q_N}{k\Delta T}\right) \& \ G = f_G \exp\left(-\frac{Q_G}{k\Delta T}\right) \tag{6}$$

The parameter k is the Boltzman constant, whereas f_N and f_G are functions representing the influence of the structure and the heating rate on the nucleation growth rates respectively.

[19] and [20] proposed the kinetics of ferrite to austenite transformation. Equation 7 shows the proposed relationship:

$$V_\gamma^\alpha = \begin{cases} V_{\alpha_0}\left[1 - \frac{5.6\dot{T}}{6\times 10^{-12} V_{\alpha_0}(T-T_C)^{5.6} + 5.6\dot{T}}\right], T_C < T < T_D \\ V_{\alpha_0} - \frac{(V_{\alpha_0} - V_D^\alpha)\dot{T}^2}{\dot{T}^2 + 1.2\times 10^{-3}(V_{\alpha_0} - V_D^\alpha)\left[(T-T_C)^2 - (T_D - T_C)^2\right]}, T \geq T_D \end{cases} \tag{7}$$

Here, V_{α_0} is the initial ferrite fraction, T_C is the starting temperature of ferrite to austenite transformation and T_D the end of this transformation. These temperatures are determined by means of dilatometric analysis.

The fraction of martensite formed below the martensite start temperature M_s is calculated by using the Koistinen-Marburger formula, Eq. 8.

$$V_m = V_\gamma\left(1 - \exp\left[-b(M_s - M_f)\right]\right) \tag{8}$$

Where V_m is the martensite volume fraction, V_γ is the existing austenite fraction. The martensite start and martensite finish temperatures are expressed by M_s and M_f respectively. The parameter b is the evolution coefficient and is usually taken as 0.011.

5 Results and Discussion

Subsequently, the results of the temperature field simulation and the phase simulation are examined and discussed.

In order to validate the simulation model, the temperature profile of the simulation and the measured values of the welding trials are compared, Fig. 4. The temperature was evaluated in identical locations for the experiments and simulations.

Table 2 shows the comparison of the maximum temperatures at the three measuring positions. Additionally the percentage of agreement is calculated. In this study, a deviation of about 10% is considered as acceptable. By comparing the measured and simulated temperature, it becomes clear that the greater the distance to the weld seam, the more accurate the temperature plotting becomes. Due to the high temperatures in the weld seam, an exact temperature recording in this area is not possible with Type K thermocouples. Since the numerical model is an approximation of the welding source, the area inside the weld is subject to a certain degree of uncertainty. This explains why a temperature comparison shows a higher deviation especially in the immediate vicinity of the welding source.

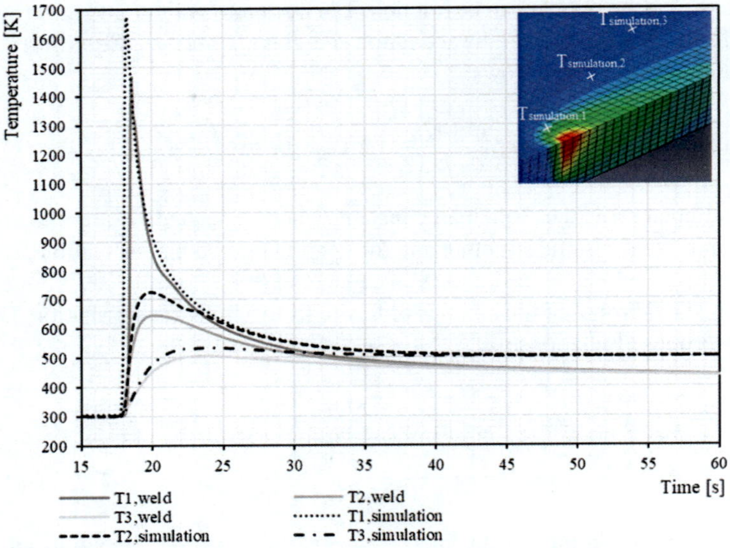

Fig. 4. Temperature profile of the simulation and the welding process

Table 2. Comparison of simulated and measured maximum temperature values

	T_1	T_2	T_3
Temperature Weld [K]	1476.15	644.95	505.35
Temperature Simulation [K]	1670.76	722.63	531.44
Percentage of Agreement [%]	88.35	89.25	95.1

In addition to the comparison of the temperature profiles, the geometry of the simulated weld seam and HAZ is verified with a cross section, Fig. 5.

With the comparison of the cross section, it becomes clear that the geometric reproduction of the melt pool also lies within an acceptable range. The HAZ begins with the austenitizing temperature A_{c1} (at 723 °C) and ends at the melting temperature T_m (at 1520 °C). The fusion zone can be read from the fusion line of the weld in the cross section. The HAZ can be identified by the changed morphology in contrast to the base material.

With the results of the thermal history the volume fraction within the fusion zone of each phase is calculated, Fig. 6.

Before the welding process the material consists of 96% ferrite and 4% pearlite. While the heating of the material starts and the temperature exceeds the austenite finish temperature A_{c3} the material consists completely of austenite. During the cooling phase, the fraction of austenite gradually decreases, so that the material consists of almost 85% martensite.

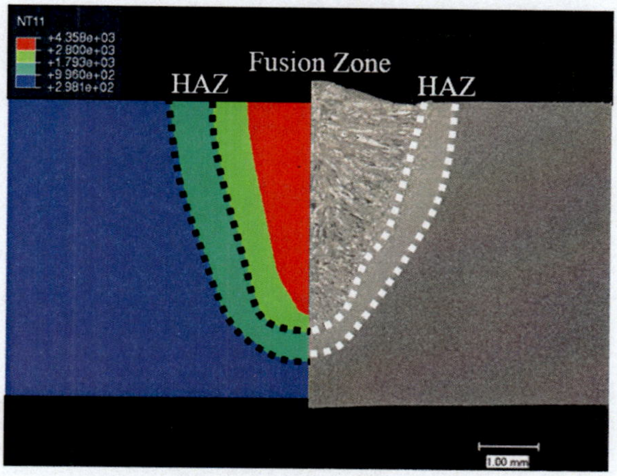

Fig. 5. Comparison of the welded cross section with the simulated thermal results

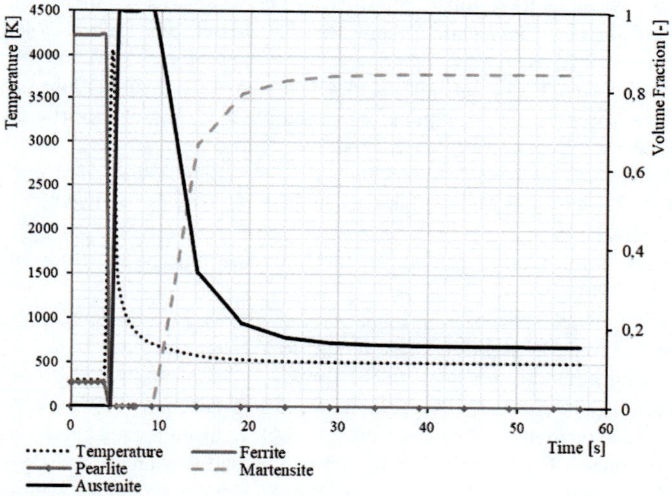

Fig. 6. Volume fraction of the phases and calculated temperature profile within the weld seam

Altogether, a numerical model for the simulation of the temperature distribution and finally the phase transformation is created with the presented numerical correlations. The model is still under development and has to be further extended by the bainite phase. Furthermore, a verification of the calculated phase components is necessary. As soon as this has been done, the next step will be the simulation of residual stresses and distortion. Ideally, the model should be extended to include hardness and grain size distribution. In addition, the aim is to incorporate the thermo-mechanical properties of an LTT material into the model so that the residual stress distribution can be represented as a function of different phases in the base material and the weld seam. Further work is done to improve the model.

Acknowledgement. The presented investigations were carried out at RWTH Aachen University Welding and Joining Institute ISF within the framework of the Collaborative Research Centre SFB1120–236616214 "Bauteilpräzision durch Beherrschung von Schmelze und Erstarrung in Produktionsprozessen" and funded by the Deutsche Forschungsgemeinschaft e.V. (DFG, German Research Foundation). The sponsorship and support is gratefully acknowledged.

Simulations were performed with computing resources granted by RWTH Aachen University under project rwth0436.

References

1. Mackerle, J.: Finite elemente analysis and simulation of welding: a bibliography (1976–1996). Modelling Simul. Mater. Sci. Eng. **4**, 501–533 (1996)
2. Mackerle, J.: Finite elemente analysis and simulation of welding: a bibliography (1996–2001). Modelling Simul. Mater. Sci. Eng. **10**, 295–318 (2002)
3. Ueda, Y.: Computational Welding Mechanics. Osaka University, Joining and Welding Research Institute (1999)
4. Webster, P., Ananthaviravakumar, N., Hughes, D., et al.: Measurement and modelling of residual stresses in a TIG Weld. J. Appl. Phys. A: Mater. Sci. Process. **74**(7), 1421–1423 (2002)
5. Dilthey, U., Reisgen, U., Kretschmer, M.: Comparison of FEM simulations to measurements of residual stresses for the example of a welded plate: a state-of-the-art report. Modelling Simul. Mater. Sci. Eng. **8**, 911–926 (2000)
6. Radaj, D.: Fachbuchreihe Schweißtechnik. Bd. 143: Eigenspannungen und Verzug beim Schweißen: Rechen- und Messverfahren. Düsseldorf: Verl. für Schweißen und verwandte Verfahren (2002)
7. Wohlfahrt, H.: Schweißeigenspannungen. In: Härterei-technische Mitteilungen. Band 31 Heft 1/2, 56–71 (1976)
8. Dilthey, U.: Schweißtechnische Fertigungsverfahren. Bd. 2: Verhalten der Werkstoffe beim Schweißen. 2. Aufl., Düsseldorf: DVS-Verlag (1995)
9. Peiter, A. (Hrsg): Handbuch Spannungsmesspraxis. Experimentelle Ermittlung mechanischer Spannungen. Braunschweig: Vieweg&Sohn Verlagsgesellschaft (1992)
10. Radaj, D.: Fachbuchreihe Schweißtechnik. Bd. 141: Schweißprozeßsimulation: Grundlagen und Anwendungen. Düsseldorf: Verl. für Schweißen und verwandte Verfahren (1999)
11. Goldak, J., Chakravarit, A., Bibby, M.: A new finite element model for welding heat source. Metallurgical Trans. B **15B**, 299–305 (1984)
12. Goldak, J.A., Akhlaghi, M.: Computational Welding Mechanics. Springer Science, New-York (2005)
13. Avrami, M.: Kinetics of phase change, 1: general theory. J. Chem. Phys. **7**, 103–112 (1939)
14. Avrami, M.: Kinetics of phase change, 2: transformation-time relations for random distribution of nuclei. J. Chem. Phys. **8**, 212–224 (1940)
15. Avrami, M.: Kinetics of phase change, 3: granulation, phase change and microstructure. J. Chem. Phys. **9**, 177–184 (1941)
16. Johnson, W.A., Mehl, R.F.: Reaction kinetics in process of nucleation and growth. Trans. Am. Inst. Min. Metall. Eng. **135**, 416–458 (1939)
17. Kolmogorov, A.N.: Statistical theory of crystallization of metals. Izv. Akad. Nauk SSSR, Ser. Mat. Bull. **1**, 355–359 (1937). (in Russian)
18. Koistinen, D.P., Marburger, R.E.: A general equation prescribing extend of austenite-martensite transformation in pure FE-C alloys and plain carbon steels. Acta Metall. **7**, 59–60 (1959)

19. Caballero, F.G., Capdevilla, C., Andres, C.G.: Modelling of kinetics of austenite formation in steels with different initial microstructures. ISIJ Int. **41**(10), 1093–1102 (2001)
20. Datta, D P., Gokhale, A.M,: Austenitization kinetics of pearlite and ferrite aggregates in a low carbon steel containing 0.15 wt pct C. In: Metallurgical Transactions A, vol. 12, pp. 443–450 (1981)

Metallographic Comparison for Laser Welding of Cu-ETP and CuSn6 with Laser Beam Sources of 515 nm and 1030 nm Wavelength

Marc Hummel[1(✉)], Christoph Schöler[2], and Arnold Gillner[1,3]

[1] Chair for Laser Technology LLT, RWTH Aachen University, Steinbachstraße 15, 52074 Aachen, Germany
marc.hummel@llt.rwth-aachen.de

[2] Nonlinear Dynamics of Laser Processes NLD, RWTH Aachen University, Steinbachstraße 15, 52074 Aachen, Germany

[3] Fraunhofer-Institute for Laser Technology ILT, Steinbachstraße 15, 52074 Aachen, Germany

Abstract. Laser micro welding is used for manufacturing of battery and fuel cell components for electrical applications. The laser micro welding process allows processing of parts in the micron range but reduces the surface quality of the parts at the same time. Laser beam sources with a wavelength in the near infrared range have established themselves for this purpose. Laser beam sources in the visible wavelength range however prove to be an alternative due to an increased absorption of the laser energy in copper-based alloys.

This paper presents the observation of laser welding on Cu-ETP and CuSn6 with a 515 nm and 1030 nm disc laser regarding surface roughness and the weld seam geometry for deep penetration welding. The results are evaluated as a function of laser power and feed rate. Both laser beam sources are operated at same focal diameters. Experimental results are used to validate a keyhole model.

Keywords: 1030 nm · 515 nm · Copper · Green laser · Laser welding

1 Introduction

In today's world, where climate change and the depletion of fossil fuels are a daily topic of conversation, the significance of switching to renewable electrical energy sources is of growing demand. On the one hand, new electrical storage technologies are required for this change, on the other hand, manufacturing processes are needed to join the electrical contacts of battery packs for example, without errors and with short process times.

Worldwide sales of laser beam sources almost tripled from US$ 5.9 billion to US$ 14.6 billion between 2006 and 2019. This capital increase in the laser industry shows the rising importance of laser sources for industry and research. [1] With the widespread use of laser beam sources in the manufacturing industry, a further increase in process reliability and reproducibility of results in laser beam welding is becoming more and more important. This is necessary to reduce costs and save resources. Furthermore, an optimization of the energy efficiency is targeted in order to further save costs. Today, laser

U. Reisgen et al. (Eds.): EMPOrIA 2020, LNME, pp. 14–28, 2021.
https://doi.org/10.1007/978-3-030-70332-5_2

beam sources with a wavelength in the near infrared range between 1030 and 1070 nm are widely used for processing metallic materials [2, 3]. Thanks to their high brilliance, they offer the possibility to focus the laser beam to a few 10 μm beam diameter. Laser beam sources with a wavelength in the visible range seem to be an ideal alternative due to the increased absorption of up to 60% of the laser radiation on copper [4].

In this work, the influence of the laser wavelength on the process is investigated. Therefore, similar setups and laser parameters are selected. Only the wavelength differs between the two lasers. In addition, the process without further influences such as extraction of the copper vapor or inert gas is examined to determine the visible physical differences. Furthermore, the results are used to validate a numerical keyhole model. This model can help to design optimized laser processes for manufacturing in the future.

2 State of the Art

2.1 Laser Micro Welding

Laser welding is with increasing demand used for joining electronic components based on aluminum and copper material. The precise energy input and the high feed rates allow short process times and high reproducibility [2]. Laser welding can be divided into two main process regimes, heat conduction welding and deep penetration welding, as shown schematically in Fig. 1.

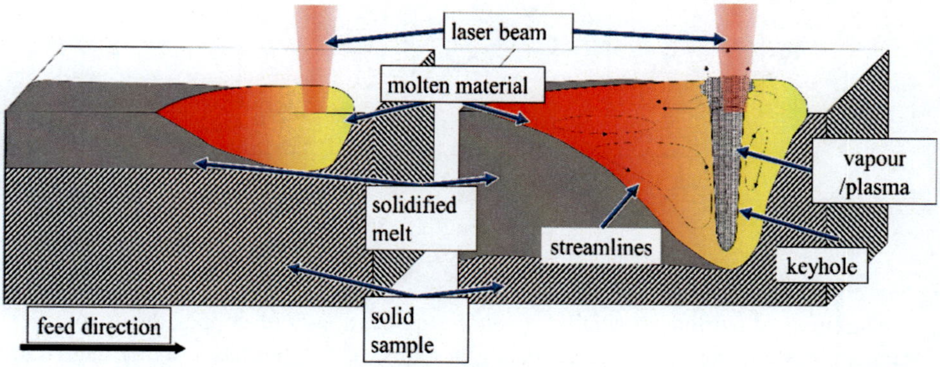

Fig. 1. Heat conduction welding (left), deep penetration welding (right) [3]

On the left side in Fig. 1 the heat conduction welding is shown. The laser beam heats the surface of the material above the melting point. A molten bath is formed. The expansion of the melt pool is determined by the heat conduction into the interior of the material. This produces a lens-like seam cross-section with an aspect ratio (weld seam width/weld seam depth) of ~2:1. By increasing the laser intensity, the vaporization temperature on the surface is locally exceeded. This causes the outflowing vapor to press the material surface inwards. In a state of equilibrium, a keyhole is created in which the laser beam can penetrate the material and is reflected and absorbed several times on the keyhole surface. This results in an enhanced aspect ratio and an increased total absorption of the laser beam. [3].

An important aspect of this process control is the precise control of the energy input into the material. Since small geometries and small material thicknesses are used in the production of electronic components, these would be welded through, if the energy input is too high. In order to implement the precise control of the welding process, knowledge of the physical processes is of great importance. However, the basic prerequisite for the deep welding process is the melting of the solid base material at room temperature. The absorption values (absorbed laser power/total laser power) here differ greatly from those in the liquid melt. The measured absorption curves with a spectrometer in solid state at room temperature are shown in Fig. 2.

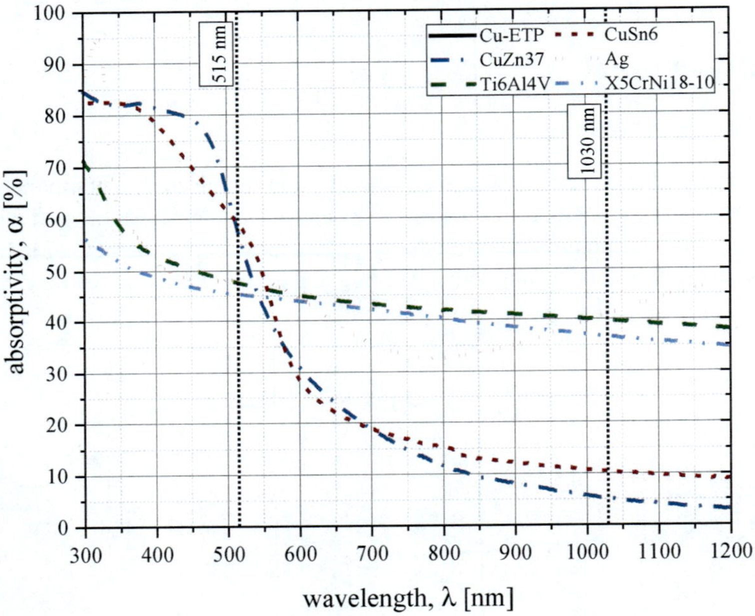

Fig. 2. Absorptivity of metallic materials depending on the wavelength (measured)

The curves show, that especially in welding applications with copper, the use of laser beam sources with shorter wavelengths in the range of 515 nm has a higher degree of absorption at ambient temperature and an increased absorption in the process can also be experimentally evaluated [5, 6]. During a deep welding process regime of copper alloys at 1064 nm wavelength, the average energy absorptivity is around 70–80% [7].

De Bono et al. us a frequency doubled pulsed disk laser with a pulse peak power of 1500 W at a wavelength of 532 nm to show the successful and reproducible laser beam micro welding of copper foil connections. The nominal spot diameter calculated for the experiments is 200 μm [8]. In the investigations of Pricking et al. the focus is on heat conduction welding. A sufficient welding depth is achieved with a spot diameter of 410 μm by using a laser beam source with a wavelength of 515 nm and a laser power of 1700 W. The results of welding tests on copper show a high reproducibility regarding welding depth and seam width. In addition, no pores are formed during the welding process [9]. Altarazi et al., Kaiser et al. and Pricking et al. also show that due to a higher degree of absorption, the welding result with a green laser beam source is independent of the condition of the material surface [9–11]. The use of combined green and infrared beam sources improve the overall energy coupling and reduces the number of ejections during the welding process. The use of green laser beam sources reduces the porosity of the weld seam and number of ejections [6]. Ramsayer et al. also shows that an increase in process stability is possible regardless of the feed rate. The tests are carried out using a frequency-doubled disk laser at a wavelength of 515 nm with 325 W output power on CuSn6 and CuFe2P with a spot diameter of 48 μm [4]. Kaiser et al. and Haubold et al. demonstrate that welding depth can be increased by extracting the process vapor from the copper during welding with 515 nm laser beam wavelength. This indicates an increased absorption of the laser beam at 515 nm by the copper vapor, generated during the deep welding process [12, 13].

3 Modeling and Experimental Setup

Modeling

For calculation of three-dimensional quasi-steady keyhole profiles, a reduced process model is applied. The model assumptions lead to a drastic reduction of the mathematical and numerical complexity and facilitate fast calculations of keyhole shapes for huge aspect ratios as a function of process parameters. To this, continuum-mechanical conservation laws across the keyhole surface and heat transfer in the solid are modelled in two-dimensional lateral layers. Each layer is parametrized by a constant radius, its relative position to the laser axis, and a constant surface temperature. Light absorption at the keyhole surface is modelled by Fresnel equations in the high conductivity limit. Light propagation inside the keyhole is described by a ray tracing algorithm that provides the contributions of multiple reflections to the absorbed energy flux. The layer parameters are determined iteratively in order to satisfy a set of interface conditions at the keyhole surface. The three-dimensional keyhole shape is then constructed layer by layer out of the geometric layer parameters. For a detailed description of the model and the keyhole calculation the reader may be referred to a previous work of the authors [14].

The simulations will be restricted to Cu-ETP. Due to model assumptions only constant material parameter values can be used. Values for density, thermal conductivity, and specific heat are chosen at room temperature. The material parameters are summarized in Table 1. Shape parameters of the laser beam are obtained from measurements shown in Fig. 3. Between simulations with 515 nm and 1030 nm laser light two parameters are changed: absorptivity and Rayleigh length. The latter influences beam propagation inside the keyhole up to the first incidence on the keyhole surface and leads to different distributions of the intensity and Poynting vectors. The Rayleigh lengths are 1.48 mm and 2.96 mm for the 1030 nm and 515 nm laser beams, respectively.

Table 1. Material parameters for Cu-ETP [15, 16].

Parameter	Value
Density [kg/m^3]	8,930
Thermal conductivity [W/m/K]	394
Specific heat [J/kg/K]	386
Ambient temperature [K]	293.15
Boiling point [K]	2,868.15
Melting enthalpy [kJ/kg]	209.30
Evaporation enthalpy [J/kg]	4,800
Surface tension [N/m]	0.95
Absorptivity at 1030 nm [1]	0.13
Absorptivity at 515 nm [1]	0.42

Experimental Setup

For the presented investigations, two Trumpf disc lasers with 1030 nm and 515 nm are used in order to show the process result differences between the two wavelengths. Both lasers are equipped with optical setups to achieve identical focal diameter of 200 μm as shown in Table 2.

Table 2. Properties of used laser beam sources and optical setups

Parameter	Trumpf TruDisc 8001	Trumpf TruDisc Pulse 421
Laser beam wavelength [nm]	1030	515
Maximum laser power [W]	8000	4000
Average laser power [W]	8000	400
max. Pulse duration [ms]	cw	50
max. Pulse energy [J]	–	40
Fiber core diameter [μm]	100	100
Focal length collimation [mm]	150	80
Focal length optics [mm]	300	160
Resulting focal diameter [μm]	200	200
Laser beam intensity profile	Top hat	Top hat

To evaluate the real laser intensity distribution in the focal plane, a caustic measurement was conducted. The exemplary result of the Trumpf TruDisc Pulse 421 setup is shown in Fig. 3.

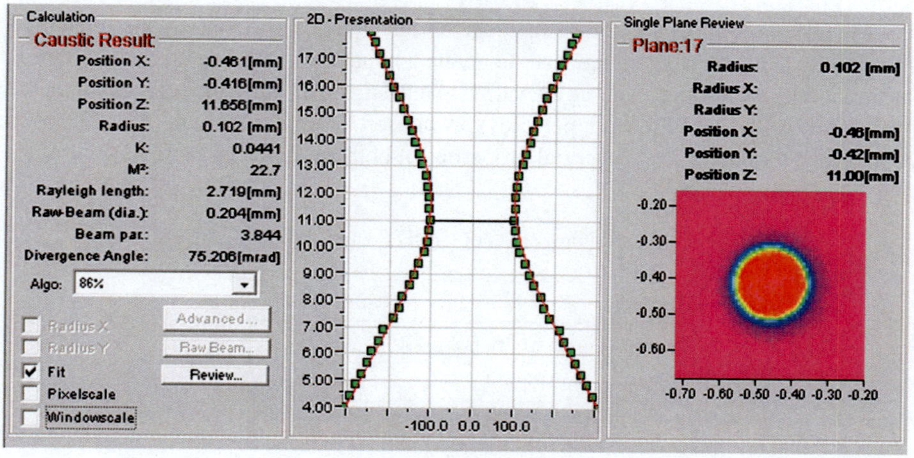

Fig. 3. Caustic measurement for Trumpf TruDisk Pulse 421

The result shows a measured focal diameter of 204 μm. With a measurement uncertainty of ±3% of the measuring instrument (Primes Micro Spot Monitor), this measurement value corresponds to the expected focus diameter.

The Trumpf TruDisk Pulse 421 is a pulsed laser. The maximum duration for which a laser pulse can be emitted in this case is 50 ms. To create a continuous weld seam, the sample material is moved under the laser beam for the duration of a pulse. This causes the weld seam to have a different length depending on the feed rate. Despite all this, a linear weld seam can be produced with a pulsed laser by this so-called SHADOW process [17]. The pulse duration of the laser with 1030 nm wavelength is adapted to the pulse duration of the Laser with 515 nm wavelength to ensure identical process parameters and boundary conditions. The schematic view of the setup is shown in Fig. 4.

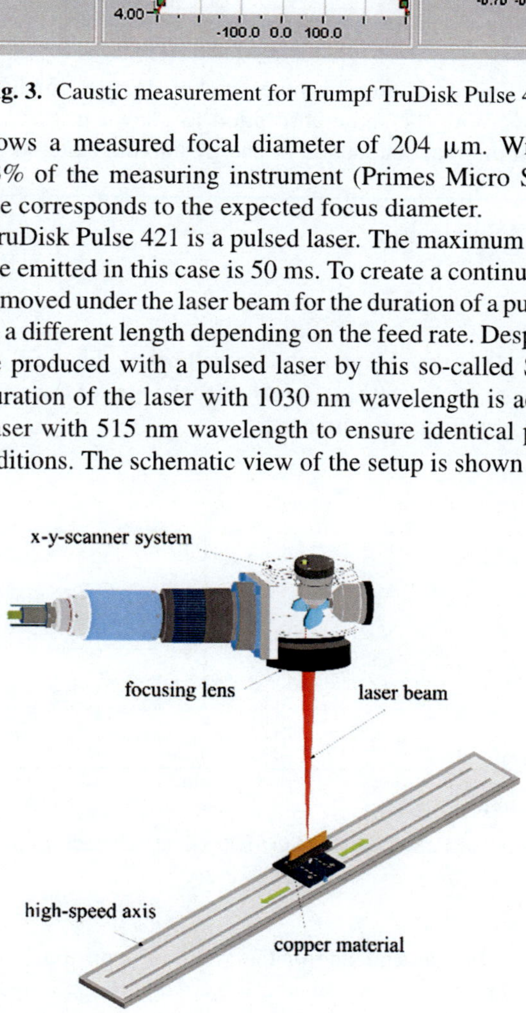

Fig. 4. Schematic representation of the experimental setup

The sample is moved under the laser on the carriage of a horizontally mounted high-speed axis. The axis triggers the laser to create a continuous weld seam on the material surface. Laser power and feed rate are varied to investigate the influence of the parameters on the process. Two different copper alloys are chosen for the experiments. Cu-ETP (pure copper) and CuSn6 (alloy). Material properties of both are listed in Table 3.

Table 3. Material properties of Cu-ETP and CuSn6 [15, 16, 18]

Parameter	Cu-ETP	CuSn6
Thermal conductivity [W/mK]	>393	75
Melting point [K]	1356	1313

The striking difference is the five times higher thermal conductivity of Cu-ETP compared to CuSn6. With the same absorption as shown in Fig. 2, this results in a completely different thermal conduction of energy into the material.

4 Results and Discussion

Experiments

The experimentally determined data are discussed below. Figure 5 shows the welding depth determined from cross sections. Here the dependence of laser power, feed rate and laser wavelength is of primary importance.

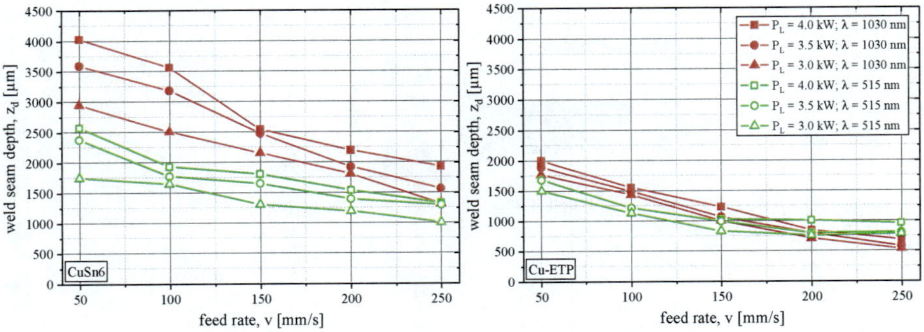

Fig. 5. Weld seam depth of CuSn6 (left) and Cu-ETP (right) depending on laser power, feed rate and laser wavelength

The graphs show the dependence of the welding depth on laser power and feed rate, whereby both laser wavelengths are compared. For CuSn6 (left), the values of the welding depth for 1030 nm wavelength are significantly higher than the values for 515 nm at a feed rate of 50 mm/s. With increasing feed rate, a different sensitivity can be seen with both lasers. If the feed rate is increased to 250 mm/s, the welding depth at 1030 nm wavelength is reduced by up to 2100 μm, whereas the welding depth at 515 nm wavelength is only reduced by a maximum of 1350 μm. It can also be seen that this causes the curves to converge. At an even higher feed rate, it can be assumed that the curves would intersect between 1030 nm and 515 nm wavelength. This behavior can be seen in the right diagram for Cu-ETP. This results in a range between 150 and 250 mm/s feed rate where the laser weld seams with 515 nm wavelength are even deeper than those of 1030 nm laser wavelength. For welding applications with a wide range of different feed rates, the laser with 515 nm wavelength provides more consistent welding depths.

Looking at the weld seam width, an opposite behavior than for the weld seam depth is visible. The graphs in Fig. 6 show the same representation, but this time with the weld seam width plotted on the y-axis.

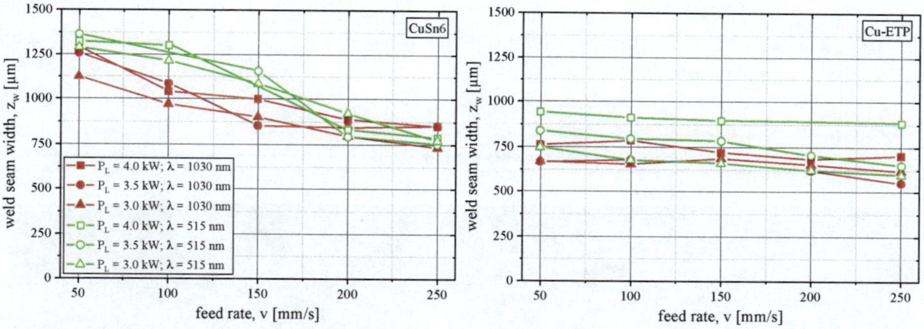

Fig. 6. Weld seam width of CuSn6 (left) and Cu-ETP (right) depending on laser power, feed rate and laser wavelength

It can be seen that for both CuSn6 and Cu-ETP the seam width at 515 nm wavelength tends to be higher than width at 1030 nm wavelength. The seam width of CuSn6 with 515 nm wavelength and increasing feed decreases more than with 1030 nm wavelength. With CuSn6 the curves tend to be constant for both wavelengths. In addition, the seam width is lower in absolute terms for Cu-ETP than for CuSn6. This can be explained by the 5 times higher thermal conductivity of Cu-ETP as listed in Table 3.

Calculating the aspect ratio (weld seam depth/weld seam width) from the graphs shown before, this results in a new image as shown in Fig. 7.

For CuSn6, the aspect ratio for 515 nm wavelength is between 1 and 2 and remains constant despite increasing feed rate. On the other hand, the aspect ratio for 1030 nm decreases from 2.5–3.5 down to about 2. For Cu-ETP, the point of intersection is again earlier at a feed rate of about 200 mm/s. At higher feed rates from 200 mm/s upwards, weld seams with greater aspect ratio are formed at 515 nm wavelength compared to 1030 nm wavelength.

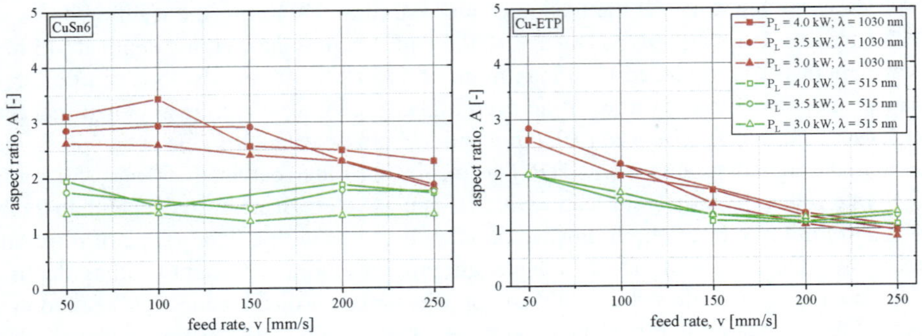

Fig. 7. Aspect ratio for CuSn6 (left) and Cu-ETP (right) depending on laser power, feed rate and laser wavelength

In Fig. 8 and Fig. 9 a representation of the weld seam cross sections for both laser wavelengths and materials are shown.

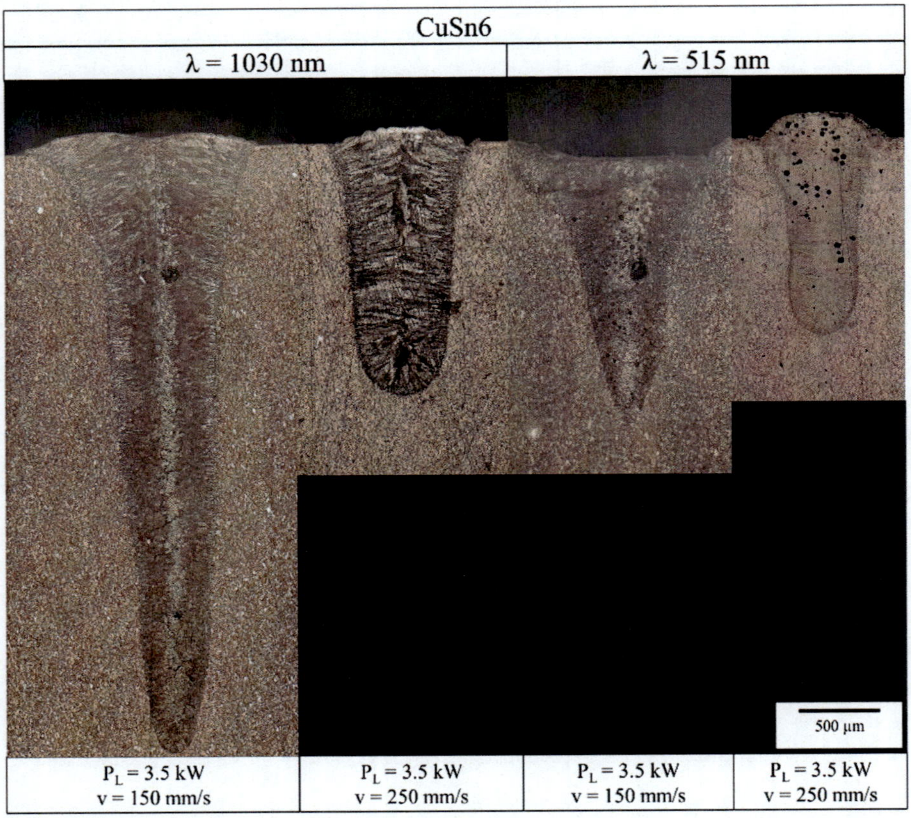

Fig. 8. Weld seam cross sections of CuSn6 for 1030 nm laser (left) and 515 nm laser (right)

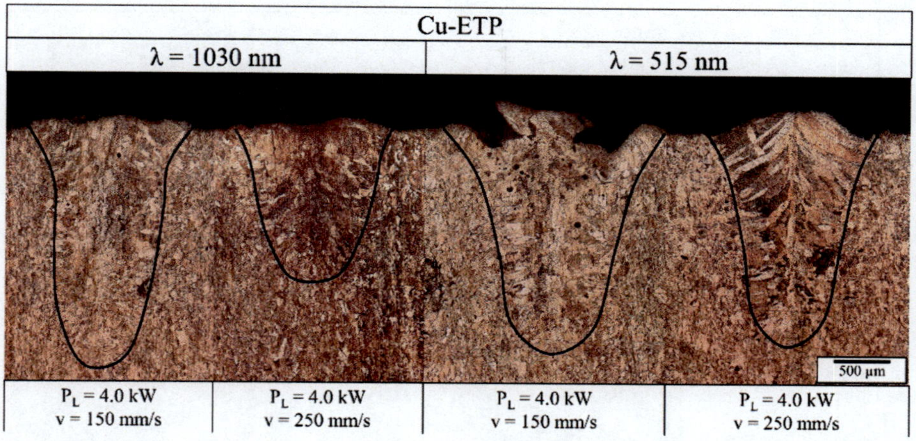

Fig. 9. Weld seam cross sections of Cu-ETP for 1030 nm laser (left) and 515 nm laser (right)

In the case of CuSn6, very sharply tapered weld seams are produced at the seam base with low feed rate. With increasing feed rate, the welding depth decreases for both. However, the weld seam becomes wider at the seam base. It can also be seen, that weld seams with the 515 nm laser beam source have stronger pore formation than with the 1030 nm wavelength. For Cu-ETP on the other hand, the weld seams show a similar geometry for both wavelengths and parameters. Once again, a stronger pore formation can be observed at 515 nm wavelength. In addition, the melt throw-up is more pronounced on the upper side of the seam with 515 nm wavelength. These results indicate a greatly increased process dynamic in the melt pool with the 515 nm wavelength process.

Measuring the weld seam surface roughness with a 3D-microscope (Keyence VHX-6000) the findings of an increased process dynamic can be confirmed. As shown in Fig. 10, the surface roughness of both materials with 515 nm wavelength is higher than with 1030 nm wavelength.

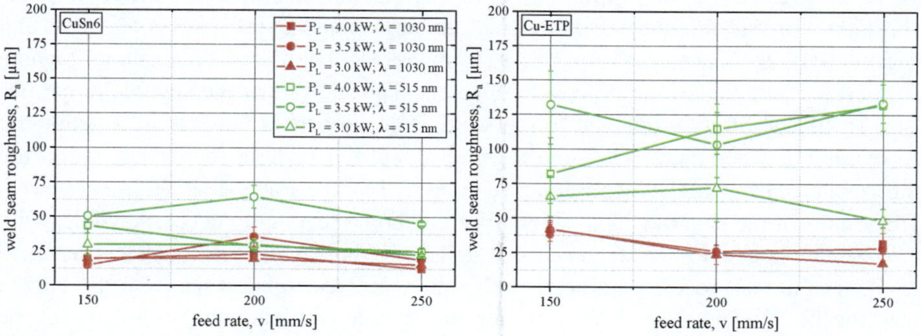

Fig. 10. Weld seam surface roughness for CuSn6 (left) and Cu-ETP (right) depending on laser power, feed rate and laser wavelength

These findings are also confirmed by the surface photographs shown in Fig. 11.

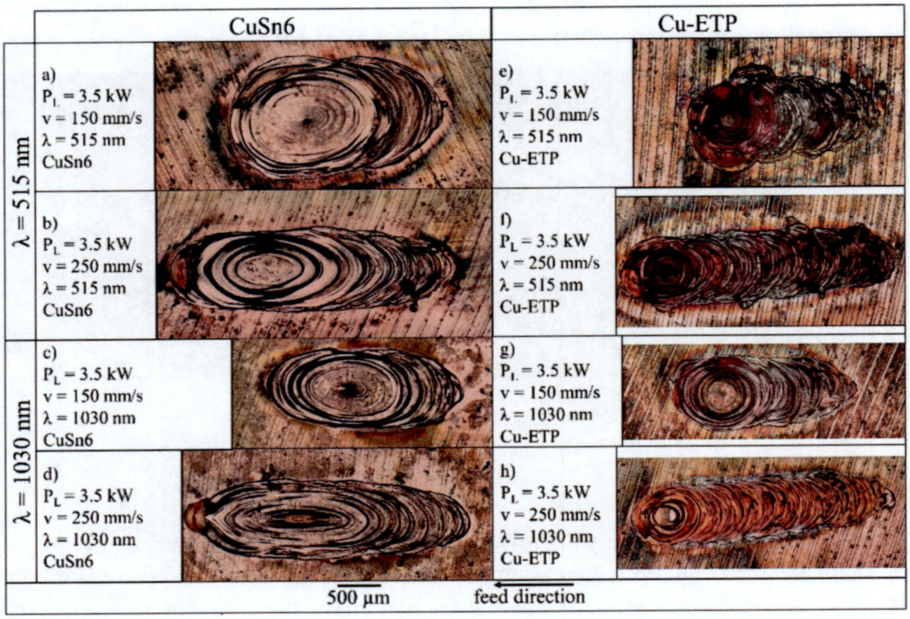

Fig. 11. Comparison of weld seam surface on CuSn6 (left) and Cu-ETP (right) between 515 nm laser (top) and 1030 nm laser (bottom)

The weld seams produced with 515 nm wavelength show more irregular structures visible due to a stronger formation of spatter around the weld seam as well as a more irregular surface and stronger cratering compared to 1030 nm for both CuSn6 and Cu-ETP.

Simulation

The experimental results from Fig. 5 and Fig. 6 are assessed using the keyhole model presented in Sect. 3 in order to identify the mechanisms leading to the different trends observed with 515 nm and 1030 nm laser radiation. The values of keyhole depths and widths (at the keyhole entry) are extracted from the calculated three-dimensional keyhole profiles as a function of the process parameters. The analysis of the keyhole properties then should be directly linkable to the weld seam properties because of a geometric correlation between them. The simulation results of the keyhole depths and widths are shown in Fig. 12. In both cases similar trends as in the experimental results are observed. The graphs of the keyhole depth show different slopes leading to deeper keyholes with the 1030 nm laser at low feed rates. In contrast the keyhole width acts much less sensitive to changes in the process parameters although a slightly higher sensitivity still can be seen for the 1030 nm laser. The values of the keyhole width are always greater for the 515 nm laser and do not reflect the "intersecting" behavior of the keyhole depth.

It should be noted that the power values used in the simulations differ from the nominal values in the experiments. They had to be manipulated in order to obtain comparable keyhole depths. For 1030 nm and 515 nm radiation 80% and 60% of the nominal values were found by trial and error calculations, respectively. The reduction to effective powers can be explained by losses due to the focusing optic (approximately 10%, the only losses considered in the calculations of the keyhole width) and by interaction with the metal vapor inside the keyhole (scattering, absorption) which is wavelength-dependent. There might be further reasons for the discrepancy like a non-stationary process in the experiments or the two-dimensional model structure (c.f. [14]). Nevertheless, it should be concluded that the attenuation of laser light by the metal vapor is much more present in case of the 515 nm beam (c.f. [12, 13]).

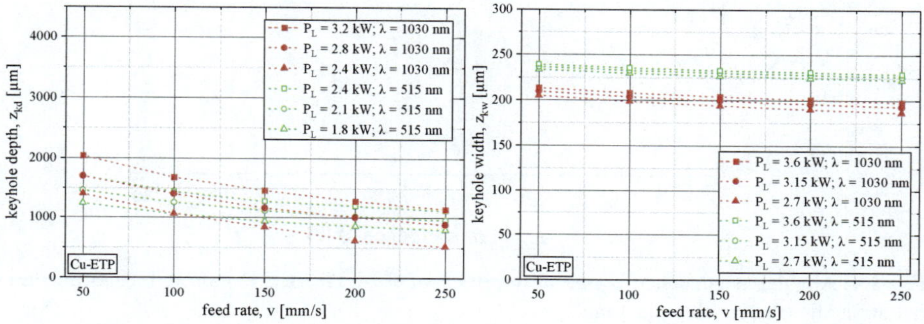

Fig. 12. Calculated keyhole depth and width from numerical model for Cu-ETP

Regarding the qualitative behavior of the keyhole depth, the model reproduces the influence of the wavelength correctly. The following hypotheses are put forward to explain the differences between a process with a wavelength of 515 nm and 1030 nm. Laser beams with a wavelength of 515 nm impinging on the horizontal material surface are absorbed to a larger extent than those at 1030 nm wavelength. Heating and melting is intensified which leads to an increased keyhole entry diameter. Due to the enlarged opening, a larger part of the total laser radiation enters the keyhole. This is particularly pronounced for high feed rates at which the weld depth reacts much more sensitive for the 1030 nm radiation. Furthermore, the radiation of 515 nm wavelength entering the keyhole is absorbed more strongly than the radiation of 1030 nm wavelength during any interaction with the keyhole surface which tends to produce steeper inclinations of the keyhole wall in this process regime.

At lower feed rates, however, there are two peculiar aspects that lead to deeper keyholes with the 1030 nm laser radiation. First, when the angle between incident radiation and surface normal exceeds a certain value, absorption according to the Fresnel equations becomes larger for 1030 nm than for 515 nm radiation (c.f. Fig. 13). This typically happens whenever the keyhole wall acquires a sufficient steepness. Second, the stronger absorption of the 515 nm light by the metal vapor becomes more noticeable the deeper the keyhole gets. In combination, both effects lead to the trends observed in the experiments and simulations.

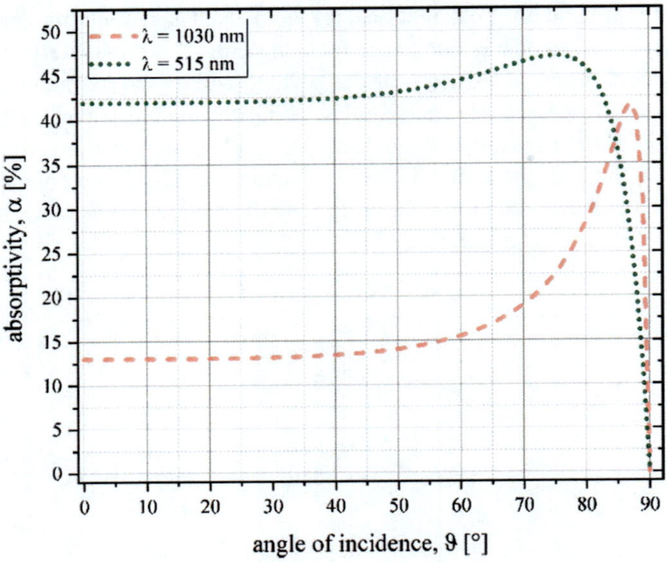

Fig. 13. Angular dependence of the absorptivity of Cu-ETP for 515 nm and 1030 nm laser radiation after the Fresnel equations

5 Conclusions and Outlook

In the present work, a parameter study on CuSn6 and Cu-ETP is conducted to compare the influence of the laser wavelength on the laser welding process. A laser beam source with 515 nm and one with 1030 nm wavelength are used with identical focal diameter. The resulting weld seams are compared based on metallographic evaluation of cross sections and surface measurements. The following conclusions can be drawn:

i. The final weld seam depth shows lower influence on feed rate in a process with 515 nm laser wavelength.
ii. For feed rates < 150 mm/s the laser beam source with 1030 nm wavelength produces weld seams with greater depth than 515 nm. At feed rates, for CuSn6 > 250 mm/s and Cu-ETP > 200 mm/s, the laser produces deeper welds at a wavelength of 515 nm.
iii. The laser beam source with 515 nm wavelength produces weld seams with greater surface roughness and increased pore formation compared to the 1030 nm laser beam source.
iv. For 515 nm laser wavelength a stronger interaction of light and copper vapor is expected.
v. The trends of the weld seam depth and width are attributed to differences in Fresnel absorption and light vapor interaction by means of simulation studies.

In future investigations the influence of local power modulation on the process with both laser wavelengths and identical optical parameters should be investigated. Due to the homogeneous distribution of the laser power in the material, it can be investigated more precisely what influence the laser beam wavelength has on the formation of pores inside the weld seam. For the simulation, a consideration of light-vapor interaction inside the keyhole and three-dimensional heat conduction losses at the keyhole bottom are expected to give more reliable results.

For the experimental proof, further in situ experiments have to be conducted to determine the geometry of the vapor capillaries, the absorption behavior of the laser radiation in the process and the position of the laser beam axis and keyhole axis. This allows to determine to which degree the laser radiation with different wavelengths actually influences the process.

Acknowledgements. The presented investigations were carried out at RWTH Aachen University within the framework of the Collaborative Research Centre SFB-1120–236616214 "Bauteilpräzision durch Beherrschung von Schmelze und Erstarrung in Produktionsprozessen" and funded by the Deutsche Forschungsgemeinschaft e.V. (DFG, German Research Foundation). The sponsorship and support is gratefully acknowledged.

References

1. Breitkopf, A.: Umsatz mit Lasern weltweit in den Jahren 2006 bis 2019, 11 May 2020]. https://de.statista.com/statistik/daten/studie/163232/umfrage/weltweiter-umsatz-mit-lasern-seit-2006/
2. Häusler, A., Mehlmann, B., Olowinsky, A., Gillner, A., Poprawe, R.: Efficient copper microwelding with fibre lasers using spatial power modulation. Lasers Eng. LIE **36**(1), 133–146 (2017)
3. Poprawe, R.: Tailored Light 2. Springer , Heidelberg (2011)
4. Ramsayer, R.M., Sebastian Engler, G.S.: New approaches for highly productive laser welding of copper materials. In: 1st International Electric Drives Production Conference (EDPC) (2011)
5. Hummel, M., Schöler, C., Häusler, A., Gillner, A., Poprawe, R.: New approaches on laser micro welding of copper by using a laser beam source with a wavelength of 450 nm. J. Adv. Join. Process. **1**, 100012 (2020). https://doi.org/10.1016/j.jajp.2020.100012
6. Engler, S., Ramsayer, R., Poprawe, R.: Process studies on laser welding of copper with brilliant green and infrared lasers. Phys. Proc. **12**, 339–346 (2011). https://doi.org/10.1016/j.phpro.2011.03.142
7. Conzen, J.H., Häusler, A., Stollenwerk, J., Gillner, A., Poprawe, R., Loosen, P.: Laserstrahl-Mikroschweißen von mikroelektronische Baugruppen unter Anwendung von örtlicher und zeitlicher Energiedeposition. Elektronische Baugruppen und Leiterplatten EBL (2017)
8. De Bono, P., Metsios, I., Blackburn, J., Hilton, P.: Laser processing of copper and aluminium thin sheets with green (532nm) and infrared (1064nm) pulsed laser beam sources. In: ICALEO (2013)
9. Pricking, S., Huber, R., Klausmann, K., Kaiser, E., Stolzenburg, C., Killi, A.: High-power CW and long-pulse lasers in the green wavelength regime for copper welding. SPIE LASE 2016:97410G. https://doi.org/10.1117/12.2213293.
10. Altarazi, S., Hijazi, L., Kaiser, E.: Process parameters optimization for multiple-inputs-multiple-outputs pulsed green laser welding via response surface methodology. In: International Conference on Industrial Engineering and Engineering Management 4–7 December 2016, Bali, Indonesia (2016)

11. Kaiser, E., Huber, R., Stolzenburg, C., Killi, A.: Sputter-free and uniform laser welding of electric or electronical copper contacts with a green laser. In: LANE, vol. 8 (2014)
12. Kaiser, E., Dold, E., Killi, A., Zaske, S., Pricking, S.: Application benefits of weling copper with a 1 kW, 515 nm continuous wave laser. In: 10th CIRP Conference on Photonic Technologies [LANE 2018], vol. 10 (2018)
13. Haubold, M., Ganser, A., Eder, T., Zäh, M.F.: Laser welding of copper using a high power disc laser at green wavelength. Proc. CIRP **74**, 446–449 (2018). https://doi.org/10.1016/j.procir.2018.08.161
14. Schöler, C., Nießen, M., Hummel, M., Olowinsky, A., Gillner, A., Schulz, W.: Modeling and simulation of laser micro welding. In: Lasers in Manufacturing Conference (2019)
15. Copper, L.A.: In: Ullmann's Encyclopedia of Industrial Chemistry. Wiley-VCH Verlag GmbH & Co KGaA, , Hoboken (2000)
16. Brillo, J., Egry, I.: Surface tension of nickel, copper, iron and their binary alloys. J. Mater. Sci. **40**(9–10), 2213–2216 (2005). https://doi.org/10.1007/s10853-005-1935-6
17. Kramer, T., Olowinsky, A., Durand, F.: SHADOW - a new welding technique. In: SPIE 2002 (4637) (2002)
18. Moeller, E. (ed.): Handbuch Konstruktionswerkstoffe: Auswahl, Eigenschaften, Anwendung, 2nd edn. Hanser, München (2014)

Numerical Investigation of Keyhole Depth Formation in Micro Welding of Copper with 1030 nm and 515 nm Laser Radiation

Christoph Schöler[1(✉)], Markus Nießen[2], and Wolfgang Schulz[1,2]

[1] Nonlinear Dynamics of Laser Processing, RWTH Aachen University, Steinbachstr. 15, 52074 Aachen, Germany
christoph.schoeler@nld.rwth-aachen.de

[2] Fraunhofer Institute for Laser Technology, Steinbachstr. 15, 52074 Aachen, Germany

Abstract. The keyhole depth formation in laser micro welding of copper is numerically investigated as a function of feed rate and laser power. The calculations of the keyhole depth are based on an approximate process model for the three-dimensional steady keyhole profile and carried out for laser beams of the wavelengths 1030 nm and 515 nm with equal intensity distributions in the beam waist. It is found that under standard process conditions deeper keyholes are established with the 1030 nm laser when the feed rate is decreased whereas under conditions that minimize the attenuation of the laser radiation by the metal vapor the 515 nm laser leads to an overall more efficient process. Furthermore, the results reveal a different sensitivity of the keyhole depth to feed rate and laser power for both laser beams. Based on model equations and absorption mechanisms, the different behavior towards 1030 nm and 515 nm laser radiation is analyzed. An instructive consideration of the keyhole front in the upper keyhole part is presented and three absorption mechanisms that affect the geometrical form of the keyhole are identified: direct absorption, absorption contributions due to multiple reflections, and radiation attenuation by the metal vapor.

Keywords: Laser welding · Copper · 1030 nm vs. 515 nm · Keyhole depth · Numerical simulation

1 Introduction

In basic research on laser micro welding of copper it has been reported that using 515 nm laser radiation a more efficient process leading to deeper welds can be achieved than with a conventional 1030 nm laser [1]. However, this assumes that the interaction between the 515 nm laser radiation and the copper vapor has to be minimized in order to prevent strong radiation attenuation. Under standard process conditions, where no measures are taken to reduce the light-vapor interaction, it has been observed that the seam depths of copper welds with 1030 nm and 515 nm lasers fall within the same range of values (c.f. Fig. 1). From Fig. 1 it furthermore becomes obvious that the seam depth is more sensitive to changes in feed rate for welds with the 1030 nm laser leading to deeper welds at small feed rates.

U. Reisgen et al. (Eds.): EMPOrIA 2020, LNME, pp. 29–39, 2021.
https://doi.org/10.1007/978-3-030-70332-5_3

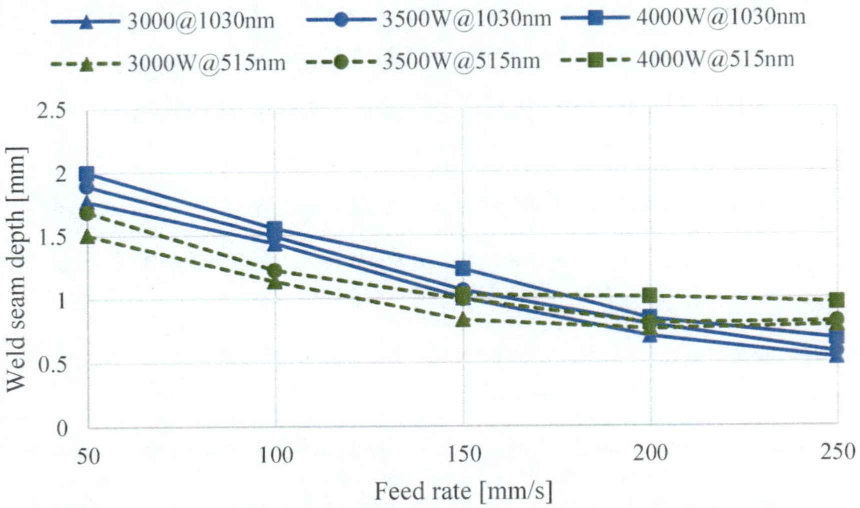

Fig. 1. Experimental values of the seam depth of bead-on-plate welds in Cu-ETP for 1030 nm and 515 nm laser radiation as a function of feed rate and laser power obtained with a Trumpf TruDisc 8001 and a Trumpf TruDisc Pulse 421 laser, respectively. Beam radius: 0.1 mm (Marc Hummel, personal communication, RWTH Aachen University, June 30, 2020)

The goal of this work is to identify mechanisms of the formation of the keyhole depth which can explain the differences in seam depth and its sensitivity to changes in feed rate based on numerical simulations of the three-dimensional keyhole profile with 1030 nm and 515 nm laser radiation.

The text is structured as follow. First, an overview of the applied keyhole model is presented. Then, numerical experiments are carried out to calculate the keyhole depth as a function of feed rate and laser power for both laser wavelengths. It is verified that the observed behavior of the weld seam is reproduced by the keyhole depth calculations. Last, a discussion regarding the keyhole formation which is based on the model equations and the considered absorption mechanisms is given.

2 Mathematical Modeling

Here, an outline of the model that is used to calculate the three-dimensional quasi-steady keyhole profiles is given. From these profiles the keyhole depths will be obtained in the further analysis. For a more detailed model description the reader may be referred to a previous work of the authors [2]. A simple scheme of the laser welding process showing the laser beam entering the keyhole is depicted in Fig. 2.

2.1 Layer Model and Model Parameters

The keyhole wall is considered steep, as is typical for high aspect ratio welds. It is assumed that the continuum mechanical quantities change only slowly with depth for the most part of the keyhole. Heat and mass transfer, including phase transitions, are

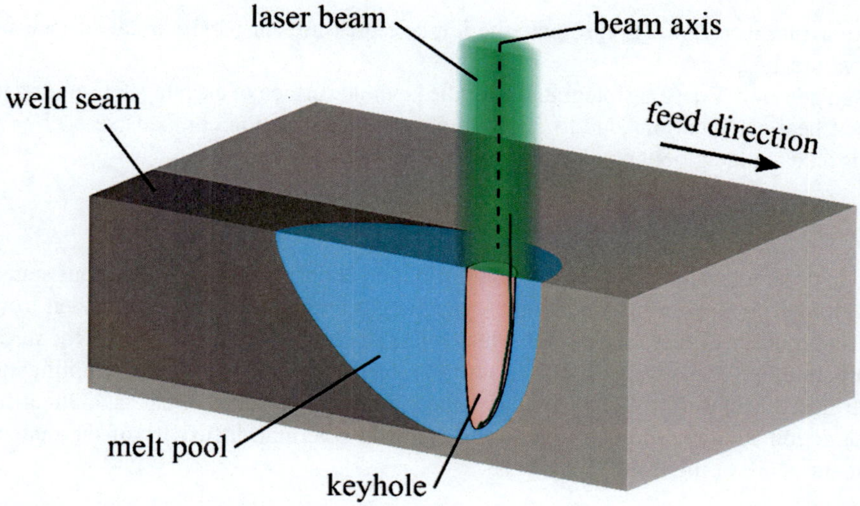

Fig. 2. Process scheme of laser welding indicating the path of a single light ray in the keyhole

modeled in two-dimensional lateral layers normal to the beam axis. Coupling between the layers is conveyed by Fresnel absorption which depends on the local shape of the three-dimensional keyhole surface. Multiple reflections inside the keyhole and their absorption at the surface are realized by a ray tracing algorithm.

The geometric shape of the keyhole surface in each layer is approximated by a circle which is characterized by its radius R and its front position A. Furthermore the layer surface temperature T_s is introduced, which takes a constant value in each layer since lateral variations of the temperature along the keyhole surface are neglected. The front position corresponds to the relative position of the keyhole front with respect to the laser beam axis. For each layer the values of the three model parameters are calculated iteratively to satisfy the set of conditional equations presented in the following.

2.2 Conditional Equations

The conditional equations result from local and global balance considerations at the keyhole surface for which a best fit by the layer parameters (R, A, T_s) is sought. At the front position $x = A$ the interface conditions Eqs. (1) and (2), expressing, respectively, the local conservation of momentum and heat during the transition from the liquid to the gaseous phase, are evaluated:

$$\rho u^2 = \gamma / R \tag{1}$$

$$\rho u H_v = q_a - q_l \tag{2}$$

where ρ denotes the mass density of the gas at the phase boundary, u the gas efflux velocity normal to the boundary, γ the surface tension, H_v the specific latent heat of vaporization, q_a the absorbed heat flux, and q_l the heat flux entering the liquid material.

The last term in Eq. (1) represents the Laplace pressure for a cylindrical surface with curvature $1/R$.

In addition, Eq. (2) is integrated over the keyhole surface of each layer assuming that latent heat contributions due to evaporation and condensation compensate each other. This leads to the global power balance.

$$\oiint (q_a - q_l) dS = 0. \tag{3}$$

In order to evaluate Eqs. (1) to (3) information about the liquid and gaseous states is required. The mass efflux ρu and the dynamic gas pressure ρu^2 are established from a Hertz-Knudsen evaporation model, which relates them to the gas pressure and the surface temperature T_s. The gas pressure is set to ambient pressure for further simplification. The heat flux into the liquid q_l is obtained from a two-dimensional, analytical heat conduction solution and is a function of feed rate, thermal diffusivity of the material, and curvature of the front.[2].

2.3 Absorbed Heat Flux

The heat flux due to absorption of laser radiation at the keyhole surface is calculated by

$$q_a = \mu A_S(\mu) I_L + q_{MR} \tag{4}$$

where $\mu = -\hat{s} \cdot \hat{n} = \cos\theta$, θ being the angle of incidence between the light's Poynting vector $\hat{s}$ and the local normal vector of the keyhole surface $\hat{n}$, A_S is the absorptance of the material's surface, I_L is the intensity of the direct laser light, and q_{MR} is the contribution of multiple reflections. The surface elements and the local normal vectors are obtained by a three-dimensional discretization of the keyhole surface (c.f. Fig. 3).

The functional dependence $A_S(\mu)$ results from Fresnel equations in the high conductivity limit [3] and is parametrized by a frequency- and temperature-dependent material parameter ϵ. In the present work it is considered for unpolarized light for which it reads:

$$A_S(\mu) = \frac{2\epsilon\mu}{\epsilon^2 + 2\epsilon\mu + 2\mu^2} + \frac{2\epsilon\mu}{2 + 2\epsilon\mu + \epsilon^2\mu^2} \tag{5}$$

With regard to Fig. 3 the contribution of multiple reflections to the total heat flux at node (i, j) is calculated by the ray tracing algorithm as

$$q_{MR}^{ij} = \sum_{k=1}^{K} \sum_{l=1}^{L_k} \prod_{m=1}^{M_{kl}} A_{n_k}^{-1} \left(1 - R(\theta_{n_k l})\right) R(\theta_{n_k lm}) P_{n_k l} \tag{6}$$

where k, l, and m are local indices for surface elements, rays, and reflections, respectively, K is the number of surface elements adjacent to node (i, j), L_k is the number of rays hitting element n_k, M_{kl} is the number of reflections of ray l prior to hitting element n_k, A_{n_k} is the area of element n_k, R is the reflectance due to the Fresnel equations ($R = 1 - A$, c.f. Eq. (5)), $\theta_{n_k l}$ is the angle of incidence of ray l of element n_k, $\theta_{n_k lm}$ is the angle of incidence of the m th reflection of ray l of element n_k, $P_{n_k l}$ is the initial power of ray l of element n_k, and n_k is a partial indexing for identification of the surface elements.

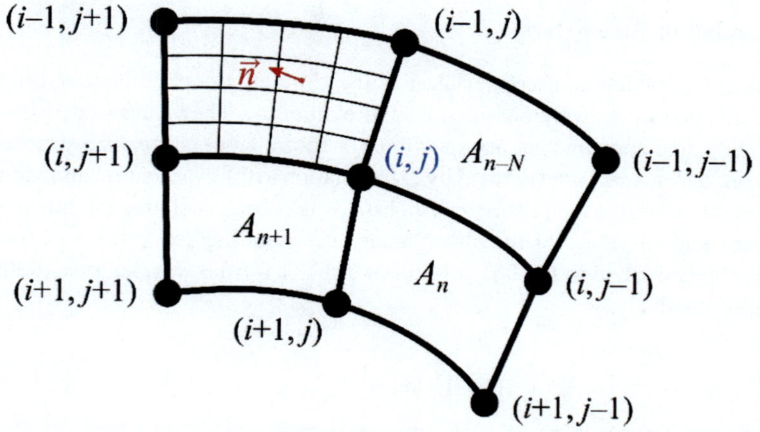

Fig. 3. Surface mesh with node and element indexing scheme around the node (i, j) for $j \geq 1$ with N nodes per layer. A partitioning of 4×4 as indicated in the upper left element is used by the ray tracer for emitting one ray per each sub-element in each iteration.

It should be noted that K, L_k, M_{kl}, and n_k are functions of the node number (i, j). (For instance, K equals 2 if the node is part of the first or last layer and 4 else.)

The initial power of a ray is calculated from the power of the raw beam taking into account a "zeroth" reflection from the sub-element which it is emitted from (c.f. Fig. 3). Intersection handling and redirection of rays is accomplished with basic geometry.

2.4 Limitations

The simplified model is bound to several restrictions which will cause deviations from the real process. It allows for constant material parameters only. The heating and cooling contributions from the workpiece top and the keyhole bottom are neglected. Solid and liquid are considered as one common phase. There is a uniform flow through the capillary, i.e., circulations in the weld pool are not taken into account. Neither light-vapor interaction inside the keyhole nor vapor heating are modeled, and only the effect of radiation attenuation by the metal vapor will be accounted for roughly by adjusting the laser intensity values globally during calculations with the 515 nm laser.

3 Simulation Results

In this section numerical results of the keyhole depth obtained with the proposed keyhole model are presented. The keyhole depth is calculated as a function of feed rate and laser power for lasers of the wavelengths 1030 nm and 515 nm under standard process conditions, including attenuation of the 515 nm radiation by the metal vapor, and under idealized conditions, neglecting the interaction between laser radiation and metal vapor. The keyhole depth results are examined with respect to the experimental results on the seam depth referred to in Sect. 1.

3.1 Simulation Parameters

The material parameters of copper used in the simulation are summarized in Table 1. The thermal properties were obtained from literature [4]. The values of density, thermal conductivity, and specific heat are specified for room temperature. The surface tension was obtained from measured data [5] by extrapolation to the evaporation temperature. For both laser wavelengths the parameter ϵ in Eq. (5) is chosen such that the function $A_S(\mu)$ reproduces known values of the absorptance at normal incidence [6], i.e., for $\mu = 1$. These values are listed in the last column of Table 1 for the wavelengths 1030 nm and 515 nm, respectively.

Table 1. Material parameters

Density [g/cm^3]	Thermal conductivity [W/m/K]	Specific heat [J/g/K]	Heat of fusion [J/g]	Heat of vaporization [J/g]	Boiling point [K]	Surface tension [N/m]	Absorptance []
8.93	394	0.385	210	4810	2868	0.95	0.10, 0.45

The investigated process parameters are listed in Table 2. They are chosen so that they correspond to typical parameters for setups with the Trumpf TruDisc 8001 and the Trumpf TruDisc Pulse 421 laser beam sources. The chosen Super-Gaussian beam profile (of fourth order) resembles a top hat beam profile with smoothed flanks and is motivated by measurements of the beam caustic (c.f. [2]). Note that the 1030 nm and 515 nm laser beams share the same distributions of intensity and Poynting vector only in the focal plane because of wavelength-dependent propagation.

Table 2. Process parameters

Laser power [W]	Feed rate [mm/s]	Focal diameter [μm]	M^2 beam quality []	Beam profile
3000–4000	50–250	200	20	Super-Gaussian

In order to account for the strong attenuation of the 515 nm radiation by the copper vapor under standard process conditions, i.e., conditions for which there are no external measures to minimize the interaction between laser radiation and metal vapor (c.f. [1]), the intensity values in simulations with the 515 nm laser are reduced by an overall factor of 20%. The value is found from numerical tests and comparisons with existing experimental data (c.f. Fig. 1). It should be understood as an average relative attenuation of the laser radiation between both wavelengths.

3.2 Calculation Results

The results of the keyhole depth calculations are shown in Fig. 4 and Fig. 5. Figure 4 shows that deeper keyholes are obtained with the 515 nm laser radiation for all considered feed rates and laser powers under idealized process conditions. Under standard conditions, including attenuation of the 515 nm laser radiation, a similar trend can be observed as in the experiments on the seam depth shown in Fig. 1. Due to their different slopes the graphs of the keyhole depth in Fig. 5 reveal deeper keyholes for the 515 nm laser at high feed rates and for the 1030 nm laser at low feed rates. This observation is compatible with the assumption that the process of keyhole depth formation is more sensitive to changes in feed rate when using 1030 nm laser radiation. It can also be seen that changing the laser power hardly influences the sensitivity to the feed rate which shows that the power calibration performed on the process with 515 nm light does not impair the conclusions of the following analysis from a qualitative point of view. It should also be noted that the absolute values of the keyhole depth in Fig. 4 and Fig. 5 will deviate from those in the real process due to the several model limitations mentioned in Sect. 2.4.

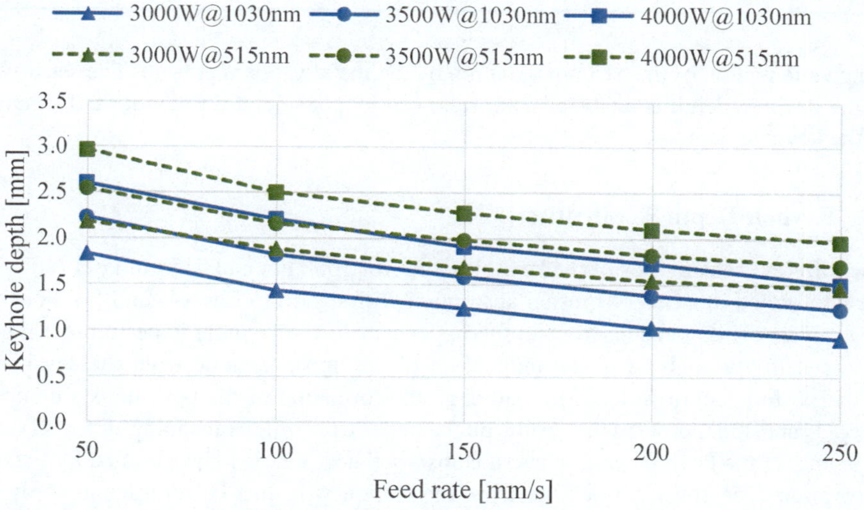

Fig. 4. Calculated values of the keyhole depth in copper for 1030 nm and 515 nm laser radiation as a function of feed rate and laser power under idealized process conditions

4 Discussion

It is rather intuitive that under idealized process conditions, neglecting attenuation by the metal vapor, deeper keyholes are obtained with 515 nm laser radiation since the absorptance of copper at 515 nm is by far greater than at 1030 nm (c.f. Table 1). Therefore, the focus of the discussion is on the observations under standard process conditions at

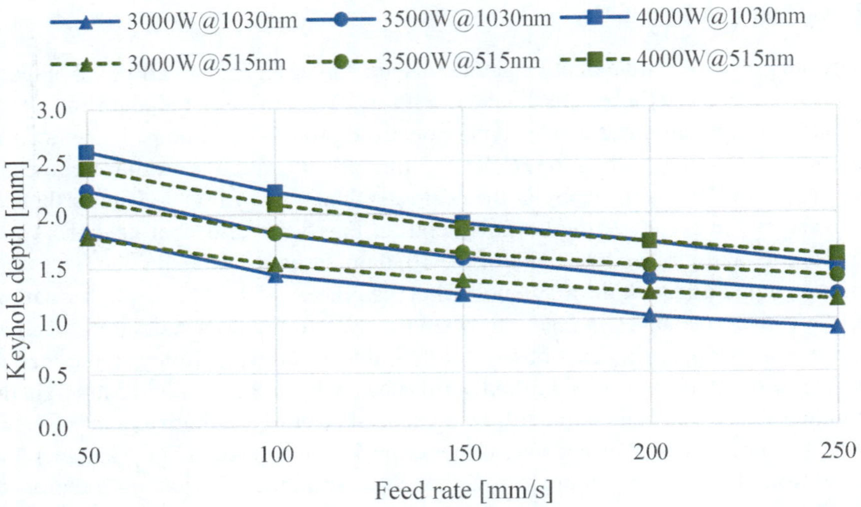

Fig. 5. Calculated values of the keyhole depth in copper for 1030 nm and 515 nm laser radiation as a function of feed rate and laser power under standard process conditions

which attenuation of the 515 nm radiation by the metal vapor is present. The sensitivity of the keyhole depth is analyzed first, followed by considerations about the formation of the keyhole depth.

4.1 Keyhole Depth Sensitivities

The different sensitivities of the keyhole depth for 1030 nm and 515 nm laser radiation are interpreted in terms of Fresnel absorption with regard to Eqs. (4) and (5). For this, an analysis of the keyhole front in the upper part of the keyhole is performed. Based on a sensitivity analysis of the inclination of the upper keyhole front the sensitivity of the keyhole depth is deduced. Although the formation of the keyhole is a complex three-dimensional process the results on the upper front will be meaningful for all other segments of the keyhole wall where the absorbed heat flux is likewise ruled by Fresnel absorption. The sensitivity of the front inclination will thus be representative of the average keyhole wall inclination and, consequently, of the keyhole depth.

From now on consider the keyhole front in a two-dimensional coordinate system containing the feed axis and the optical axis. Near the workpiece top the intensity distribution on the keyhole surface is primarily established by direct laser radiation, i.e., contributions of reflections are secondary [2]. Neglecting the term q_{MR} in Eq. (4), the factor $\mu A_S(\mu)$ then gives the percentage of the incident intensity that is absorbed by the front as a function of the cosine of the angle of incidence μ. From Fig. 6 it becomes clear that for the majority of μ-values $\mu A_S(\mu)$ is larger for 515 nm laser radiation, with a maximum difference at $\mu = 1$. There it corresponds to the reference value of the absorptance listed in Table 1. For smaller μ-values the differences in $\mu A_S(\mu)$ decrease and for $\mu < 0.083$ the 1030 nm laser even yields greater absorption. It should be noted that the monotonous graphs of $\mu A_S(\mu)$ show different slopes which will become important in the later discussion.

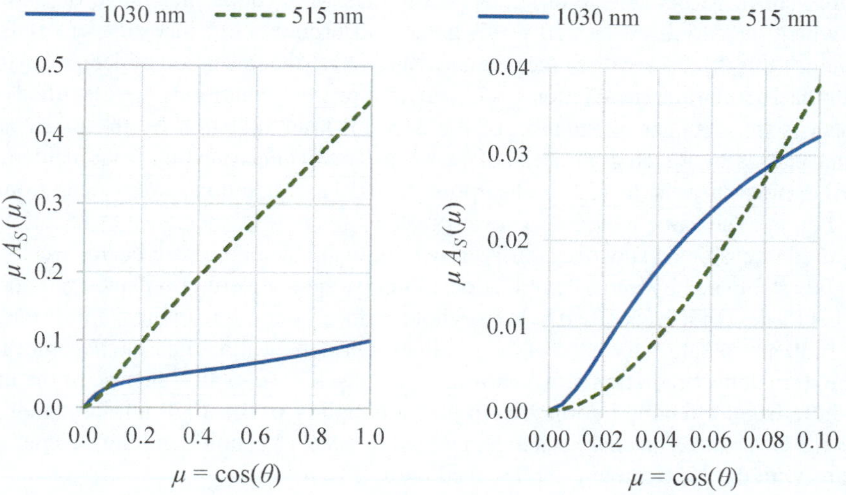

Fig. 6. Intensity absorption factor $\mu A_S(\mu)$ from Eq. (4) for 1030 nm and 515 nm laser radiation as a function of the cosine of the angle of incidence θ

By definition, μ is related to the inclination of the keyhole front. Near the workpiece top propagation distances are small compared to the Rayleigh lengths of both laser beams so that the Poynting vector is directed mainly parallel to the optical axis. In this case μ corresponds to the cosine of the front inclination angle with respect to the feed axis. That is, for large values of μ the front is flat, for small values it is steep. In order to maintain a steady keyhole front a certain value of the absorbed intensity q_a is required according to Eq. (2). This value depends on process and material parameters and local properties at the keyhole front (e.g., the heat flux into the material, curvature of the front). In general, the higher the feed rate, the higher the value is. For a given intensity I_L of the incident laser radiation the front will then incline to absorb the required value of q_a according to Eq. (4). Referring to Fig. 6, the quotient of q_a and I_L defines a target value of $\mu A_S(\mu)$ which can be incorporated into the plot as a horizontal line. The intersections of this line with the graphs of $\mu A_S(\mu)$ define the local values of μ for both lasers.

Considering a top-hat-like beam profile, there is a typical intensity value I_L everywhere along the upper keyhole front. Following the construction above there will be a representative front inclination for each laser power and feed rate. Increasing the laser power or reducing the feed rate will lower the quotient of q_a and I_L and thus lower the value of μ according to Fig. 6. However in the range $0.1 \leq \mu \leq 1$ the decrease will be much faster for the 1030 nm laser because of the lower slope of the graph of $\mu A_S(\mu)$. Hence there is a higher sensitivity of the front inclination to feed rate and laser power for 1030 nm radiation. Under the initial assumptions this sensitivity is reflected by the keyhole depth as observed in Fig. 5.

4.2 Mechanisms of Keyhole Depth Formation

So far only the sensitivity of the keyhole depth but not its absolute values has been addressed. In particular, the different sensitivities were found in a range of μ-values for which the 515 nm laser still yields greater absorption and hence steeper fronts. To conclude this discussion three mechanisms that lead to the formation of deeper keyholes with the 1030 nm laser radiation under standard process conditions are identified. One aspect is the stronger attenuation of the 515 nm laser radiation by the metal vapor which causes an overall diminution of the intensity and hence smaller front inclinations. Furthermore, there is the higher absorption of 1030 nm radiation at very steep fronts as per Fig. 6. (The value $\mu = 0.083$ corresponds to an angle of incidence of 85.24°.) The required intensities do not necessarily have to be available in the laser beam directly, but can be established by multiple reflections. Finally, the intensity of reflected radiation is higher the less of it is absorbed by the keyhole surface. According to Fig. 7 this is the case for 1030 nm radiation for most of the μ-values. Once the wall inclination becomes large enough to deflect radiation downwards, i.e., at least 45° ($\mu = 0.7$) in case of the upper keyhole front, effective into-the-depth power coupling of the 1030 nm laser radiation sets in. These three mechanisms together lead to deeper keyholes and thus deeper weld seams with the 1030 nm laser at low feed rates.

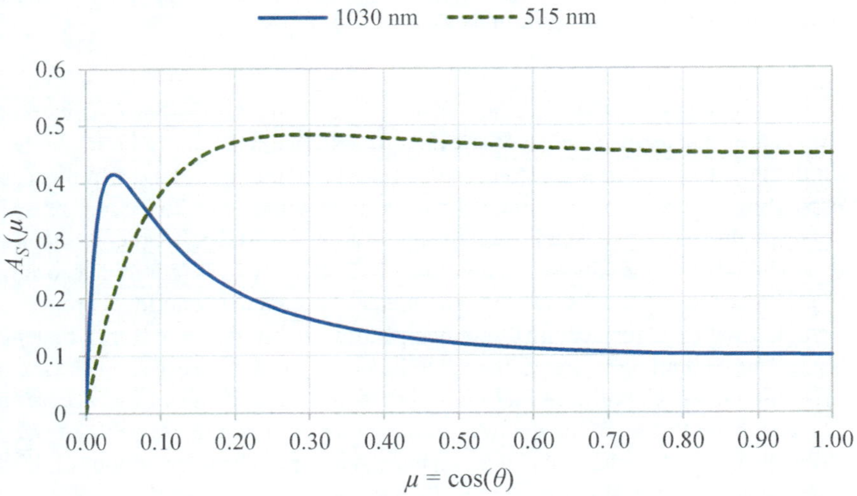

Fig. 7. Fresnel absorptance $A_S(\mu)$ from Eq. (5) for 1030 nm and 515 nm laser radiation as a function of the cosine of the angle of incidence θ

5 Conclusion

Using an approximate process model the keyhole depth was investigated as a function of feed rate and laser power for copper welds with 1030 nm and 515 nm laser beams. A more efficient process was found for 515 nm laser radiation under idealized conditions where there is no radiation attenuation by the metal vapor. Under standard conditions, including attenuation of the 515 nm radiation, the calculations revealed deeper keyholes at low feed rates and a higher sensitivity of the keyhole depth to feed rate for the 1030 nm laser as observed in experiments. This behavior could be explained by a consideration of the upper keyhole front in terms of Fresnel absorption and a local energy balance, and by the identification of three absorption mechanisms that affect the formation of the keyhole. In further investigations the effects of preheating and polarization state will be investigated, and laser beam sources in the blue visible spectrum [7] as well as other copper materials will be addressed.

Acknowledgement. The presented investigations were carried out at RWTH Aachen University within the framework of the Collaborative Research Centre SFB-1120–236616214 and funded by the Deutsche Forschungsgemeinschaft e.V. (DFG, German Research Foundation). The sponsorship and support is gratefully acknowledged.

References

1. Kaiser, E., Dold, E.-M., Killi, A., Zaske, S., Pricking, S.: Application benefits of welding copper with a 1 kW, 515 nm continuous wave laser. In: 10th CIRP Conference on Photonic Technologies, LANE 2018. Bayerisches Laserzentrum GmbH, Erlangen (2018)
2. Schöler, C., Nießen, M., Hummel, M., Olowinsky, A., Gillner, A., Schulz, W.: Modeling and simulation of laser micro welding. In: Lasers in Manufacturing 2019, World of Photonics Congress, Munich, Germany, 24–27 June 2019. WLT e.V, Erlangen (2019)
3. Schulz, W., Simon, G., Urbassek, H.M., Decker, I.: On laser fusion cutting of metals. J. Phys. D Appl. Phys. **20**, 481–488 (1987)
4. Lossin, A.: Copper. In: Ullmann's Encyclopedia of Industrial Chemistry, vol. 1, p. 15. Wiley-VCH Verlag GmbH & Co. KGaA, Weinheim, Germany (2000)
5. Brillo, J., Egry, I.: Surface tension of nickel, copper, iron and their binary alloys. J. Mater. Sci. **40**, 2213–2216 (2005)
6. Hess, A., Schuster, R., Heider, A., Weber, R., Graf, T.: Continuous wave laser welding of copper with combined beams at wavelengths of 1030nm and of 515nm. Phys. Proc. **12**, 88–94 (2011)
7. Hummel, M., Schöler, C., Häusler, A., Gillner, A., Poprawe, R.: New approaches on laser micro welding of copper by using a laser beam source with a wave-length of 450 nm. J. Adv. Join. Process. **1**, 100012 (2020)

Reduction of Hot Cracks During Electron Beam Welding of Alloy-247 LC

Aleksej Senger[1](✉), Torsten Jokisch[2], Simon Olschok[1], and Uwe Reisgen[1]

[1] Welding and Joining Institute, RWTH Aachen University, Pontstr. 49, 52062 Aachen, Germany
senger@isf.rwth-aachen.de

[2] Gas and Power, Siemens AG, Nonnendamm 55, 13629 Berlin, Germany

Abstract. Alloy-247 LC belongs to the group of precipitation hardening materials and is characterized by good creep resistance at higher temperatures. Although the material offers good cast workability, the weldability of the material is very limited due to its high crack tendency. This paper identifies and optimizes electron beam welding parameters with regard to hot crack reduction when welding conventionally cast components. The crack evaluation is carried out using scanning electron microscopy and light microscopy of the upper beads and the cross sections. In order to statistically verify the results, crack identification was carried out by micro-CT measurements. A welding speed dependent dendrite arm distance measurement additionally supports the crack investigation. Furthermore, a crack-optimized welding parameter was successfully transferred to a dissimilar joint weld (conventionally solidified to directionally solidified) and the potential for welding the high-performance material Alloy-247 LC with the electron beam was demonstrated by creep tests.

Keywords: Electron beam welding · Nickel-based alloy · Alloy-247 LC · Dendrite · Hot crack · Creep test

1 Introduction

The improvements in the efficiency of gas turbines are achieved by various combinations of technological measures. These include improved cooling, development of thermal insulation layers, aerodynamic optimization of the blades, reduction of losses and the further development of materials with better high-temperature characteristics [1]. The last-named has experienced an enormous development in metallurgy over the last decades. The specific mechanical and chemical properties have been improved with increasing quantities of different alloying elements. In addition, the casting processing of the superalloy has been expanded. With the innovative alloys and the possibility of directional solidification (DS) and single crystalline solidification (SX), the efficiency and lifetime of turbine materials have been further optimized [2–4]. Alloy 247 LC is a chemically modified superalloy specially developed for directional solidified blades and guide vanes. This enables the production of complex, core-containing, thin-walled blades. From an economic point of view, a repair of the guide vanes is opportune [5]. A

U. Reisgen et al. (Eds.): EMPOrIA 2020, LNME, pp. 40–50, 2021.
https://doi.org/10.1007/978-3-030-70332-5_4

hybrid design of the guide vanes, using different solidification states, would also have a positive influence on the economic efficiency.

The broad spectrum of alloying elements constitutes a great challenge for welding technology. In particular, the weldability of the age-hardening nickel-base alloy is correlated with the aluminium and titanium content [6]. To increase strength, aluminium and titanium are withdrawn from the γ-matrix after solution heat treatment during the ageing process and a γ' phase Ni3(Al, Ti) is precipitated. This decreases the lattice parameter and increases the lattice distortion. The heat treatment is usually carried out after the welding process. Welding after heat treatment would soften the material in the area of the weld seam and the HAZ. However, the strengthening measure after the welding process can lead to crack formation. Due to the lower aging temperature range compared to the solution annealing temperature, aging processes already occur during the heating phase to solution annealing temperature and thus γ' precipitation. A superposition of the stresses resulting from the lattice distortion processes and the residual welding stresses in the HAZ that have not yet been removed can lead to the maximum material yield strength being exceeded and to failure of the component [7]. For this reason the residual welding stress should be reduced to a minimum.

Furthermore, hot cracking is to be considered as critical, especially during welding. When welding nickel-base alloys, a differentiation is essentially made between two types of hot cracking. Intercrystalline liquation cracks in the HAZ, according to the mechanism of constitutional liquefaction of the low-melting phases, as well as solidification cracks in the seam. For both crack types, crack initiation is due to thermally induced stresses. Cracks in the weld metal occur mainly during solidification. Due to the different solubility of the alloying elements in the melt, the concentration changes continuously in the solidification interval. The degree of segregation depends on the solidification rate. This is generally influenced by the welding speed during welding. In addition, the solidification rate changes with the temperature gradient G at each location of the weld pool isotherms. Both the weld pool shape and the microstructure are determined by these two values. During the solidification interval, the solidification type changes continuously from planar to equiaxial dendritic [7, 8]. Solidification starts in the area of the highest cooling rate or the highest temperature gradient at the solidification edge to the base material. The crystals grow perpendicular to the solidification isotherms. With increasing welding speed, the weld pool shape changes from elliptical to drop-shaped with crystal growth almost perpendicular to the welding direction. The segregation front proceeds in the remaining melt with the interdendritic distance unfavourable to the shrinkage stresses resulting from the process. The consequence is crack initiation in the melt.

With the electron beam, a precise and versatile tool is available. On the one hand, critical residual welding stresses can be reduced by specific heat control and on the other hand, the solidification morphology can be positively influenced. The welding process thus offers the ideal conditions for improving the seam quality in terms of crack reduction when welding Alloy 247 LC. The welding speed is investigated as a primary parameter of the process for modifying the heat control.

2 Experimental Setup

2.1 Material

For the investigations, the precipitation-hardening nickel-based alloy Alloy-247 LC was examined in the conventionally cast (CC) state and later also in the directionally solidified state. After the casting process, the material was solution annealed at 1232 °C for 2 h 15 min under reduced atmospheric pressure. Subsequently, the specimens for the welding tests were eroded out of the ingots and milled to final dimension ($100 \times 60 \times 6$). Both the CC and the DS base material have a coarse-grained structure with macroscopic dendrite spacing. In the interdendritic areas of the base material a multi-phase structure with γ' precipitations of Ni3(Al,Ti) in the γ matrix is present. The size and shape of the γ' precipitates is inhomogeneously distributed due to the missing aging process after the solution heat treatment. This chemical composition is presented in Table 1.

Table 1. Composition of Alloy-247 LC (wt. %). Examined with atomic emission spectroscopy

	Ni	C	Si	Cr	Co	Mo	W	Ti	Al	Ta	Hf	Fe
Alloy-247 LC	bal.	0.07	0.1	8.2	9.0	0.5	8.9	0.8	5.6	3.5	1.6	0.4

2.2 Process

The electron beam welding machine ProBeam K-7 with 120 kV acceleration voltage was used as a test facility. The welding was performed at a pressure of 10^{-4} mbar and with a working distance of 400 mm to the focus lens. The focus position was left on the component surface at all beam currents. As in practice a certain welding depth is required, all tests were welded with a constant welding depth of 5 mm $\pm$ 0.5 mm. This also provides comparability of the results. For this purpose, a power required for the welding depth was determined for each welding speed. An overview of the parameters investigated is shown in Table 2. It can be seen that with decreasing welding speed a higher energy per unit length is required to achieve the same weld depths. Before welding, the plates were cleaned in an acetone ultrasonic bath for 10 min and then clamped with a vice perpendicular to the welding direction, over the entire length. A 90 mm long seam was welded in the middle of the 100 mm long sample.

Table 2. Overview of the examined parameters

Alloy-247 LC		Specimen thickness [mm]	E [J/mm]	U_b [kV]	I_b [mA]	V_s [mm/s]	Focus position
CC	BoP	6.0	960	120	4.0	0.5	surface
CC	BoP	6.0	600	120	5.0	1.0	surface
CC	BoP	6.0	450	120	7.5	2.0	surface
CC	BoP	6.0	270	120	9.0	4.0	surface
CC	BoP	6.0	210	120	10.5	6.0	surface
CC	BoP	6.0	164	120	10.9	8.0	surface
CC	BoP	6.0	138	120	11.5	10.0	surface
CC + DS	tack welding	12.0	78	120	6.5	10.0	surface
CC + DS	joint welding	12.0	1680	120	14	1.0	surface

2.3 Evaluation Methods

In order to be able to evaluate the welding results and create a sufficient basis for interpretation, both metallographic and radiographic examinations were carried out. For this purpose, all samples were cut out of the seam according to Fig. 1. The first and last 5 mm of the seam were not considered in the evaluation.

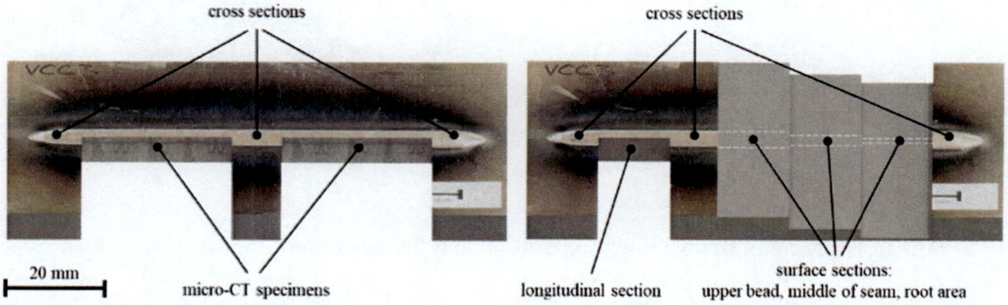

Fig. 1. Drawing of the mechanical sample preparation for three transverse sections and two micro-CT images (left) and of the sections for the dendrite investigation (right)

In the first step, all samples were examined for surface cracks with the scanning electron microsope. The crack position and the total crack length were determined. These results are used as a first indication for crack evaluation. The crack identification and localization inside the weld was performed with the assistance of microCT scans. For this purpose, samples of the same size were taken from the weld samples. Due to the high density of the material, the sample geometry was limited to a maximum thickness of 3.3 mm. One side was milled to the centre of the seam and subsequently the 3.3 mm thick sample was cut out. A total of 2 specimens per examined parameter were prepared for the Micro CT examination. The crack positions and crack lengths were also determined. Due to the lack of information about the types of crack, metallographic examinations of transverse, longitudinal and surface sections were carried out in addition to the microCT. For this purpose the samples were etched according to Kalling-II (100 mL ethanol, 100 mL hydrochloric acid (32%) and 5 g copper(II)-chloride) at 21 °C for 45 s and then magnified with a microscope. Simultaneously to crack identification, the secondary dendrite arm spacing (SDAS) was measured using the metallographically prepared samples according to [9]. Due to the continuously changing welding isotherm and the associated dependence of the dendrite growth direction, the sample preparation for the determination of the primary dendrite spacing is very complex. In addition, the direction of dendrite growth changes with the welding speed and thus the effort for an adequate measurement. For this reason, the dendrite spacing was not measured within the scope of this work.

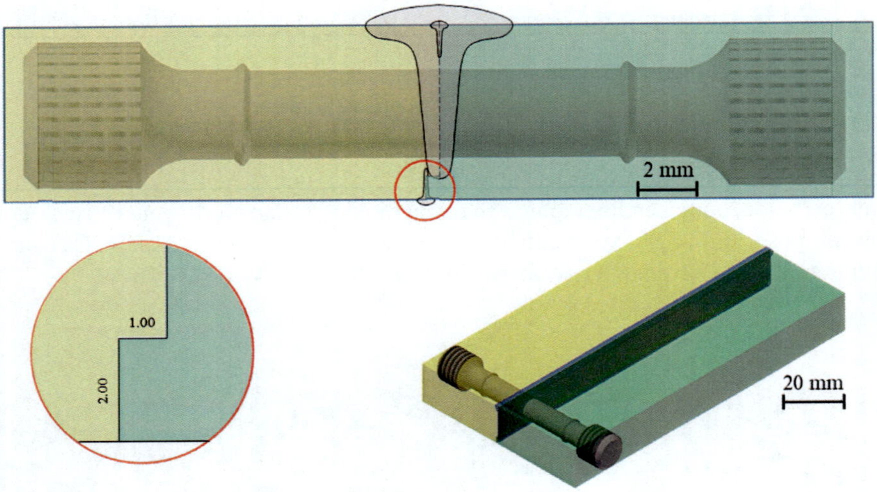

Fig. 2. Drawing of the welding preparation and the position of the creep specimen

Furthermore, specimens for creep tests were welded with the crack-reduced 1.0 mm/s parameter. For this purpose, the parameter for a 12 mm thick joint weld (Alloy-247 CC to Alloy-247 DS) was adjusted, Table 2. Due to the low viscosity of the melt and the resulting unacceptable weld sinkage welding in PA-position, the specimen preparation was selected according to Fig. 2. Like the preliminary tests before, the specimens were heat-treated, mechanically prepared and then cleaned with acetone. Before the joint welding, the samples were tack welded on both sides by an 80 mm long linear seam. The parameters are shown in Table 2. By choosing a narrow tack weld geometry, the subsequent weld ensures that the cracked tack weld is completely over-welded in the welding area.

The samples for the creep tests were extracted according to Fig. 2. A total of 10 creep specimens were prepared from two joint welds. Subsequently, the samples were subjected to a three-stage heat treatment. Solution annealed at 1243 °C for 4h, in the stabilization annealed at 1080 °C for 2 h and ageing at 870 °C for 20 h.

3 Results

3.1 Influence of the Welding Speed on the Welding Depth

The weld geometry has a significant influence on the heat distribution over the welding depth and thus on the stress distribution in the component. Figure 3 shows the seam geometries in relation of the welding speed. At welding speeds of 1 mm/s and 2 mm/s it can be seen that the weld depth of 5 mm ± 0.5 mm is not achieved. Due to the low welding speed, the leading heat flow changes continuously the welding depth over the entire weld seam length. In addition, nail head formation increases and the ratio of weld depth to weld width decreases significantly. Due to the low thermal conductivity of the alloy, this effect can be observed up to a welding speed of 4 mm/s.

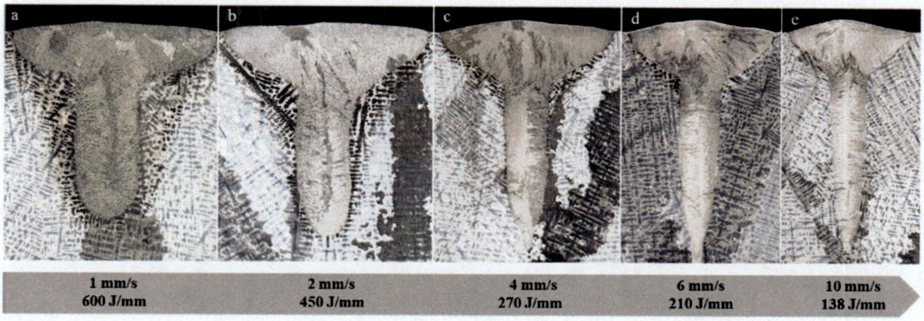

Fig. 3. Light microscopy images of cross-sections at different welding speeds

3.2 Influence of the Welding Speed on the Dendrite Morphology

The welding speed as a direct parameter has a significant influence on the temperature gradient G and the crystallization rate R. Figure 4 illustrates the influence of the welding speed on SDAS. The SDAS decreases with increasing welding speed due to the high solidification rate. Furthermore, the SDAS is inhomogeneous over the seam depth. Towards lower welding speeds, the SDAS increases in areas of the pronounced nail head. Regardless of the welding speed, the solidification rate is highest in the root zone. As a result, the dendrite arm distance in the root area is closer than in the other areas of the weld.

Furthermore, the inhomogeneity of the SDAS increases over the seam depth to higher welding speeds. At 1.0 mm/s welding speed, the SDAS varies over the seam depth by only 1.2% of the maximum value. In contrast, the variation at higher welding speeds

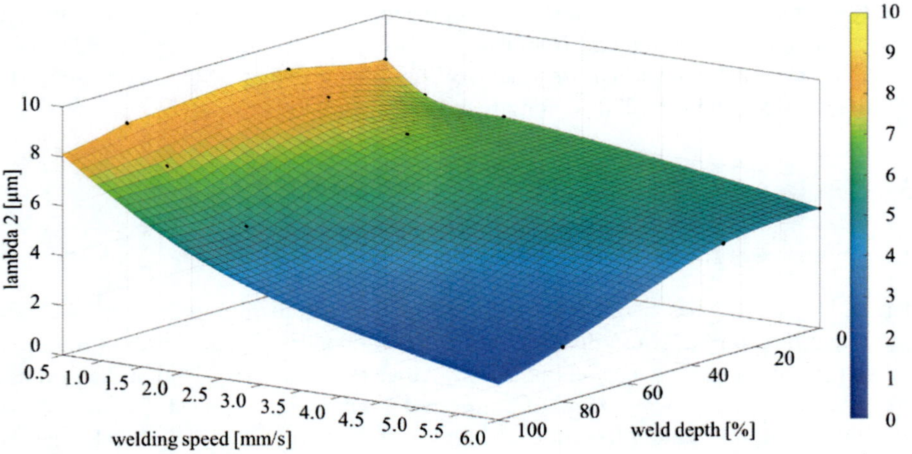

Fig. 4. Dendrite arm distance over the welding depth in percent depending on the welding speed. Evaluated by means of transverse, longitudinal and surface sections in three areas of the seam (upper bead, seam centre and seam root).

(6.0 mm/s) increases to approximately 50% of the maximum value. According to Whitesell [10], the SDAS is dependent on both the temperature gradient and the solidification rate. In welding, however, these two indirect parameters cannot be separated from the welding speed. The solidification rate is inevitably dependent on the temperature gradient. This allows a conclusion to be drawn about the qualitative course of the temperature gradient.

3.3 Crack Formation Depending on the Welding Speed

The first evaluation of a weld seam is always performed visually on the surface. The result gives a first indication of the seam quality. Figure 5 shows an example of three SEM images of seam surfaces that were welded with different speeds. The energy per unit length was kept constant at 100 J/mm.

It can be seen that the crack length increases with increasing welding speed. In addition, the crack position shifts from the base material in the direction of the weld seam. At high welding speeds, the crack progresses along the solidification line perpendicular to the welding isotherm. In this area, thermal stresses occur perpendicular to the dendrite growth direction and tear open the still liquid interdendritic area. With decreasing welding speed, more time is available for solidification. On the one hand, a microcracks can be filled due to the larger dendrite spacing and the subsequent melt flow, on the other hand, the thermal stress (transverse to the welding direction) only develops when the critical area (which runs transverse to the stress) has sufficient strength. According to Rappaz [11], the hot crack tendency decreases with higher SDAS and lower thermal stress. However, the residual stress states in Fig. 5 cannot be compared due to the different stiffness states. An investigation with constant welding depth minimizes the possible influence of stiffness to a minimum. Figure 6 shows the total crack length on the upper

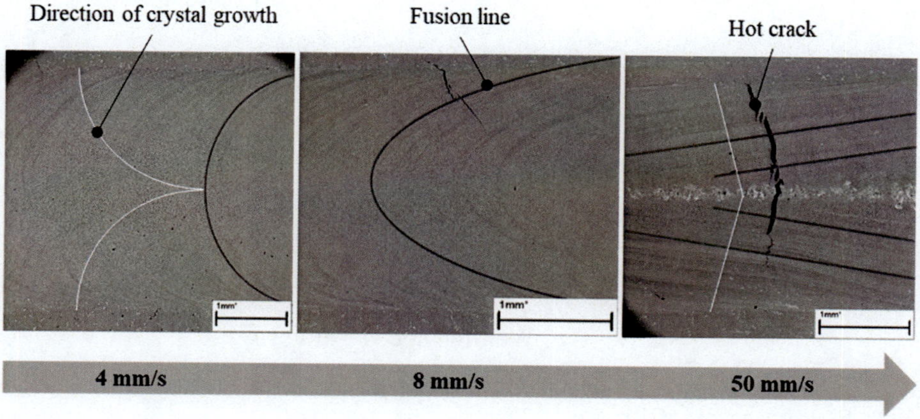

Fig. 5. SEM Images of weld pool geometry, dendrite growth direction and crack initiation direction depending on the welding speed

bead over the welding speed at constant welding depth. For this purpose all cracks were cumulated to a total crack length.

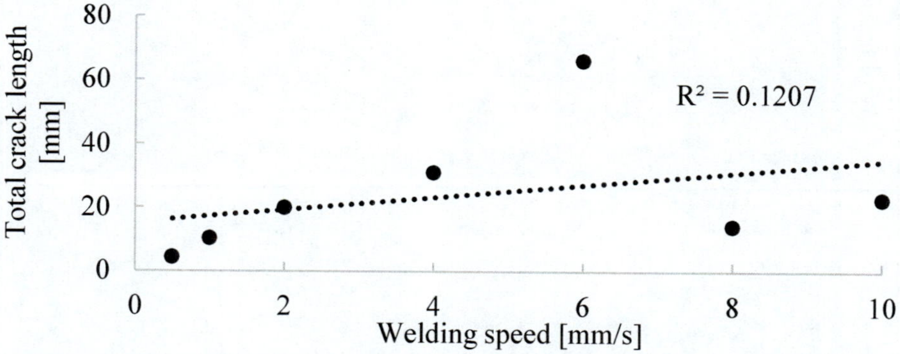

Fig. 6. Correlation between welding speed and total crack length on the upper bead at constant welding depth. Evaluated on SEM images.

The total crack length increases with increasing welding speed. At higher speeds, the maximum crack length is limited by the seam width and thus the total crack length on the upper bead does not increase significantly. In addition, a differentiation is made in the following between two types of hot cracking. At higher welding speeds, the solidification crack mainly occurs in the seam. Where, on the other hand, the liquation cracks occurs in the HAZ at lower welding speeds. In the range between 2 mm/s and 8 mm/s welding speed, both types of hot crack are observed. For this reason, a maximum of the total crack length at 6 mm/s welding speed is shown in Fig. 6.

A microCT evaluation of the weld seams in Fig. 7 shows a similar correlation between the welding speed and the total crack length.

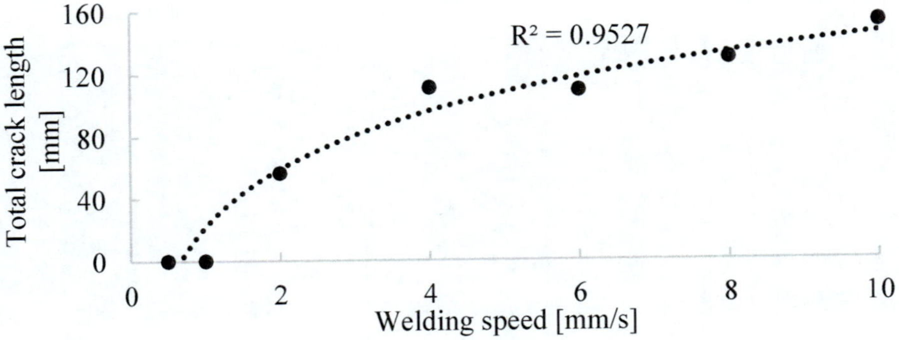

Fig. 7. Correlation between welding speed and total crack length in the seam at constant welding depth. Evaluated on microCT images.

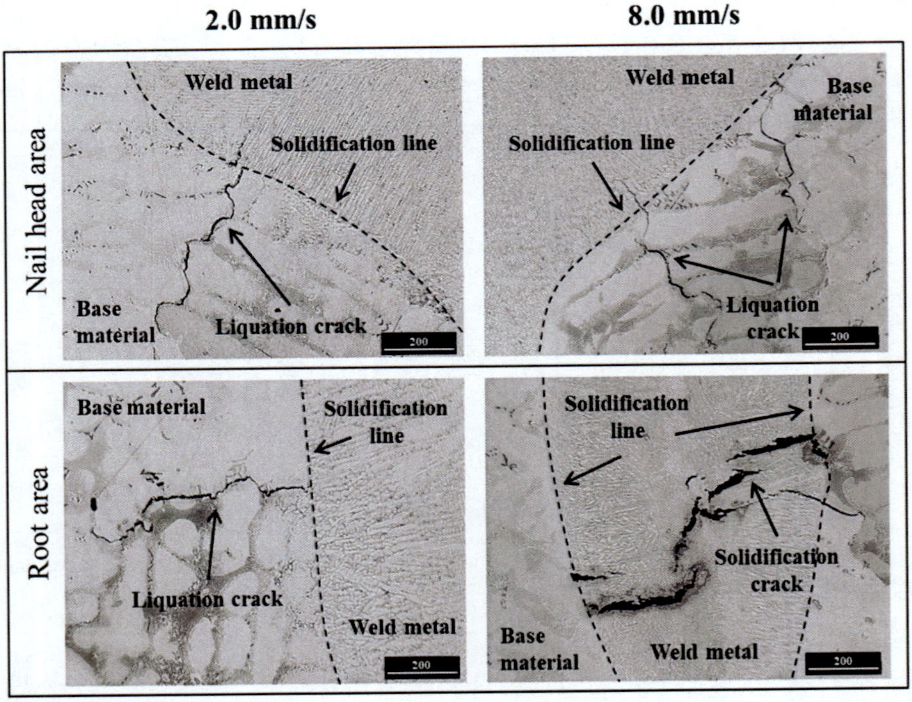

Fig. 8. Light microscopy images showing a parameter and position dependent crack formation near the nail head and in the root area of a weld for 2.0 mm/s and 8.0 mm/s welding speed

In contrast to the evaluation of the upper bead, the total crack length in the weld seam increases linearly with increasing welding speed. A metallographic examination was carried out to identify the types of cracks. Figure 8 shows an example of the typical hot cracks when welding Alloy-247 LC. Comparable to the crack investigations on the upper bead, only predominantly hot cracks occur in the HAZ at lower welding speeds.

In contrast, at higher welding speeds the solidification crack in the deep welding area cannot be avoided. According to Fig. 4, a higher temperature gradient can be assumed due to the smaller dendrite arm distance in this area. This in turn has a negative effect on the thermally induced residual stresses and the solidification rate, which are mainly responsible for the solidification cracks in the weld metal.

3.4 Welding of Creep Specimens

The previous results indicate that at lower welding speeds, the total crack length decreases to a minimum when welding Alloy-247 LC in the CC and solution heat treated condition. A weld joint between a CC and a DS (transverse to the direction of weld) specimen was investigated in the following. Heat treatment subsequent to the welding process increases the γ‘precipitation and thus the creep resistance to higher temperatures. For this purpose, the samples were solution annealed at 1243 °C for 4 h after welding, stabilised at 1080 °C for 2 h and aged at 870 °C for 20 h.

For the creep tests the 1.0 mm/s welding speed parameter was preferred to the 0.5 mm/s parameter due to the lower leading heat flow. The beam current was adjusted to the sheet thickness. Despite clamping perpendicular to the welding direction, the specimens were tack welded with a welding depth of 2.1 mm on both sides over a length of 80 mm. The creep test specimens were mechanically prepared as shown in Fig. 2. The results of the creep test thus reflect the weld quality in the middle area of the weld.

Figure 9 shows the results of the creep tests of welded and aged specimens, at a test temperature of 850 °C and 950 °C in comparison with the base materials. The individual measuring points of the welded specimens are on the extrapolated minimum curve of the CC base material.

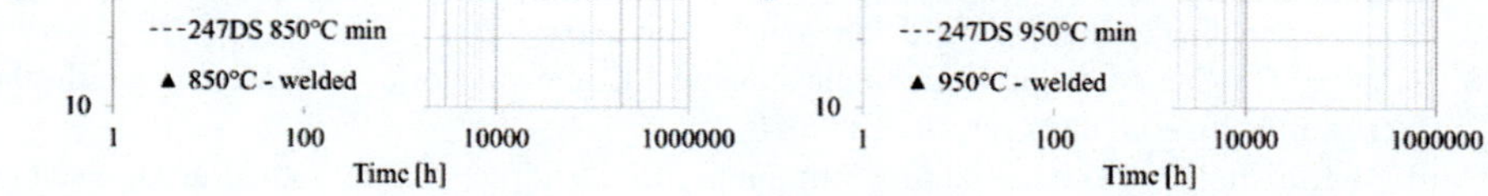

Fig. 9. Creep rupture strengths of the hybrid weld (CC with DS) at two different test temperatures (850 °C and 950 °C). Mean strength of the base materials (solid lines) and the specified minima according to Siemens (dashed lines).

This shows that the liquation cracks occurring parallel to the direction of the load or the cracks that could occur during post-heat treatment have no significant influence on the creep strength.

4 Conclusion

In this study it could be shown that the hot cracking tendency could be reduced with a reduced temperature gradient. As a direct parameter in electron beam welding, the welding speed has a significant influence on the temperature gradient and thus on the solidification rate and weld pool geometry while the power density distribution of the beam remains constant. At lower welding speeds the solidification crack in the seam and on the upper bead could be completely suppressed due to the lower temperature gradient. As Rappaz [11] has already shown with the hot cracking criterion, the SDAS and the temperature gradient correlate with the hot cracking probability (solidification crack). In contrast, the liquation crack in the HAZ could not be suppressed either in the area of the nail head or in the area of the deep welding. However, at lower welding speeds, the total length decreased significantly. Furthermore, creep tests have shown that possible liquation cracks in the area of the deep welding area that occurring parallel to the direction of the load (transverse to the welding direction) have no influence on the high-temperature strength. Even a heat treatment to increase strength after the welding process has no negative influence on crack formation with the optimized welding parameter.

References

1. Mom, A.J.A.: Introduction to Gas Turbines. Woodhead Publishing, Sawston (2013). https://doi.org/10.1533/9780857096067.1.3
2. Betteridge, W., Shaw, S.W.K.: Development of superalloys. Mater. Sci. Technol. **3**(9), 682–694 (1987)
3. Sims, C. T.: A history of superalloy metallurgy for superalloy metallurgists. Superalloys 1984. The Metallurgical Society of AIME, Warandale (1984)
4. Harris, K., Erickson, G.L., Schwer, R.E.: Directionally solidified and single-crystal superalloys, properties and selection: irons, steels, and high-performance alloys. ASM Handbook Committee (1990)
5. Harris, K., Erickson, G.L., Schwer, R.E.: MAR M 247 derivations - CM 247 LC DS alloy and CMSX single crystal alloys: properties & performance. Superalloys 1984. The Metallurgical Society of AIME, Warandale (1984)
6. Kelly,T. J.: Welding metallurgy of investment cast nickel-based superalloys. Weldability of Materials, pp. 151–157. ASM International, Materials Park, OH (1990)
7. Kou, S.: Welding Metallurgy, 2nd edn. Wiley, New Jersey (2003)
8. Schulze, G.: Die Metallurgie des Schweißens – Eisenwerkstoffe - Nichteisenmetallische Werkstoffe, 4th edn. Springer, Berlin (2010)
9. BDG-Richtlinie VDG-Merkblatt: Bestimmung des Dendritenarmabstandes für Gussstücke aus Aluminium-Gusslegierungen, P220, BDG-Informationszentrum Giesserrei, Düsseldorf (2011)
10. Whitesell, H.S., Li, L., Overfelt, R.A.: Influence of solidification variables on the dendrite arm spacings of ni-based superalloys. Metall. Mater. Trans. B. **31B**, 546–551 (2000)
11. Rappaz, M., Drezet, J.-M., Gremaud, M.: A new hot-tearing criterion. Metall. Mater. Trans. A. **30A**, 449–455 (1999)

Validation of the EDACC Model for GMAW Process Simulation by Weld Pool Dimension Comparison

Oleg Mokrov, Marek Simon(✉), Ivan Shvartc, Rahul Sharma, and Uwe Reisgen

Welding and Joining Institute, RWTH Aachen University, Pontstrasse 49, 52062 Aachen, Germany
simon@isf.rwth-aachen.de

Abstract. The EDACC model (evaporation-determined arc cathode coupling) was developed to represent a more physically in depth description of the heat flux and current density distribution on the surface of the weld pool in gas metal arc welding (GMAW). To validate the model, geometry parameters available from experiments from GMAW processes were compared to the corresponding geometry parameters from simulations. The process simulation model was supplied with an approximation for the surface deformation and simulations were performed without the surface deformation and with the surface deformation for a common Gaussian heat flux and current density distribution, as well as for the EDACC model. The EDACC model parameter of the bulk plasma temperature was calibrated to match the total current. The results of the validation are presented and their fidelity is discussed.

1 Introduction

The gas metal arc welding (GMAW) process remains the most widely used fusion welding process among the others, when considering the amount of filler material sold in Germany [1]. Its simple use, high deposition rates and, easy automation made it widely known and usable. However, the simulation of the GMAW process remains a field of current engineering research. It consists of modeling the complex interaction of physical phenomena which occur during the welding process and which have not always been fully understood in sufficient depth. The formation of the droplets, the molten pool behavior and, the metal transfer during the welding process have remained relevant research topics throughout the years. Therefore, further research and a continuing development of new and more precise mathematical models are still relevant.

The key factor, which determines the formation of the weld seam, is the flow of the molten metal in the weld pool with its visible and invisible complex physical processes [2]. The hydrodynamics, electromagnetics, heat and current distribution, momentum and mass transfer of the droplets into the molten pool as well as the free surface deformation are some of these processes [3]. The cathode region determines the distribution of the heat flux and the current density on the weld pool surface and is therefore a sensitive boundary

U. Reisgen et al. (Eds.): EMPOrIA 2020, LNME, pp. 51–59, 2021.
https://doi.org/10.1007/978-3-030-70332-5_5

condition for the process. Until now, mainly a Gaussian distribution was assumed, but the latest development was presented with the evaporation determined model of arc cathode coupling (EDACC) [4], see Fig. 1.

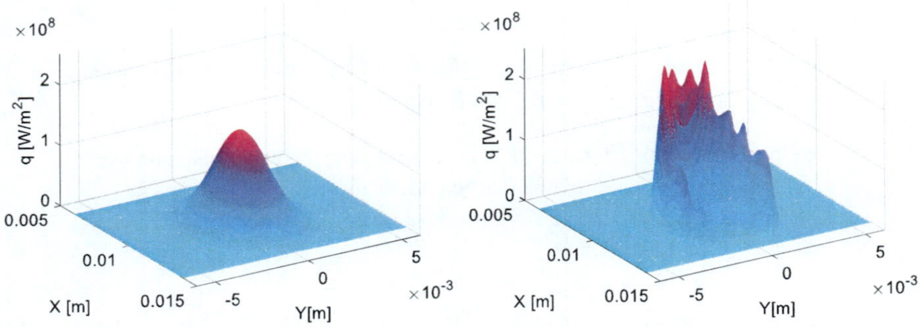

Fig. 1. Exemplary heat flux distribution for the Gaussian heat source model (left) and the EDACC model (right), adapted from [4]

While the assumption of the Gaussian distribution was dependent of the actual conditions of the arc only, the EDACC model is based on the surface temperature of the weld pool as well as the plasma temperature and their interaction via the evaporation. This represents a deeper consideration of the underlying physics. Although a qualitative comparison of the models has been performed in [5], the results have not been compared against experimental data, yet.

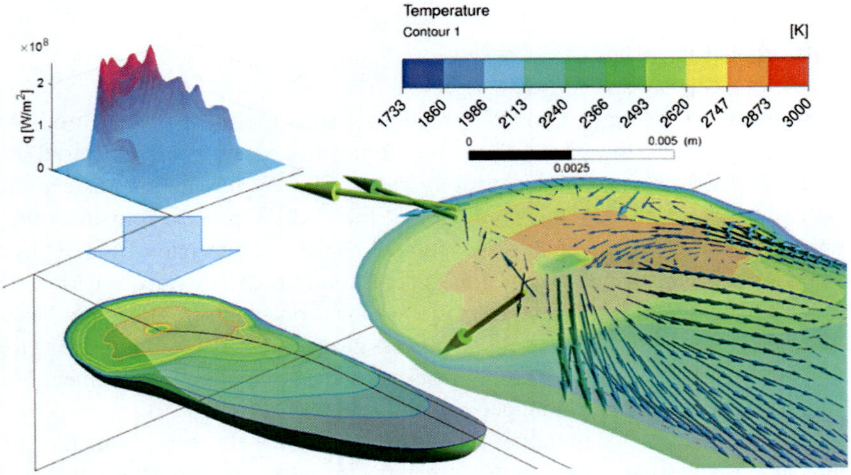

Fig. 2. Symbolized application of the EDACC model to the fully hydrodynamic weld pool calculation

The model used in [5] was extended with an approximation of the free surface deformation (FSD) in ANSYS CFX. This allows to study the effects of the changed current density distribution on the Lorentz-forces, which are considered as main drivers for the hydrodynamics in the weld pool, see Fig. 2. This allows a validation of the EDACC model in comparison with geometrical weld pool data (reinforcement, depth, length) from real welding experiments. Therefore, three different cases are examined during the investigation: the application of a Gaussian heat source model (GHSM) without FSD, the application of the GHSM with FSD and application of the EDACC model with FSD.

2 Methods

2.1 Geometrical Data

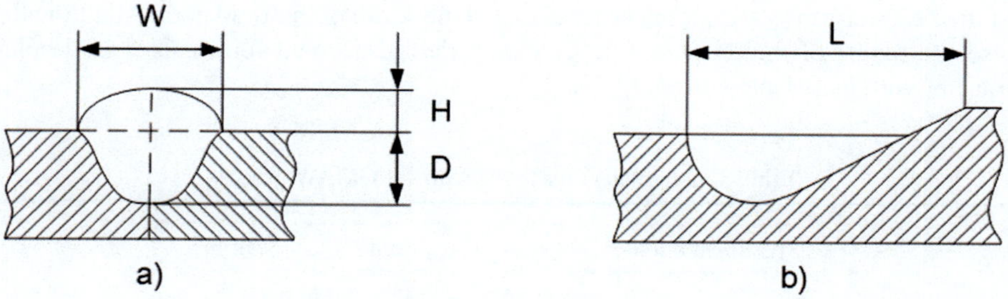

Fig. 3. Section views of the weld seam from [6] a) Cross-section b) Longitudinal section

For the validation of the models (i.e. [5] with extension to the FSD) and further simulation results, the experimental data from previously generated experiments were used, see Table 1. The main geometry parameters under consideration in the present work can be seen in Fig. 3. Those are the weld bead reinforcement (*H*), the weld bead width (*W*), the depth of the penetration (*D*) and the length of the molten pool (*L*). The welding was performed on steel S235JR and the main process parameters are current (*I*), voltage (*U*), welding speed ($v_{w.s.}$) and wire feed rate ($v_{w.f.r.}$) and they can be seen in Table 1. The geometrical parameters resulting from the experiments are shown in Table 2.

Table 1. GMAW process parameters

Experiment number, №	I, [A]	U, [V]	$V_{w.s}$, [*m*m/min]	$V_{w.f.r.}$, [m/min]
1	200	26.9	300	6
2	250	29.4	300	8
3	300	38.7	300	10

Table 2. Geometrical weld pool parameters. (H), (W), (D) from blown-out experiments, (L) from tracer experiments.

Experiment number, №	H, [mm]	W, [mm]	D, [mm]	L, [mm]
1	3.4 ± 0.09	10.9 ± 0.06	2.8 ± 0.2	26.6 ± 2.6
2	3.8 ± 0.2	12.8 ± 0.3	3.6 ± 0.2	34.4 ± 0.9
3	3.6 ± 0.1	17.9 ± 0.2	5.1 ± 0.2	42.5 ± 0.2

2.2 Heat Source Calibration Data

Gaussian Heat Source Model. The GHSM model needed to be calibrated for the radius of the heat source $\mathbf{r_{heat\_source}}$, in order to match the width of the weld pool. Additionally also the radius of the droplets $\mathbf{r_{droplet}}$ was varied to account for changes in droplet transfer with increasing current (Table 3).

Table 3. Parameters for the GHSM for GMAW process

Experiment number, №	r_{heat_source}, [mm]	$r_{droplet}$, [mm]
1	9.3	0.6
2	10	0.5
3	14.6	0.4

EDACC Heat Source Model The EDACC heat source, as defined in [4] and [5], also needed to be calibrated for the cathode voltage drop $\boldsymbol{U_d}$, the plasma temperature $\boldsymbol{T_{plasma}}$ and the radius of the heat source $\mathbf{r_{heat\_source}}$, see Table 1, in order to match the total current and the power of the real process. The radius was also chosen to match the width of the resulting weld pool, while the voltage and the temperature were adjusted to give a reasonable ratio between power and current. The resulting values for the plasma temperature can be considered reasonable, based on the scientific works of [7] and [8], where it is stated that the plasma temperature in GMAW can vary between 6000 [K] and 9000 [K]. Also, for the voltage drop, the results from [9] were used for orientation, considering a small negative contribution by the anode voltage drop (Table 4).

3 Results

3.1 Quantitative Comparison

Figure 4 presents a comparison of the simulation results with the experimental data for the weld bead reinforcement, the depth and the width. In the comparison of the weld bead reinforcement a constant underestimation can be noticed. The model for the weld bead

Table 4. Parameters for the EDACC for GMAW process

Experiment number, №	U_d, [V]	T_{plasma}[K]	r_{heat_source}, [mm]
1	16	6850	5.2
2	18	6670	6.2
3	20	6080	9.0

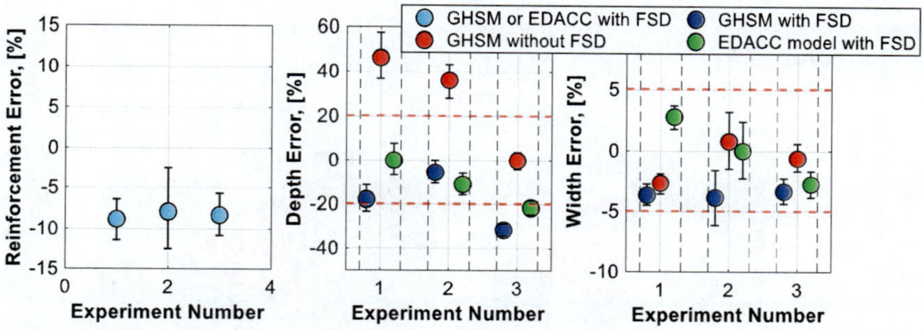

Fig. 4. Reinforcement (left), depth (center) and width (right) error comparison with GMAW process experimental data.

reinforcement was only applied in the cases where the free surface deformation was taken into account, so for the GHSM and the EDACC, the same value for the reinforcement was used. The depth seems to be very much overestimated in the cases, where the free surface was not considered. However for the comparison of the EDACC with the Gauss model, the results almost lie within 0 and -20%, i.e. slightly underestimating the depth in the simulation compared to the real experimental weld pool results. The error of the width for all cases lies within $\pm5\%$, which is to be expected as the width was one of the parameters, the model was calibrated for.

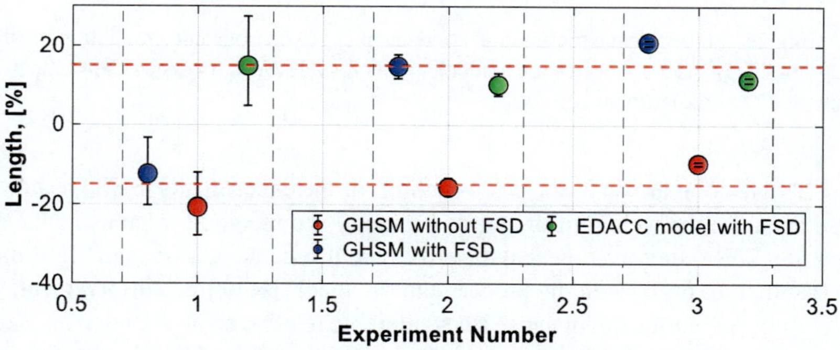

Fig. 5. Length comparison with experimental data for GMAW.

For the length comparison as can be seen in Fig. 5, it can be seen that the length is always overestimated by the EDACC model, by ~10–15%. Only in Experiment 1, the low current case, the GSHM with FSD underestimates the length. It should be noted however, that the length for the experimental data was taken from tracer experiments, which rather tends to overestimate the real weld pool length. Therefore an overestimation by the simulation needs to be judged especially critical. In this perspective, the GHSM without the FSD seems to yield the most accurate results. However, since it lacks the consideration of the weld seam deformation and it gives very large errors for the depth, this agreement should not be taken as a proof of accuracy.

3.2 Qualitative Comparison

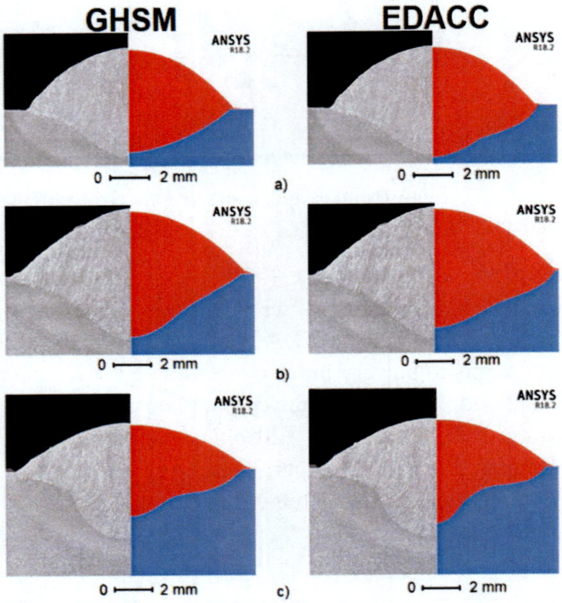

Fig. 6. Comparison between experimental weld seam cross-sections and simulation results performed with GHSM (left) and EDACC model (right) for GMAW process a) Experiment №1 b) Experiment №2 c) Experiment №3.

The qualitative comparison to the cross-sections of the experiments in Fig. 6 shows that the agreement for the weld reinforcement geometry is quite good when using the model for the FSD. This seems to show that the error in reinforcement, as seen in Fig. 4 must be due to fluctuations of the welding process and the small size of the reinforcement, where already small deviations might cause a considerable relative error. However, it is clearly noticeable that the depth is clearly deviating quite strongly. Overall, the differences between the GHSM (left) and the EDACC (right) are subtle, and they both seem to capture the shape of the cross section equally well.

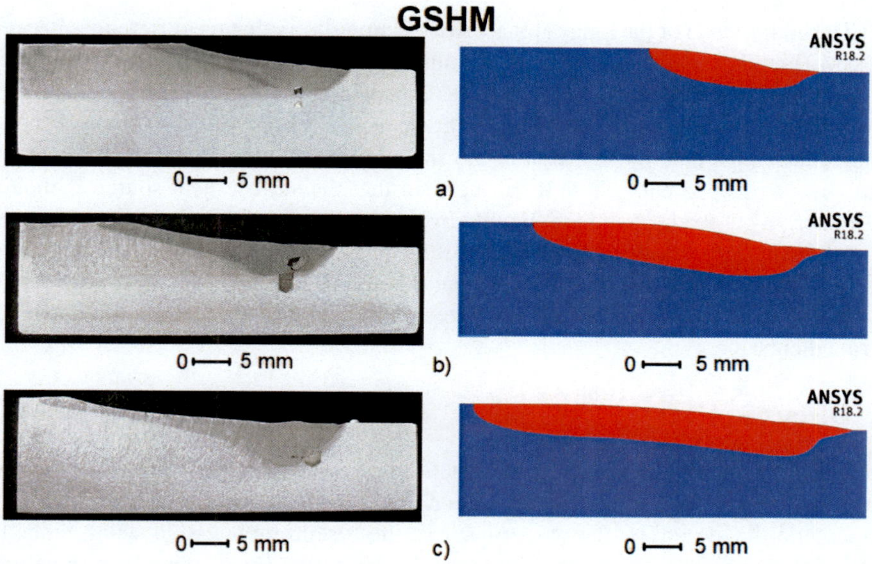

Fig. 7. Comparison between experimental weld seam longitudinal sections and simulation results performed with GHSM (a, b, c) for GMAW process a) Experiment №1 b) Experiment №2 c) Experiment №3.

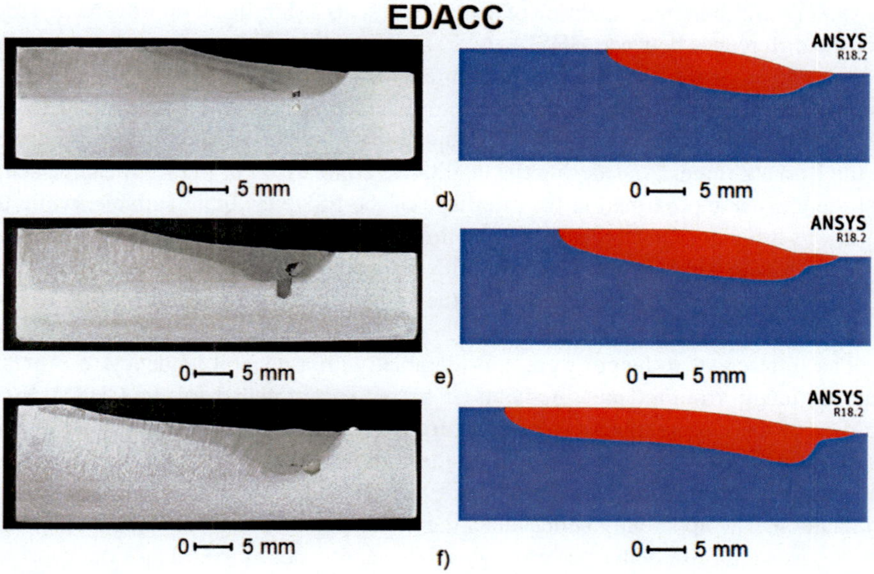

Fig. 8. Comparison between experimental weld seam longitudinal sections and simulation results performed with EDACC model (d, e, f) for GMAW process d) Experiment №1 e) Experiment №2 f) Experiment №3.

The comparison of the longitudinal cross-section shows that the differences between GHSM (Fig. 7) and EDACC (Fig. 8) are subtle again. Only for the first experiment in a) and d), a noticeable difference becomes apparent. Here the GHSM is considerable shorter, which can be also seen in Fig. 5. The main differences between the experimental results and the simulations lie first in the shape of the melting front, as in all cases, a "tip" becomes apparent, where it can be seen that the surface heat source is melting the material, but the heat and momentum by the droplets have not yet affected the weld pool shape. Another noticeable difference is the angle and curvature of the solidification front. In the experiments, the shape is much more concave and the angle with the top surface of the weld seam is much more steep, while in the simulation it seems almost perpendicular.

4 Discussion

The reinforcement seems to show a constant underestimation. The model for the reinforcement was based on mass conservation, i.e. it is based on the fact, that whatever comes into the system by the wire, must exit the system by the weld seam. Since the approximation for the shape seems quite good as seen from Fig. 6, it has to be concluded that more mass has entered the system than could be accounted for by the process parameters. Possibly the welding velocity was not well calibrated in the experiment. The differences in width were marginal, as was expected due to the radius of the heat source being calibrated for the resulting weld pool width. However, the error in depth was considerable, although the addition of the FSD presents a significant improvement. However the differences between GHSM and EDACC were not as pronounced as expected. This might be due to the mesh resolution and due to the omission of the consideration of the arc column, which might have even more effects when accounting for the non-axisymmetric distribution of the current density vector field and the resulting effect on the Lorentz force. Considering the qualitative comparison in Fig. 6, it can be stated that the chosen approximation for the shape of the FSD was quite suitable, while the depth was difficult to obtain by the simulation. However, as the droplet transfer was not considered with high accuracy and with the suspicion of inaccurate setting of the wire feed velocity on the wire feeding device, the agreement in depth could be improved by a more accurate treatment in these areas.

The differences in length were considerable, with a general tendency to overestimate the length from the measurements, in all but Experiment 1 for the GHSM. Since the data of the measurements were taken from trace experiments, which tend to even more overestimate the weld pool length, these differences are even more concerning. The seeming agreement for the length for Experiment 1 with GHSM still lacks a clear explanation. The apparently better matching of the simulation without the FSD should not be mistaken for a better performance of this approach. Since it lacks the consideration of the weld seam reinforcement and it gives very large errors for the depth, this agreement should rather be taken as an indication that some parts in the underlying model, still need improvement. In particular, the shape of the melting front in Fig. 7 and Fig. 8 could be improved by a more accurate positioning of the droplet mass source. For improvement of the shape of the solidification front in Fig. 7 and Fig. 8 a more accurate solidification model and possibly a finer meshing will be investigated in the future.

5 Conclusion

The presented validation of the EDACC model shows that the EDACC model does perform as well as the GHSM when it comes to the determination of the weld pool shape. However, the EDACC model yields the advantage of giving more realistic surface temperatures of the weld pool, i.e. below boiling temperature, as was discussed in [5]. The EDACC model will also give rise to a modified distribution of current density and heat flux, as was discussed in [4] and [5] and can be seen from Fig. 1. This proves the additional benefit of the EDACC model when performing numerical GMAW process simulations. As a next step, the very simplified solidification model will be improved.

Acknowledgement. The presented investigations were carried out at RWTH Aachen University within the framework of the Collaborative Research Centre SFB1120-236616214 "Bauteilpräzision durch Beherrschung von Schmelze und Erstarrung in Produktionsprozessen" and funded by the Deutsche Forschungsgemeinschaft e.V. (DFG, German Research Foundation). The sponsorship and support is gratefully acknowledged.

References

1. Reisgen, U., Stein, L.: Fundamentals of joining technology: Welding, Brazing and Adhesive Bonding, DVS Media GmbH, Düsseldorf (2016)
2. Cho, M.H., Lim, Y.C., Farson, D.F.: Simulation of the weld pool dynamics in the stationary pulsed gas metal arc welding process and final weld shape. Weld. J. **85**(12), 271–283 (2006)
3. Mokrov, O., et al.: Numerical investigation of droplet impact on the weld pool in gas metal arc welding. Mater. Sci. Eng. Technol. **48**(12), 1206–1212 (2017). https://doi.org/10.1002/mawe.201700147
4. Mokrov, O., Simon, M., Sharma, R., Reisgen, U.: Arc-cathode attachment in GMA welding. J. Phys. D Appl. Phys. **52**(36), 364003 (2019). https://doi.org/10.1088/1361-6463/ab2bd9
5. Mokrov, O., Simon, M., Sharma, R., Reisgen, U.: Effects of evaporation-determined model of arc-cathode coupling on weld pool formation in GMAW simulation. Weld. World **64**, 847–856 (2020). https://doi.org/10.1007/s40194-020-00878-3
6. Reisgen, U., Schiebahn, A., Mokrov, O., Lisnyi, O., Sharma, R.: The experimental analysis of the influence of submerged arc welding parameters on weld bead geometry formation. CWA J. (Journal de l'ACS) **11**, 90–94 (2015)
7. Kozakov, R., et al.: Spatial structure of the arc in a pulsed GMAW process. J. Phys. D Appl. Phys. **46**(22), 224001 (2013). https://doi.org/10.1088/0022-3727/46/22/224001
8. Valensi, F., et al.: Plasma diagnostics in gas metal arc welding by optical emission spectroscopy. J. Phys. D Appl. Phys. **43**(43), 434002 (2010). https://doi.org/10.1088/0022-3727/43/43/434002
9. Zhang, G., et al.: Study of the arc voltage in gas metal arc welding. J. Phys. D Appl. Phys. **52**, 085202 (2019). https://doi.org/10.1088/1361-6463/aaf588

New Opportunities for Brazing Research by *in situ* Experiments in a Large Chamber Scanning Electron Microscope

Anke Aretz[1](✉), Riza Iskandar[1], Joachim Mayer[1], Kirsten Bobzin[2], Alexander Schmidt[2], and Thomas E. Weirich[1]

[1] RWTH Aachen University, Central Facility for Electron Microscopy (GFE), 52074 Aachen, Germany
aretz@gfe.rwth-aachen.de

[2] RWTH Aachen University, Surface Engineering Institute (IOT), 52056 Aachen, Germany

Abstract. The present study demonstrates that *in situ* heating experiments in a LC-SEM are capable to provide hitherto not accessible insights into the kinetics of the brazing process. Continuous video documentation and temperature logging of the entire experiment yielded the on-set time of the wetting of the base material and the upper time limit that should not be exceeded to avoid quality losses of the joint. For the here investigated system of 75Sn20Cu5Ge filler metal with aluminum alloy EN AW-42100 base metal, the maximum holding time has been determined as 3.5 min at 450 °C. The wetting point at this temperature is reached after one minute, which proves the effectiveness of the here investigated system.

Keywords: Scanning electron microscopy · SEM · Large chamber · LC-SEM · *In situ* · Brazing · Wetting

1 Introduction

The usual workflow of materials characterization by electron microscopy puts main emphasis on the pre- and post-mortem state of the sample under investigation. This traditional approach fails however, if information about intermediate stages during mechanical, chemical and/or temperature treatment of the sample are crucial to gain in detail understanding of the process. In consequence, questions related to the mechanism and how the material transforms into a certain morphology or microstructure remain frequently unanswered. To fill this knowledge-gap the development of *in situ* materials characterization techniques and equipment became one of the top priorities among the past two decades. Nevertheless, while the implementation of *in situ* capabilities is rather straightforward and became readily available for some characterization techniques, this is unfortunately not the case in conjunction with electron microscopes. For the latter, a much more elaborate instrument design is required in order to manage the high vacuum demands at elevated temperatures and/or upon varying pressure [1, 2].

About 15 years ago one of the unique large-chamber scanning electron microscopes (LC-SEM) was installed at the main authors institute [3]. The main benefit of this type

U. Reisgen et al. (Eds.): EMPOrIA 2020, LNME, pp. 63–69, 2021.
https://doi.org/10.1007/978-3-030-70332-5_6

of LC-SEM is its huge chamber that was designed to overcome the common restrictions of conventional SEMs, which do not allow implementation of *in situ* experiments on the larger scale. To reach this goal, the large-chamber hosts the whole column of the SEM, which enables to align the electron optics to almost any position with regard to the sample [4]. Hence, this setup allows investigation of samples with dimensions of up to 0.7 m in diameter and 300 kg in weight. Aside imaging and high-frame rate video capabilities with secondary electrons (SE), the LC-SEM is equipped with a backscattering electron (BSE) detector and a silicon-drift energy dispersive X-ray (EDX) analysis system from BRUKER. Experiments can be operated under variable pressure (VP) of up to 3000 Pa. Hence, this LC-SEM facilitates integration of even very bulky setups necessary for *in situ* tensile testing, laser beam micro welding or larger sample heating equipment [5]. For the here presented *in situ* brazing experiment we employed a heating module from Kammrath & Weiss GmbH, Dortmund, Germany (Fig. 1).

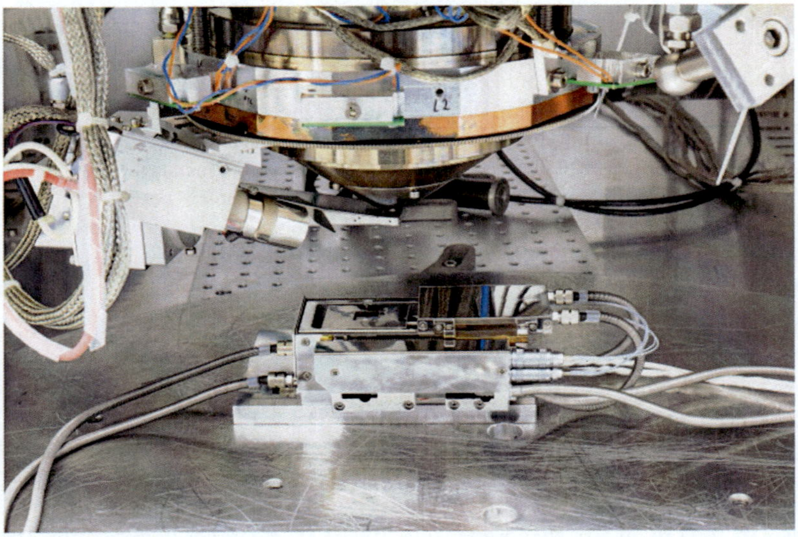

Fig. 1. Heating module in the LC-SEM (bottom). The electron optical column with an SE and BSE detector can be seen at the top.

2 Focus Application: *in situ* Brazing

Throughout the joining process via brazing, locally melted phases are generated along the interface of the work pieces which form the brazing joint between the filler metal and the base material after solidification. However, melting and solidification processes as well as segregation effects might occur in the process affected zone that can lead to severe shrinkage, mismatch and even distortion of the components. Melting, solidification and segregation are temperature dependent, so the heating and cooling rate as well as the holding time of a brazing process are crucial parameters for the phase evolution and microstructure along the joining zone [6]. For further improvement of the joining process, it is therefore necessary to gain a more thorough understanding of the involved time and temperature frame and to determine the onset temperature of the wetting, the ideal maximum temperature and the period of holding time. To determine all these crucial parameters a LC-SEM outfitted with a suitable *in situ* setup is the perfect tool that assists development of optimized brazing processes with even more precise results for long-lasting and well-functioning joints. For demonstration of the capabilities of the described approach we used the brazing system 75Sn20Cu5Ge (as filler metal) with aluminum alloy EN AW-42100 (as base material) that has been investigated by our group in an earlier study [7].

2.1 Experimental

A small 0.1 mm × 0.2 mm chunk of 75Sn20Cu5Ge filler metal on a sheet of aluminum alloy EN AW-42100 was inserted into the heating module. The sample was then heated up under vacuum to 450 °C using a heating rate of about 2.4 degrees per sec. Due to the inertia of the temperature control loop, the temperature may overshoot the targeted temperature for a limited period of time. The holding time of 33 min was chosen to be relatively long compared to conventional brazing processes, in order to be able to investigate any effects even after the brazing joint has been established. The subsequent cooling took place under vacuum conditions in the LC-SEM. The temperature curve of the experiment, together with SE images recorded after certain time periods are shown in Fig. 2. The corresponding event-log of the experiment is listed in Table 1. After cooling to ambient temperature, the sample was characterized by EDX element mapping from top (Fig. 3) and in cross-section geometry (Fig. 4) for characterizing interface of the joint. A specimen for subsequent STEM-EDX element mapping in a transmission electron microscope (TEM) was prepared by focused ion beam (FIB) milling using a Dualbeam FIB FEI STRATA400 (Fig. 5) [8]. The TEM investigation was carried out with a Zeiss LIBRA200FE operated at 200 kV.

Table 1. Event-log of the LC-SEM *in situ* experiment.

#	Time [sec]	Temperature [ºC]	Remarks
1	0	RT	Launch of heat ramp (~2.4 °C/sec)
2	90	380	First visible shrinkage of the filler metal
3	150	470	Initial signs of wetting
4	210	450	Complete wetting
5	270	450	Grain structure of base becomes visible
6	400	450	Grain structure/white rim around molten filler metal
7	1640	450	Pronounced grain structure visible
8	2700	200	Termination of the experiment

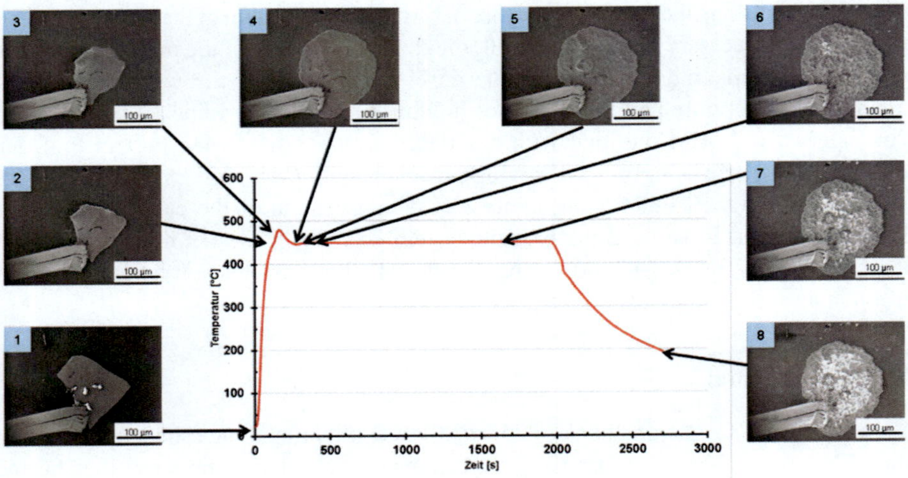

Fig. 2. Temperature curve of the LC-SEM *in situ* experiment and SE images recorded after certain time periods. The number of the images refers to the event-log in Table 1.

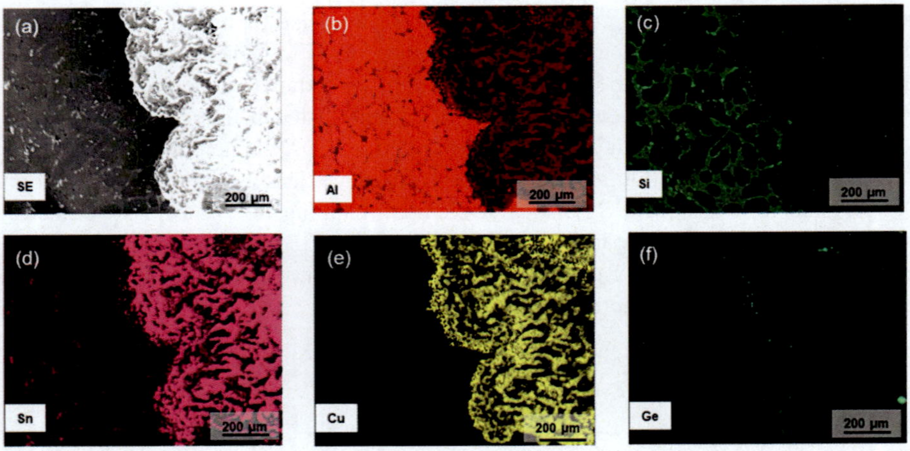

Fig. 3. SEM image and corresponding EDX element distribution maps recorded after the *in situ* experiment in the LC-SEM. (a) image from the secondary electron (SE) detector (b) aluminum, (c) silicon (d) tin, (e) copper (f) germanium.

2.2 Results

The event-log listed in Table 1 reports the first visible changes at a nominal sample temperature of 380 °C (#2). This is the point when the native oxide layer of the base material is locally destroyed and becomes dissolved by the filler metal. After another minute, the until then rather structureless molten filler metal started to develop some grain structure contrast in the SE image (#5). This must be interpreted as result of an increased surface roughness. Further annealing at 450 °C caused formation of a white rim after additional 2 min (#6), which might go along with an increased segregation of the initial filler metal. The latter process is again characterized by a further increase of surface roughness. As seen by comparison of the SE images #6 to #8 in Fig. 2, these images represent the final stage of the experiment since the shape of the filler metal remains unchanged. The main difference between these three images is the contrast gain with annealing time.

Inspection of the EDX element distribution maps recorded from the solidified sample proves the great inhomogeneity over the sample. As expected, the grains of the base material contain still the largest amounts of aluminum with formation of aluminum-copper phases in the central contact zone (Fig. 4b and 4e). However, the grain boundaries of the aluminum rich phase are found thoroughly decorated with silicon and it seems that the spread of tin took place along these tracks along the grain boundaries as well. As a consequence, the center of the re-solidified filler metal is significant depleted in silicon. In contrast to the latter, copper appears much less mobile and could not be detected outside the central wetting area. Moreover, low amounts of magnesium and germanium are distributed along the rim of the center part of the filler metal. Based on the SEM-EDX results from the cross-section sample, it appears that the formation of copper-tin phases at the interface can be ruled out due to the missing overlap of the corresponding element maps for copper (Fig. 4e) and tin (Fig. 4f). However, the STEM-EDX element distribution maps in Fig. 6 give a more detailed view of the microstructural changes directly at the interface. These distribution maps recorded at higher magnification show that copper is also dissolved to some extend in the tin matrix. Moreover, the maps show that copper mixes also with aluminum, but not with silicon. The latter element however, binds to germanium, a finding that is also supported by the element overlaps in Fig. 4c and 4d.

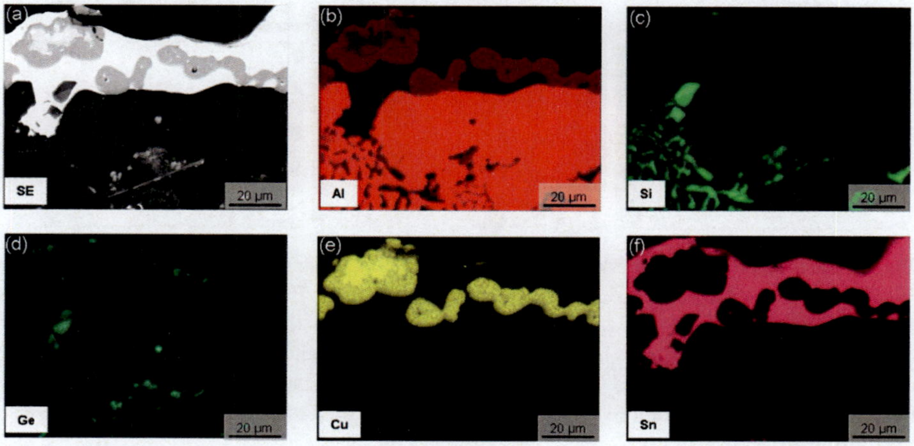

Fig. 4. SEM secondary electron (SE) image and corresponding EDX element distribution maps of a cross-section from the center part of the sample after the LC-SEM *in situ* experiment. The elemental maps in (b) and (e) prove formation of aluminum-copper phases at the interface. The latter are embedded in a probably pure tin matrix as seen in (f). At this magnification the formation of copper-tin phases at the interface cannot be justified due to the non-overlapping element maps for copper (e) and tin (f). However, as proven by Figs. 6e and 6f, some amounts of copper are indeed dissolved in the tin matrix.

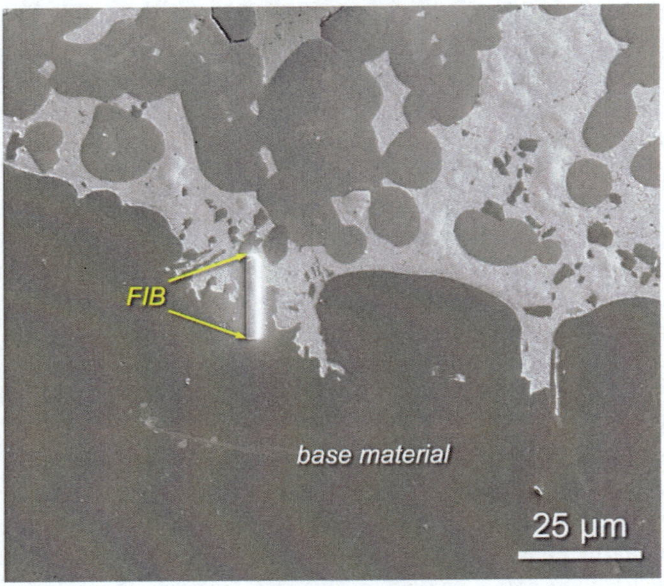

Fig. 5. Secondary electron (SE) image recorded at the interface of the cross-section sample. The length of the extracted FIB lamella for STEM-EDX analysis is indicated by the yellow arrows.

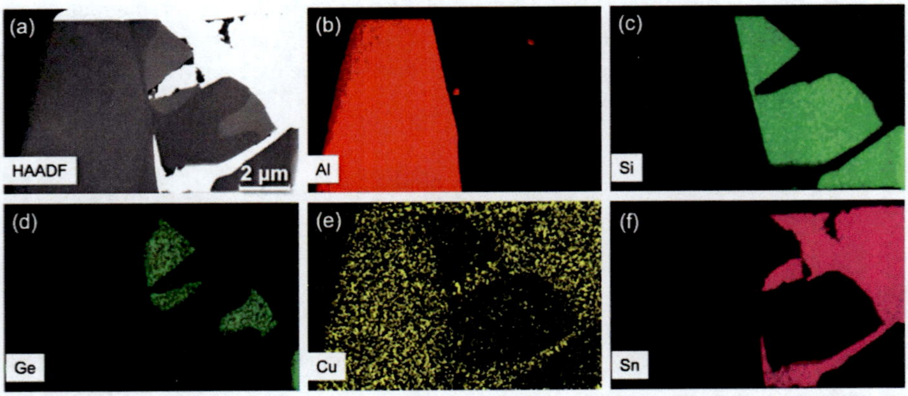

Fig. 6. High-angle annular dark-field (HAADF) scanning transmission electron microscopy (STEM) image of the FIB specimen in (a) and corresponding EDX element distribution maps for the key-elements in (b) to (f). This close-up demonstrates that the interface between base material and filler metal reveals highly in-homogenously and quite complex. Whereas copper (e) mixes with aluminum in (b) and tin in (f), this appears not to be the case for silicon (c). The latter however, binds to germanium (d), which is a rather unexpected finding (for comparison see also Fig. 4c and 4d).

2.3 Discussion and Conclusions

As demonstrated in this study,*in situ* brazing experiments in a LC-SEM are capable to provide hitherto not accessible insights into the kinetics of the initial wetting of the base material and the evolution of intermetallic phases formed along the contact zone. Due to continuous video documentation and temperature logging of the entire experiment, it is possible to identify the optimum processing parameters at the later stage of data evaluation. Although each *in situ* experiment is in fact a single shot experiment that requires several repetitions for statistical reasons, it does provide already good first estimates for the nearly optimum brazing conditions. Hence, we were also able within the present study to estimate the most critical parameter for brazing - the holding time - for a particular maximum processing temperature from a single *in situ* experiment in the LC-SEM. Thus, for the here investigated system of 75Sn20Cu5Ge filler metal with aluminium alloy EN AW-42100 base metal, the holding time should not exceed 3.5 minutes at 450 °C in order to avoid any quality losses of the joint. Moreover, the great efficiency of the here investigated brazing system could be confirmed by determining the wetting point at one minute after reaching the target temperature.

Acknowledgements. The authors gratefully thank Mr. Kevin Kistermann (GFE) for FIB sample preparation and evaluation of TEM data. This work was performed within the German Research Foundation (DFG) supported Collaborative Research Centre SFB 1120 "Precision Melt Engineering". Moreover, the DFG is acknowledged for financial support within the recent Project 163323790.

References

1. Stokes, D.J.: Principles and Practice of Variable Pressure/Environmental Scanning Electron Microscopy (VP-ESEM). Wiley, Chichester (2008). https://doi.org/10.1002/9780470758731
2. Dehm, G., Howe, J.M., Zweck, J.: In-Situ Electron Microscopy: Applications in Physics, Chemistry and Materials Science. Wiley-VCH (2012). https://doi.org/10.1002/9783527652167
3. Aretz, A., Ehle, L., Haeusler, A., Bobzin, K., Öte, M., Wiesner, S., Schmidt, A., Gillner, A., Poprawe, R., Mayer, J.: In situ investigation of production processes in a large chamber scanning electron microscope. Ultramicroscopy **193**, 151–158 (2018)
4. Klein, M., Klein, S.: Adapting human behaviour for the development of a new scanning electron microscope. In: Proceedings of International Conference on Micromechatronics for Information and Precision Equipment MIPE 1997, Tokyo, Japan, p. 324–329 (1997)
5. Ramazani, A., Schwedt, A., Aretz, A., Prahl, U.: Digital Image Correlation (DIC), Dual Phase Steel, EBSD, Extended Finite Element Method (XFEM), In Situ Bending Test, Kernel Average Misorientation (KAM) Map, Representative Volume Element (RVE). Key Eng. Mater. **586**, 67–71 (2014)
6. Schmitz, G.J., Böttger, B., Apel, M.: On the role of solidification modelling in integrated computational materials engineering "ICME". In: IOP Conference Series: Materials Science and Engineering, vol. 117, p. 012041 (2016). https://doi.org/10.1088/1757-899X/117/1/012041
7. Iskandar, R., Schwedt, A., Mayer, J., Rochala, P., Wiesner, S., Oete, M., Bobzin, K., Weirich, T.E.: Microstructural analysis of germanium modified tin-copper brazing filler metals for transient liquid phase bonding of aluminium. Materialwiss. Werkstofftech. **48**, 1257–1263 (2017). https://doi.org/10.1002/mawe.201700155
8. Giannuzzia, L.A., Stevieb, F.A.: A review of focused ion beam milling techniques for TEM specimen preparation. Micron **30**, 197–204 (1999). https://doi.org/10.1016/S0968-4328(99)00005-0

Phase-Field Modeling of Precipitation Microstructure Evolution in Multicomponent Alloys During Industrial Heat Treatments

Michael Fleck(✉), Felix Schleifer, Markus Holzinger, Yueh-Yu Lin, and Uwe Glatzel

Metals and Alloys, University of Bayreuth, Prof.-Rüdiger-Bormann-Straße 1, 95447 Bayreuth, Germany
michael.fleck@uni-bayreuth.de

Abstract. We develop a phase-field model for the simulation of chemical diffusion limited microstructure evolution. The model is applied to γ'-precipitation under the influence of realistic multi-step aging treatments in multi-component nickel-based superalloys with industrially relevant chemical complexity. The temperature-dependent thermodynamic and kinetic input parameters are obtained from CALPHAD calculations using ThermoCalc. Further, the model accounts for the lattice-misfit between the precipitate- and the matrix-phase. The required temperature-dependent elastic stiffness and lattice-misfit can be measured using resonance ultrasound spectroscopy and high temperature X-ray diffraction, respectively. This allows to account for realistic shaping of γ'-particles in the simulation. The comparison to shapes of γ'-particles in experimental microstructures serves as an important cross validation of the model. The application of the model to investigate the effect of the subsequent aging treatment on the precipitation microstructure after a brazing process is discussed.

Keywords: Precipitation hardening · Aging heat treatment · Phase-field modeling · Brazing

1 Introduction

Nickel-base superalloys have various applications at elevated temperatures, especially in stationary gas turbines and airplane engines. The main strengthening mechanism, which makes these alloys applicable for high temperatures is by coherent precipitation of ordered fcc γ'-phase ($L1_2$ structure) [1]. The aim of industrial aging heat treatments is to achieve optimum precipitate sizes, a narrow size distribution and high volume fractions up to 70% of the strengthening phase [2, 3]. This is achieved by nonisothermal multi-stage aging heat treatment, being specifically optimized for the individual alloy [4–6].

U. Reisgen et al. (Eds.): EMPOrIA 2020, LNME, pp. 70–78, 2021.
https://doi.org/10.1007/978-3-030-70332-5_7

During brazing the composition of the base material is strongly altered in the extended joining zone by the imposed braze material. This poses the material science question of how to design optimal heat treatments for the joined material [7, 8]. In this work, we discuss the precipitation microstructure evolution during multi-stage, nonisothermal aging heat treatments in joining zones based on simulations.

1.1 Precipitation Microstructure Evolution

To obtain optimal mechanical properties, it is crucial to tailor the mean precipitate size and size distribution as well as the precipitate volume fraction as a result of the aging heat treatment. Different physical mechanisms lead to a temporal evolution of the mean size and size distribution of the precipitates as well as the overall precipitate volume fraction. A general increase of the mean precipitate size is referred to as "coarsening". An individual particle can either grow due to a local supersaturation of the matrix (i.e. nonequilibrium growth) or on the expense of the surrounding, smaller particles, which are simultaneously dissolving. The latter coarsening mechanism is called "precipitate ripening", according to the commonly known Lifshitz-Slyosov-Wagner (LSW) theory of Ostwald ripening [9–11]. Which proceeds very close to the local thermodynamic equilibrium [12]. Ostwald ripening is the growth of large particles at the expense of dissolving smaller ones. The driving force is the overall reduction of interface area and thus interfacial energy. The size distribution during ripening is self-similar with respect to the mean precipitate size. Furthermore, new particles can nucleate from the supersaturated matrix, and two neighboring precipitates can coagulate, i.e. merge into a single, larger precipitate. In comparison to ripening, nonequilibrium nucleation and growth results in an increasing overall precipitate volume fraction. In an industrial aging heat treatment this typically occurs during and right after the cooling stages.

The shaping and the arrangement of the precipitate particles is predominantly influenced by elastic effects, which result from the temperature dependent misfit between the γ'-precipitates and the matrix-phase. Misfitting coherent precipitates cause an inhomogeneous strain field around the precipitates. The precipitate shape is determined by the minimum of the sum of interfacial and elastic energy [13]. The effect of elasticity on the precipitate shape rises with increasing precipitate size as bulk elastic effects then dominate the interfacial contributions [14]. With increasing misfit, the precipitate shapes become increasingly more cuboidal due to the cubic anisotropy of the elasticity [3, 15].

1.2 Phase-Field Simulations of Precipitate Microstructure Evolution

The focus of this work is on the qualification of a phase-field model for the simulation of diffusion limited precipitate microstructure evolution in nickel-base superalloys. The aim is to realize the simulation of nonisothermal multi-stage aging heat treatments including cooling and heating stages for integrated computational materials engineering. The phase-field method is frequently used to study the temporal evolution of precipitation microstructures [16–21]. The required thermodynamic and kinetic input information is obtained from CALPHAD calculations using the commercial software-package ThermoCalc [22]. Considering different nickel-base superalloys, we perform simulations of precipitation microstructure formation with the explicit consideration of up to ten different chemical components. Furthermore, an elastic term is included in the model to account for the misfit stresses as well as inhomogeneous elastic constants. The details concerning the present phase-field model are given basically in [12]. This model has been recently extended by the so-called "sharp phase-field method" [23, 24]. Respective extensions to the model are discussed in [25].

The temperature dependent thermodynamic and kinetic input parameters can be calculated from thermodynamic and thermo-kinetic CALPHAD databases as described in [12, 26]. The elastic parameters have been extracted from metallurgic experiments at Metals and Alloys at the University of Bayreuth. The temperature-dependent lattice misfit between the γ'-phase and the matrix has been measured using high-temperature X-ray diffraction. The temperature dependent anisotropic and inhomogeneous elastic constants have been measured using resonance ultrasound spectroscopy. Highly important information about the stiffness contrast between the γ-matrix and the γ'-precipitates [21] can be derived from carefully prepared singles-phase single-crystals [2, 12].

2 Results and Discussion

In this paper, we focus on the following two alloys from the 1st generation of nickel base superalloys: René 80 and Alloy 247 together with the braze material AMS4782. The compositions of the different alloys are given in Table 1.

Table 1. Composition of the different alloys in at. % considered in this work.

	Co	Mo	Cr	W	Ta	Al	Ti	Si	Hf	C	Ni
Alloy 247	10	0.7	8.3	10	3	5.5	1.0	–	1.4	0.1	Bal.
AMS4782	–	–	19	–	–	–	–	10	–	0.1	Bal.
René 80	9.2	2.4	15	1.2	-	6.4	5.7	–	–	0.8	Bal.

2.1 Simulation of Aging in René 80

After homogenization and quenching of René 80, a mean γ' precipitate size of 136 ± 30 nm was experimentally determined. Primary precipitates are observed with a volume fraction of $12 \pm 3\%$ [6]. This is the initial configuration for a one-dimensional simulation of the coarsening during the aging heat treatment of René 80. The initial setup consists of 2000 initial precipitates with randomly distributed sizes $4\xi < r < 4.5\xi$ (to avoid overlapping of the diffuse interfaces) at equilibrium volume fraction. The numerical discretization Δx has been chosen such that the smallest initial precipitate is still discretized by at least 4 grid points. The phase-field simulations were carried out assuming homogeneous temperature and initially homogeneous equilibrium solute concentration in the phases. The heat treatment for René 80 is illustrated in Fig. 1.

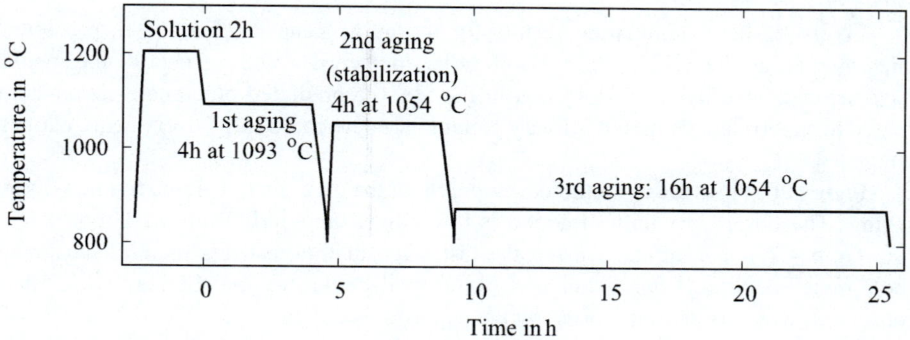

Fig. 1. Schematic illustration of the GE class A standard heat treatment for the alloy René 80.

For the René 80 the phase-field mobility M is set $M = 0.1D_{max}$, where D_{max} is the highest diffusivity amongst the alloying elements at that temperature. The nucleation of secondary precipitates leads to a decreased supersaturation of the matrix, which results in a decelerated ripening process. During the last stage of the aging heat treatment the experimentally observed volume fraction of the primary precipitates shows a significant deviation from the equilibrium content. Secondary precipitates form with a volume fraction of $10 \pm 3\%$ during cooling from 1054 °C to 873 °C [6]. Nucleation and secondary precipitates are not included in the phase-field model. We artificially decrease the phase-field mobility during the last stage of the aging heat treatment of René 80 to $M = 0.01D_{max}$ to account for the reduced precipitation kinetics due to the reduced supersaturation of the matrix through the nucleation of secondary precipitates. The γ/γ' interfacial energy density Γ used for these simulations is 4 mJ m^{-2} [27].

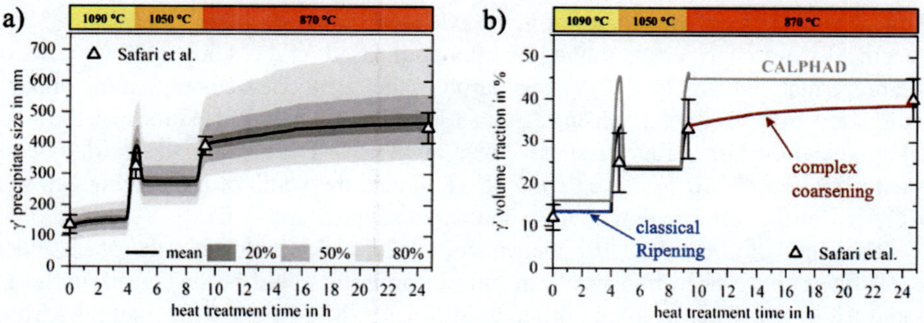

Fig. 2. Time development of the phase fraction and the mean particle radius during the simulation of the three-stage standard exposure of René 80. The mean particle size and the percentiles of $\pm 10\%$, $\pm 25\%$ and $\pm 40\%$ in relation to the mean size are indicated by gray shades.

Figure 2 shows simulation results for the alloy René 80. The mean precipitate size as well as the ±10%, ±25% and ±40% percentiles with respect to the median size are indicated by shades of gray. In Fig. 2a, the predicted precipitate size is compared to experimental sizes of primary precipitates determined after every heat treatment stage [6].

Figure 2b shows the temporal development of the precipitate volume fraction during aging. The observed volume fraction is lower than the equilibrium prediction due to the Gibbs-Thomson effect. During the last stage of the heat treatment at the lowest holding temperature a significant deviation from the equilibrium volume fraction can be observed, which is also supported by the experimental data.

Different mechanisms of precipitate coarsening can be identified over the heat treatment time. During the first stage the simulated volume fraction remains constant and the average particle radius increases over time. The increase of the mean precipitate size can be attributed exclusively to precipitate ripening, which relates to a coarsening in the sense of classic Ostwald ripening. If instead the mean particle size and the volume fraction increase simultaneously, the primary precipitates are subjected to nonequilibrium growth from a supersaturated matrix. Then, further precipitation by nucleation of secondary precipitates is also likely to happen, which is only implicitly represented in the model, by the reduction of the transformation kinetics. This is the case during the third heat treatment step at the relatively low temperature of 870 °C, where the precipitate volume fraction does not reach the CALPHAD equilibrium. Instead, it gradually changes slowly over the whole time of the heat treatment step. This is due to a comparably slow diffusion at this relatively low temperature. This indicates a complex coarsening of the particles far from the thermodynamic equilibrium.

2.2 Precipitation Within an Extended Braze Gap

Simulations of the γ' precipitation in an extended braze gap were carried out. More specifically, we consider the joining of a material based on the Alloy 247 by means of the braze material AMS4782. For this purpose, the results of a corresponding brazing simulation from Böttger et al. have been used [7, 8]. The compositions of the base material and the brazing material are shown in Table 1. The brazing was carried out at a temperature of 1210 °C for a period of 2 min, and the width of the brazing gap was 100 μm. Further information on this simulation can be found in [7, 8].

The simulation data (all 10 independent concentration fields and the phase index at all times) have been transferred in binary form and could be displayed and evaluated using the freely available program DISPLAY_MICRESS. The resulting virtual structure after the end of the soldering simulation is shown in Fig. 3a). Further, the DISPLAY_MICRESS program was used to extract the local concentrations of all 10 independent components along the horizontal center line (see green line in Fig. 3a)) in the form of an ASCII list.

Figure 3 a) Representation of the simulated microstructure after brazing [7, 8]. The coloring donates the phase. Alloy 247 was brazed with the braze material AMS4782 for the duration of 2 min at 1210 °C; The initial brazing gap width is 100 μm, as indicated by the dashed white line. b) Selected positions with their local element concentrations for subsequent γ' precipitation simulations.

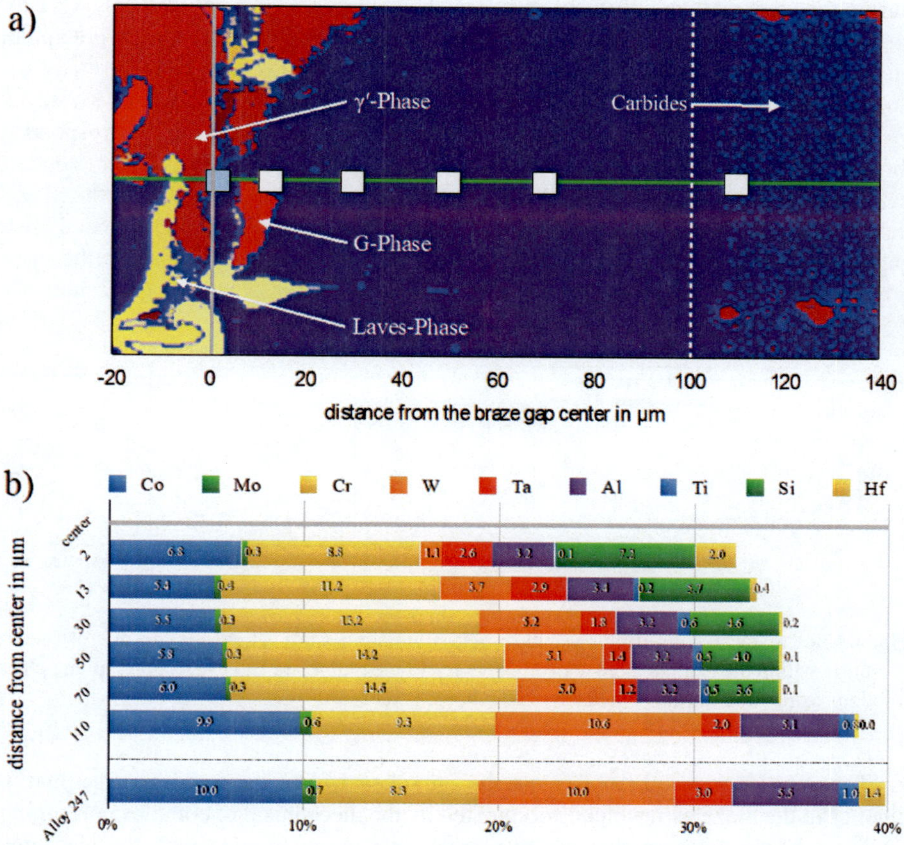

Fig. 3. a) Representation of the simulated microstructure after brazing [7, 8]. The coloring donates the phase. Alloy 247 was brazed with the braze material AMS4782 for the duration of 2 min at 1210 °C; The initial brazing gap width is 100 μm, as indicated by the dashed white line. b) Selected positions with their local element concentrations for subsequent γ' precipitation simulations.

Beside the base material, the alloy 247, 6 different locations were selected for half of the braze gap. At these locations local concentrations from the preceding brazing simulation have been extracted. With one exception, the corresponding local concentrations from the virtual EDX data of DISPLAY_MICRESS could be used directly as local alloy compositions. In the center of the braze gap, right at the position marked by the light blue square in Fig. 3 a), we find the primary γ'-phase, which has formed during the solidification of the brazing filler. Therefore, in this case, an average of the local concentration in the γ-matrix and in the adjacent γ'-phase was used as local alloy composition. For the 6 different local alloy compositions generated, the necessary, temperature-dependent thermo-kinetic input data for the phase field model were generated in a temperature range between 630 °C and 1280 °C with the help of ThermoCalc and DICTRA [12]. The input data together with an interfacial energy of $\Gamma = 10$ mJ m^{-2}, and a kinetic coefficient of $M = 0.01\ D_{\text{max}}$ is the starting point for the following microscopic precipitation simulations during the heat treatment of the brazing gap. The interface energy as well as the kinetic coefficient have been calibrated such that the simulation meets the correct precipitate phase fraction and size for the base material.

The imposed aging heat treatment of the brazing gap was chosen to have two stages: First 4 h at 1080 °C, then 8 h at 870 °C. The two aging temperatures correspond to those of the standard heat treatment for alloy 247 only the holding times differ (standard heat treatment Alloy 247: 2 h 1080 °C + 20 h 870 °C). The results from the phase-field simulation study of precipitate microstructure evolution within the extended braze gap are shown in Fig. 4. The simulation captures the expected decrease in the mean precipitate size and the local precipitate phase fraction within the braze gap center.

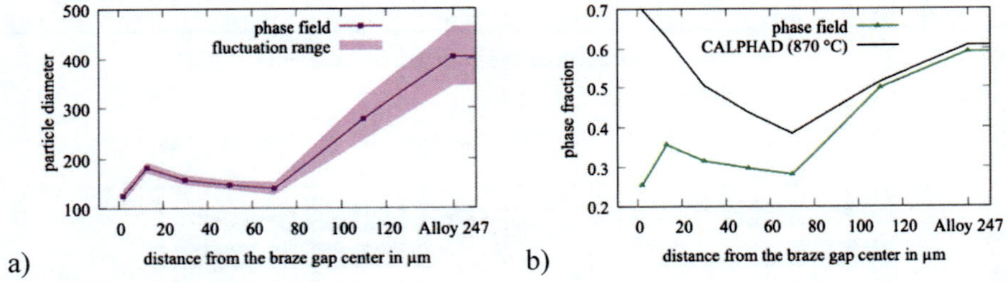

Fig. 4. Plot of the results from the phase-field simulation study of precipitate microstructure evolution within the extended braze gap. a) mean particle diameters and b) local precipitate phase fractions as function of the distance from the braze gap center.

The imposed nonisothermal two-stage aging heat treatment provides a precipitation kinetics in the braze gap, which is dominated by the mechanism of complex coarsening. In Fig. 4 b) we compare the resulting precipitate phase fraction with the respective equilibrium CALPHAD-phase fraction at 870 °C. The discrepancy between these two values indicate material stats, which are quite far away from thermodynamic equilibrium.

3 Conclusion

In this work we apply a phase-field model for the simulation of primary precipitate sizes and size distribution during the aging heat treatment of nickel-based superalloys. The model considers temperature dependent multi-component thermodynamics for non-isothermal aging including heating and cooling stages. To describe ripening in 1D, we include an artificial Gibbs-Thomson driving force [12].

For the γ'-strengthened alloy René 80, the phase-field model predicts both primary precipitate size and simultaneous evolution of the volume fraction correctly. We identify stages of the aging heat treatment during which classical ripening or complex coarsening is observed.

Finally, we discuss local variations in the γ'-precipitation kinetics within an extended braze gap. A dual-scale simulation study has been set up to predict the precipitate sizes and phase fractions as function of the distance from the braze gap center after imposing a nonisothermal two step aging heat treatment. It turns out that especially in the center of the braze gap the precipitation kinetics is dominated by the mechanism of complex coarsening, which involves simultaneous ripening and growth far from thermodynamic equilibrium.

Acknowledgements. We thank the German Federal Ministry for Economics and Energy (BMWi) for financially supporting the project COORETEC: ISar 03ET7047D and the German Federal Ministry of Education and Research (BMBF) for financial support in the project ParaPhase 01IH15005B. Further, this work is funded by the Deutsche Forschungsgemeinschaft (DFG) within the projects 387117768 and 431968427. Finally, we are very thankful to our collaboration partners at ACCESS, MTU Aero Engines AG and Siemens AG for the fruit-full discussions within the joint research project COORETEC: ISar.

References

1. Reed, R.C.: The Superalloys Fundamentals and Applications, 1st edn. Cambridge University Press, New York (2006)
2. Fleischmann, E., Konrad, C.H., Preußner, J., Völkl, R., Affeldt, E., Glatzel, U.: Influence of solid solution hardening on creep properties of single-crystal nickel-based superalloys. Metal. Mater. Trans. A **46**, 1125 (2015)
3. Holzinger, M., Schleifer, F., Glatzel, U., Fleck, M.: Phase-field modeling of γ' -precipitate shapes in nickel-base superalloys and their classification by moment invariants. Eur. Phys. J. B **92**, 208–217 (2019)
4. Völkl, R., Glatzel, U., Feller-Kniepmeier, M.: Measurement of the unconstrained misfit in the nickel-base superalloy CMSX-4 with CBED. Scripta Mater. **38**, 893–900 (1998)
5. Slama, C., Servant, C., Cizeron, G.: Aging of the Inconel 718 alloy between 500 and 750°C. J. Mater. Res. **12**, 2298–2316 (1997)
6. Safari, J., Nategh, S.: On the heat treatment of Rene-80 nickel-base superalloy. J. Mater. Process. Tech. **176**, 240–250 (2006)
7. Böttger, B., Altenfeld, R., Laschet, G., Schmitz, G.J., Stöhr, B., Burbaum, B.: An ICME process chain for diffusion brazing of alloy 247. Integr. Mater. Manufac. Innov. **7**, 70–85 (2018)

8. Böttger, B., Apel, M., Daniels, B., Dankl, L., Göhler, T., Jokisch, T.: Systematic phase-field study on microstructure formation during brazing of Mar-M247 with a Si-based AMS4782 filler. Metal. Mater. Trans. A **50**, 1732–1747 (2019)
9. Lifshitz, I.M., Slyosov, V.V.: The kinetics of precipitation from supersaturated solid solutions. J. Phys. Chem. Sol. **19**, 35 (1961)
10. Wagner, C.: Theory of the aging of precipitates by dissolution-reprecipitation (Ostwald ripening). Zeitsch. f. Elektrochem. **65**, 581 (1961)
11. Mushongera, L.T., Fleck, M., Kundin, J., Querfurth, F., Emmerich, H.: Phase-field study of anisotropic γ'-coarsening kinetics in Ni-base superalloys with varying Re and Ru contents. Adv. Eng. Mater. **17**, 1149–1157 (2015)
12. Fleck, M., Schleifer, F., Holzinger, M., Glatzel, U.: Phase-field modeling of precipitation growth and ripening during industrial heat treatments in Ni-base superalloys. Metall. Mater. Trans. A **49**, 4146–4157 (2018)
13. Thompson, M.E., Su, C.S., Voorhees, P.W.: The equilibrium shape of a misfitting precipitate. Acta Metall. Mater. **42**, 2107–2122 (1994)
14. Jokisaari, A.M., Naghavi, S.S., Wolverton, C., Voorhees, P.W., Heinonen, O.G.: Predicting the morphologies of γ' precipitates in cobalt-based superalloys. Acta Mater. **141**, 273–284 (2017)
15. Siebörger, D., Knake, H., Glatzel, U.: Temperature dependence of the elastic moduli of the nickel-base superalloy CMSX-4 and its isolated phases. Mater. Sci. Eng. A **298**, 26–33 (2001)
16. Bhadak, B., Sankarasubramanian, R., Choudhury, A.: Phase-field modeling of equilibrium precipitate shapes under the influence of coherency stresses. Metall. Mater. Trans. A **49**, 5705–5726 (2018)
17. Wang, Y., Banerjee, D., Su, C.C., Khachaturyan, A.G.: Field kinetic model and computer simulation of precipitation of L12 ordered intermetallics from FCC solid solution. Acta Mater. **46**, 2983–3001 (1998)
18. Zhu, J.Z., Wang, T., Ardell, A.J., Zhou, S.H., Liu, Z.K., Chen, L.Q.: Three-dimensional phase-field simulations of coarsening kinetics of γ' particles in binary Ni-Al alloys. Acta Mater. **52**, 2837–2845 (2004)
19. Pang, Y., Li, Y.S., Wu, X., Liu, W., Hou, Z.: Phase-field simulation of diffusion-controlled coarsening kinetics of γ' phase in Ni-Al alloy. Int. J. Mater. Res. **106**, 108–113 (2015)
20. Bhaskar, M.S.: Quantitative phase field modelling of precipitate coarsening in Ni-Al-Mo alloys. Comp. Mater. Sci. **146**, 102–111 (2018)
21. Cottura, M., Le Bouar, Y., Appolaire, B., Finel, A.: On the role of elastic inhomogeneity in the development of cuboidal microstructures in Ni-based superalloys. Acta Mater. **94**, 15–25 (2015)
22. Andersson, J.O., Helander, T., Höglund, L., Shi, P., Sundman, B.: Thermo-Calc & DICTRA, computational tools for materials science. Calphad **26**, 273–312 (2002)
23. Finel, A., Le Bouar, Y., Dabas, B., Appolaire, B., Yamada, Y., Mohri, T.: Sharp phase field method. Phys. Rev. Lett. **121**, 25501 (2018)
24. Fleck, M., Schleifer, F., Glatzel, U.: Frictionless motion of marginally resolved diffuse interfaces in phase-field modeling. https://arxiv.org/abs/1910.05180. Accessed 25 Jan 2020
25. Schleifer, F., Holzinger, M., Lin, Y.-Y., Glatzel, U., Fleck, M.: Phase-field modeling of γ/γ'' microstructure formation in Ni-based superalloys with high γ'' volume fraction. Intermetallics **120**, 106745 (2020)
26. Fleck, M., Querfurth, F., Glatzel, U.: Phase field modeling of solidification in multi-component alloys with a case study on the Inconel 718 alloy. J. Mater. Res. **32**, 4606–4615 (2017)
27. Ardell, A.J., Ozolins, V.: Trans-interface diffusion-controlled coarsening. Nature Mater. **4**, 309 (2005)

Coating

TiO_x/Cr_2O_3 Heating Coatings for Injection Molding of Polyamide

Kirsten Bobzin, Wolfgang Wietheger, Hendrik Heinemann, and Andreas Schacht(✉)

RWTH Aachen University, Surface Engineering Institute (IOT), Kackertstr. 15, 52072 Aachen, Germany
schacht@iot.rwth-aachen.de

Abstract. Variotherm injection molding processes are used to increase the quality of molded parts. Currently, tempering systems exhibit low temperature gradients or require additional process steps. A heating system applied directly on the cavity surface of the mold tool combines two benefits: high temperature gradients without additional process steps. In this study, a heating coating made of TiO_x/Cr_2O_3 was applied on the surface of a tooling steel to control the cavity temperature by Joule heating. The heating coating is electrically insulated from the substrate by an Al_2O_3 coating. Homogeneity of temperature distribution was investigated as well as thermal durability for processing polyamide in the temperature range from 70 °C to 220 °C. The coating system withstood 5,000 thermal cycles at high heating rates and without degradation.

Keywords: Functional coating · Atmospheric plasma spraying (APS) · Ceramic feedstock · Semiconductor

1 Introduction

Injection molding is the main manufacturing process in plastics production [1]. It is characterized by short process cycle times resulting in high economic efficiency. To solidify the polymer melt the mold is tempered to a temperature below the glass transition temperature of the polymer. For the commonly used polymer polyamide 6 (PA6) the mold is typically tempered at around $T_{mold} = 70$ °C, while the temperature at the injection is $T_{inj} = 220$ °C [2]. In some cases, such as thin walled parts or long flow paths for the polymer, premature solidification may result in defective parts [3, 4]. To improve the flowability of the polymer melt and its capability to fill the mold cavity, a variotherm molding process is used. The mold temperature is increased during the injection and the mold filling period and is decreased afterward. Fluid-based tempering systems consist of two circulation systems at two different temperatures. As the whole mold needs to be tempered, these systems result in a relatively low tempering dynamic [5]. A higher tempering dynamic can be achieved using induction or radiation to heat the cavity surface directly. However, those heat sources have to be inserted into the cavity of the opened mold requiring an additional process step [5]. Therefore, dynamic tempering systems are required to improve part quality without sacrificing efficiency.

U. Reisgen et al. (Eds.): EMPOrIA 2020, LNME, pp. 81–90, 2021.
https://doi.org/10.1007/978-3-030-70332-5_8

Heating coatings, which are applied directly on the mold cavity surface, are a promising option as the heat is generated in close proximity of the polymer melt. The effect of Joule heating in conducting and semiconducting materials is exploited to generate the necessary heat. When using metallic materials thin coatings as well as long conduction paths are required to achieve a sufficient heating performance due to their inherent low resistivity [6]. By utilization of the suspension spraying process, thin metallic heaters can be applied, as recently reported in literature [7]. The innately higher resistivity of semiconducting materials such as TiO_x enables remarkable heating using thicker coatings. Several studies can be found on the semiconducting properties of titanium suboxides [8–10]. While stoichiometric TiO_2 is an electrical insulator, the oxygen deficiency in substoichiometric TiO_x enables electrical conductance. Even slight changes in oxygen content may result in a change of conductivity of multiple orders of magnitude [11]. Utilizing atmospheric plasma spraying (APS) with a mixture of Ar and H2 as process gases offers reducing conditions during the spraying process to improve the conductivity of the deposited TiO_x coating [10]. Application of TiO_x as heater at elevated temperatures is challenging as re-oxidation can occur [12]. However, addition of Cr_2O_3 to the TiO_x feedstock improves the stability at elevated temperatures [8].

In semiconducting materials, the resistivity usually has a negative correlation with the temperature known as negative temperature coefficient (NTC), while conductors exhibit a positive temperature coefficient (PTC). This correlation can be described by the Arrhenius Eq. (1) with the resistivity ρ, the pre-exponential coefficient ρ_0, the activation energy E_a, the universal gas constant R and the temperature T [13].

$$\rho_{coat} = \rho_0 e^{-E_a/RT} \tag{1}$$

In previous investigations by the authors, TiO_x/Cr_2O_3 coatings were characterized for heating cycles up to a temperature of T = 150 °C [14]. In this work, however, TiO_x/Cr_2O_3 based heating coatings were investigated in regards of its potential application in injection molding of PA6, which requires an injection temperature of T = 220 °C. The temperature distribution as well as the reachable heating rate on the coating surface were characterized. Further, the NTC behavior of TiO_x/Cr_2O_3 was quantified.

2 Materials and Methods

Atmospheric plasma spraying system TriplexPro™-210 (Oerlikon Metco, Pfäffikon, Switzerland) was employed to produce the heating coating system investigated in this study. The heating coating is made of TiO_x/Cr_2O_3 and is separated from the substrates by an insulating Al_2O_3 coating. A mild steel (WERTZ GmbH, Aachen, Germany) with the dimensions of $100 \times 50 \times 5$ mm^3 was used as substrate, as it is a common steel for manufacturing of injection molds. The feedstock for the insulating coating was Al_2O_3 powder (Amdry™ 6062, Oerlikon Metco, Pfäffikon, Switzerland), which has a crushed morphology and a particle size of $-45 + 22$ μm. The experimental TiO_x/Cr_2O_3 powder (ceram, Albbruck-Birndorf, Germany) is composed of 80 wt.-% TiO_x and 20 wt.-% Cr_2O_3, which has a jointly fused and crushed morphology. Its particle fraction is $-25 + 5$ μm. The spray parameters are given in Table 1. After the deposition of the coatings, the specimens were

Table 1. Deposition parameters using TriplexPro™-210

Feedstock; fraction [μm]	Al_2O_3	TiO_x/Cr_2O_3
Fraction [μm]	−45 + 22	−25 + 5
Nozzle diameter [mm]	9	9
Process gases [SLPM]	60 Ar	60 Ar
	6 H_2	6 H_2
Current [A]	450	450
Stand-off distance [mm]	130	120
Meander width [mm]	5	5
Robot velocity [mm/s]	800	1,000
Injection nozzle diameter [mm]	2	2
Carrier gas [SLPM]	5.5 Ar	7.5 Ar
Powder feed rate [g/min]	21.1	23.6
Coating thickness [μm]	~ 100	~ 50

heat treated in atmosphere at T = 300 °C for t = 24 h to prevent a running-in phase during the first heating cycles, which was described in literature [8].

The heating function as well as the durability of the coating against cyclic thermal loads of injection mold processes was evaluated by conducting thermal cycle test in a self-developed test bench, which is shown in Fig. 1. An alternating current (AC) power supply (Chroma 61604, Datatec, Reutlingen, Germany) is controlled by a test protocol using LabVIEW (National Instruments, Texas, USA), while the actual surface temperature T_{act} and voltage U_{act} are recorded at 8 Hz and the current I_{act} at 5 Hz. An infrared sensor (CT Laser 3ML, Optris, Berlin, Germany) is used to detect T_{act}. I_{act} was recorded using an oscilloscope (MSOX3024A, Keysight Technologies, California, USA) together with a current probe and amplifier (TCP404XL and TCPA400, Tektronix, Beaverton, USA). The test protocol is designed to control the surface temperature with a PID controller. The voltage, which is supplied to the specimen during the heating time t_{heat}, is limited by the maximum target voltage $U_{max,tar}$. For the remainder of the total cycle time t_{cycle}, the voltage output is set to U = 0 V. This cycle is repeated automatically for an amount of n cycles. During the thermal cycle test, the specimen is constantly cooled on the backside by a heat exchanging plate tempered by water with a temperature of T_{cool} = 18 °C. Silver conductive paste was applied between clamped contact plates and the heating coating to reduce the contact resistance.

In addition to this test setup, thermographic recordings of the coating were made with an infrared camera (ThermaCam SC 500, FLIR, Oregon, USA) to investigate the temperature distribution on the surface. The emission coefficient ε of the specimen surface was set to ε_{sensor} = 0.48 for the infrared sensor and to ε_{camera} = 0.64 for the camera, as both infrared systems are operating at different wave lengths. The thermal cycle test parameters are given in Table 2. The target temperature is set to T_{tar} = 220 °C which corresponds to the typical injection temperature of PA6.

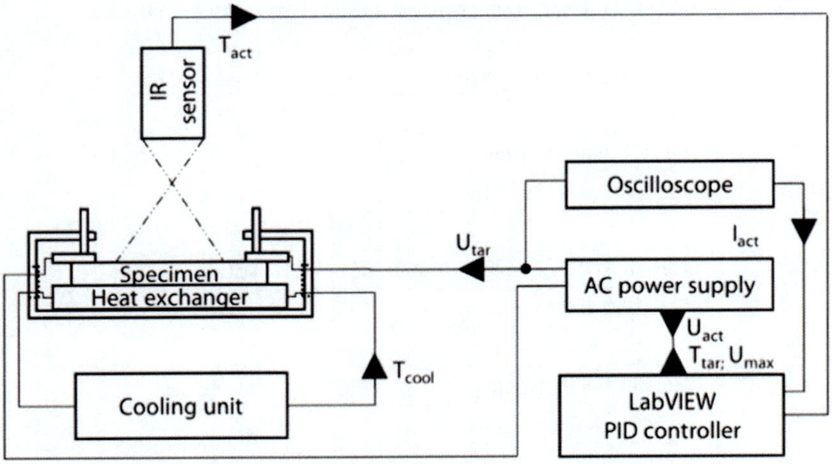

Fig. 1. Schematic of the developed thermal cycle test bench [14]

Table 2. Parameters for thermal cycle test

T_{tar} [°C]	$U_{max,tar}$ [V]	t_{cycle} [s]	t_{heat} [s]	N [-]	T_{cool} [°C]
220	95	45	14	5,000	18

As chemical changes or changes in the microstructure may occur, scanning electron microscopy (SEM) with energy-dispersive X-ray spectroscopy (EDX) were conducted before and after the thermal cycle tests. The NTC behavior of the heating coating was determined by four-terminal sensing (TM-508A, Isothermal Technology, Southport, United Kingdom) during the cooling of the specimen from 320 °C to 80 °C. This method was used before and after the thermal cycle tests to observe the coatings resistance. Additionally, the recorded values for current I and voltage U during the thermal cycle tests allow to calculate the resistance R using Ohm's law. However, this calculated resistance is the resistance of the total system, which includes contact and wire resistances.

3 Results and Discussion

The temperature distribution on the coating surface of the 10th thermal cycle as well as of the $5{,}000^{th}$ thermal cycle is represented in Fig. 2. Due to improved heat dissipation at the outer edges of the specimen the temperature in these locations is slightly lower. Both thermographic recordings show a generally homogeneous temperature distribution. Thus, the manufactured TiO_x/Cr_2O_3 heater withstood n = 5,000 thermal cycles without significant change to its heating performance.

As shown in Fig. 3, the coating temperature increases rapidly until it reaches the target value. After the heating period, a significant temperature drop initiates the cooling period. As expected the cooling is not linear, because the cooling rate decreases along

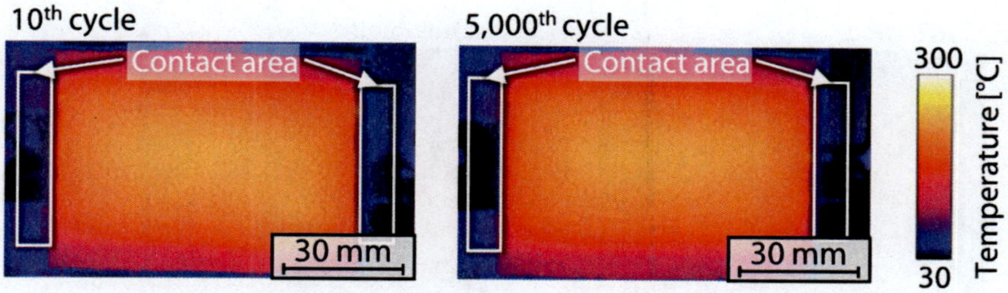

Fig. 2. Comparison of the temperature distribution on the coating surface during the heating period

the temperature difference between T_{act} and T_{cool}. Throughout the thermal cycle test quick heating from 70 °C to 220 °C was observed. However, the initial heating rate of over 13 K/s gradually dropped to 11 K/s over the course of 5,000 cycles. Nevertheless, the target temperature could be achieved during the set heating time.

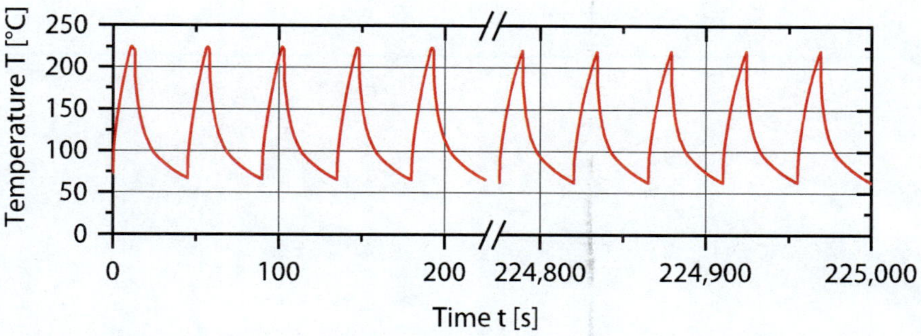

Fig. 3. Temperature curve at the beginning and the end of the thermal cycle test

To investigate the operational limitations of the heater system heating rates above 25 K/s were achieved using a higher voltage of $U = 110$ V compared to the previously used $U_{max,tar} = 95$ V. Paired with the rather slow PID controller, the actual temperature of the coating overshot the target value resulting in a maximum temperature of $T_{max} = 240$ °C. This temperature is beyond the range of application for the conductive paste in between the heating coating and the electrical contacts. In the thermographic recordings shown in Fig. 4 local hotspots at the contact areas were detected after just a few hundred thermal cycles. After continuous thermal cycles testing, the locally increased temperature leads to a complete failure of the conductive paste reducing the effective contact area. This contacting design was used as an auxiliary solution for laboratory tests only. As the heating coating itself remains undamaged, the development of a practicable contacting design is necessary in the future.

SEM images of the coating surface were taken before and after the heating cycles to detect any changes in its structure. As depicted in Fig. 5, micro cracks formed within the splats. Considering the homogeneous temperature distribution shown in Fig. 2,

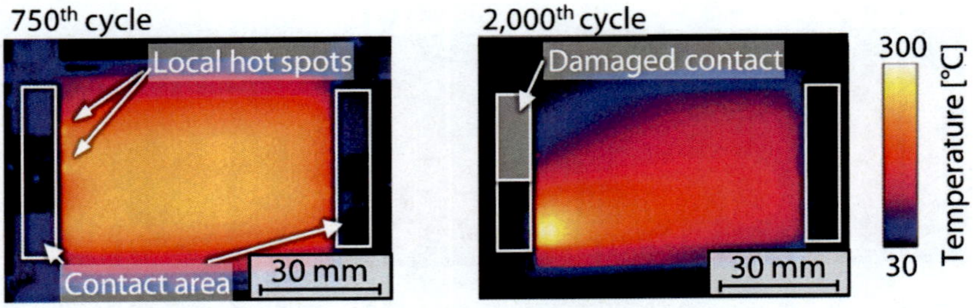

Fig. 4. Thermographic recordings of malfunctioning electrical contacts due to degenerated conductive paste at increased voltages

it is assumed that these additional micro cracks do not affect the coating's heating performance.

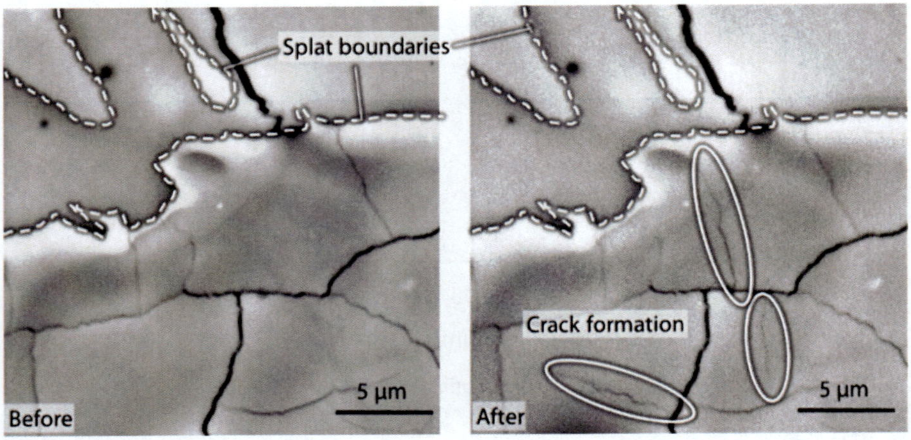

Fig. 5. Top view of the heating coating by SEM before (left) and after (right) the thermal cycle test

Furthermore, the chemical composition determined by EDX of the powder feedstock and of the heating coating before and after the thermal cycle test are listed in Table 3. The chemical composition on the surface of the specimen remained constant showing no measureable re-oxidation of the heating coating. Comparing the Cr concentration in the powder feedstock and the deposited coating a significant reduction is observed. This may be caused by a lower deposition efficiency of Cr_2O_3 than TiO_x. This results in a weight composition closer to 90/10 for TiO_x/Cr_2O_3 than the initial weight composition of the powder feedstock of 80/20.

As re-oxidation of TiO_x results in an increase of its resistivity, the resistivity of the heating coating was determined over the temperature range of 80 °C to 320 °C (353.15 K to 593.15 K) before and after the 5,000th conducted thermal cycle. The resistivity was found to be comparable and showed no significant difference as seen

Table 3. Chemical composition of the powder feedstock material and the heating coating determined by EDX in wt.-%

	Ti	Cr	O
TiO_x/Cr_2O_3 powder	46.92	10.56	42.52
Before thermal cycles	48.45	5.92	45.63
After thermal cycles	48.18	5.88	45.94

in Fig. 6. The plot fitted using the Arrhenius Eq. (1) to describe the NTC of the resistivity complies with the measured data with an R^2 of 0.99. The calculated value of the activation energy $|E_a| = 0.064$ eV for the investigated TiO_x/Cr_2O_3 coating is within the range of 0.01 eV to 1.67 eV, which is reported for titanium suboxides in literature [15]. The pre-exponential coefficient ρ_0 increased from $\rho_{0,1} = 38.6\ \Omega\ \mu m$ before the thermal cycle test to $\rho_{0,2} = 38.8\ \Omega\ \mu m$ after the thermal cycle test.

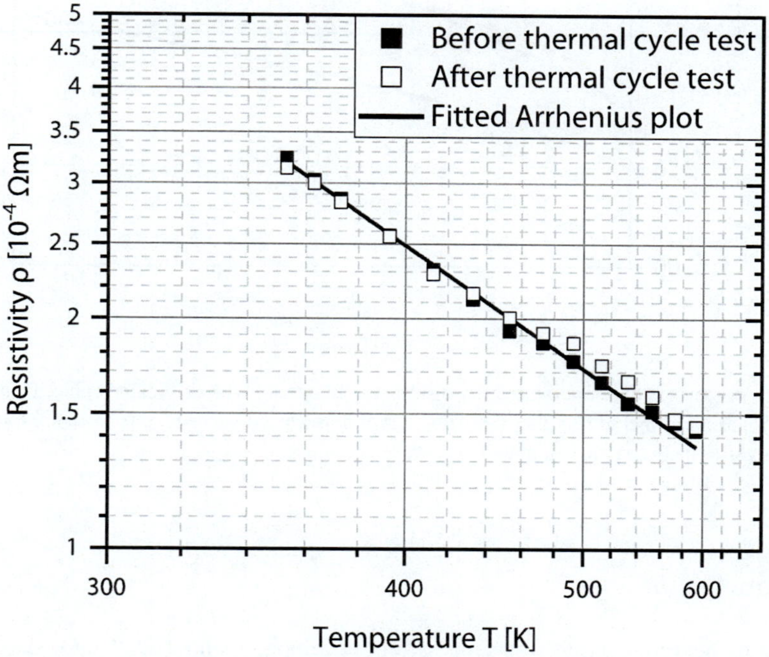

Fig. 6. Resistivity (logarithmic) of the heating coating over temperature (reciprocal). Values were measured by four-terminal sensing during cooling before and after conducting the thermal cycle test.

In Fig. 7 the coating temperature recorded by the IR-sensor is given in red. Furthermore, the calculated temperature of the heating coating corresponding to its resistivity according to the Arrhenius Eq. (1) is given in black. As the voltage and current recordings include the contact resistances, a direct calculation leads to the visible temperature offset. Considering the resistivity of the heating coating, which was determined by four-terminal sensing, the contact resistance was calculated to a value of $R_{contact} = 1.8\ \Omega$ and subtracted from the recorded total resistance. This results in the adjusted blue curve given in the diagram, which complies well with the recorded coating temperature by the IR-sensor. The correlation between the resistance and the temperature of the heating coating enables its use as a control element at the same time. This is of great interest, especially in view of the fact that such a coating shall be used in a mold where there is only limited possibility to use a temperature sensor.

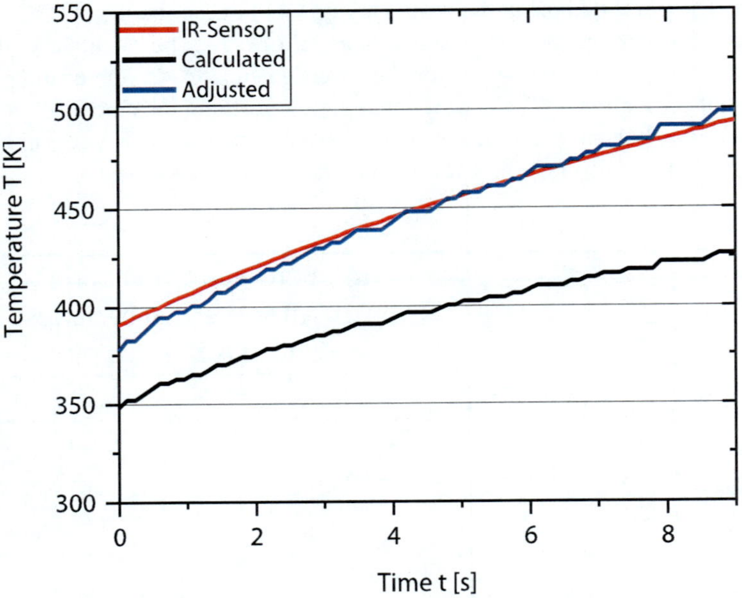

Fig. 7. Comparison of temperature measured by the IR-sensor with calculated values from voltage and current recordings during thermal cycle test. The adjusted curve results from subtracting the contact resistance from the calculated system resistance.

4 Conclusion

A ceramic heating coating was successfully developed using TiO_x/Cr_2O_3 based feed-stock material in the plasma spraying process. The coating's potential for mold tempering in injection molding was demonstrated with thermal cycles emulating the process conditions for polyamide fabrication. The main conclusions of this study are:

- The heating coating is durable for at least 5,000 thermal cycles between 70 °C and 220 °C.
- Rapid heating rates over 25 K/s were achieved in the temperature range from 70 °C to 220 °C.

- A homogeneous temperature distribution on the specimen surface was observed throughout the thermal cycles test.
- The resistivity as well as the chemical composition of the heating coating show no change during the heating cycles.
- The feasibility of controlling the heating function of the coating using the temperature-dependent resistivity of TiO_x/Cr_2O_3 was demonstrated.
- Electrical contacts with conductive paste are insufficient for a long-term application at high temperatures.

In future work, the design of the electrical contacts has to be improved. Electrical contacts with conductive paste were shown to be the limiting factor in achieving rapid heating over significant amounts of thermal cycles. Additional research has to be conducted to investigate on the limitations and further potential of heating coatings.

Acknowledgement. The presented investigations were carried out at RWTH Aachen University within the framework of the Collaborative Research Centre SFB1120-236616214 "Precision Melt Engineering" and funded by the Deutsche Forschungsgemeinschaft e.V. (DFG, German Research Foundation). The sponsorship and support is gratefully acknowledged.

References

1. Bonten, C.: Kunststofftechnik: Einführung und Grundlagen (Plastics Technology: Introduction and Basics). Einführung und Grundlagen, 1st edn. Carl Hanser Fachbuchverlag, München (2014)
2. Johannaber, F., Michaeli, W.: Handbuch Spritzgießen. Carl Hanser Verlag GmbH & Co. KG, München (2004)
3. Moayyedian, M., Abhary, K., Marian, R.: Gate design and filling process analysis of the cavity in injection molding process. Adv. Manuf. (2016). https://doi.org/10.1007/s40436-016-0138-5
4. Tredoux, L., Satoh, I.: Investigation of wave-like flow marks in injection molding: flow visualization and micro-geometry. Polymer Eng. Sci. **39**, 2233–2241 (1999)
5. Giessauf, J., Pillwein, G., Steinbichler, G.: Werkzeugtemperierung: die variotherme temperierung wird produktionstauglich (mold temperature control: variotherm temperature control is fit for production). Kunststoffe **98**, 87–92 (2008)
6. Prudenziati, M.: Development and the Implementation of High-Temperature Reliable Heaters in Plasma Spray Technology. J. Therm. Spray Technol. (2008). https://doi.org/10.1007/s11666-008-9164-6
7. Luth, J., Winkelmann, M., Wüst, F., Hartmann, S., Trenkle, F., Krieg, P., Killinger, A.: Thin, metallic heating coatings, manufactured by means of suspension spraying. Thermal Spray Bull. **12**, 72–76 (2019)
8. Scheitz, S., Toma, F.-L., Berger, L.-M., Puschmann, R., Sauchuck, V., Kusnezoff, M.: Thermally sprayed multilayer ceramic heating elements. Thermal Spray Bull. **11**, 88–92 (2011)
9. Berger, L.-M., Toma, F.-L., Scheitz, S., Trache, R., Börner, T.: Thermisch gespritzte Schichten im System Al_2O_3-Cr_2O_3-TiO_2 - ein Update. Mat.-wiss. u. Werkstofftech (2014). https://doi.org/10.1002/mawe.201400260

10. Trache, R., Berger, L.-M., Toma, F.-L., Saaro, S., Lima, R., Marple, B.: Electrical resistivity of thermally sprayed Cr_2O_3-TiO_2 coatings. In: Conference Proceedings: Int. Thermal Spray Conference, 27–29 September 2011, Hamburg (2011)
11. Hayfield, P.C.S.: Electrode material, electrode and electrochemical cell Patent US4422917, 27 December 1983
12. Berger, L.-M., Stahr, C.C., Saaro, S., Thiele, S., Woydt, M., Kelling, N.: Dry sliding up to 7.5 m/s and 800 °C of thermally sprayed coatings of the TiO_2–Cr_2O_3 system and (Ti,Mo)(C,N)–Ni(Co). Wear (2009). https://doi.org/https://doi.org/10.1016/j.wear.2008.12.105
13. Wang, C.C., Akbar, S.A., Madou, M.J.: Ceramic Based Resistive Sensors. J. Electroceram. **2**, 273–282 (1998)
14. Bobzin, K., Wietheger, W., Knoch, M.A., Schacht, A.: Heating behaviour of plasma sprayed TiOx/Cr_2O_3 coatings for injection moulding. Surf. Coat. Technol. (2020). https://doi.org/10.1016/j.surfcoat.2020.126199
15. Liu, Y., West, A.R.: Semiconductor-Insulator Transition in Undoped Rutile, TiO_2 Ceramics. J. Am. Ceram. Soc. (2013). https://doi.org/10.1111/jace.12025

Simulation of Multiple Particle Impacts in Plasma Spraying

Kirsten Bobzin, Wolfgang Wietheger, Hendrik Heinemann, and Ilkin Alkhasli(✉)

Surface Engineering Institute at the RWTH Aachen University, Kackertstr. 15, 52072 Aachen, Germany
alkhasli@iot.rwth-aachen.de

Abstract. Plasma spraying is a versatile coating process, which allows processing of nearly any feedstock material. The feedstock material is injected in powder form into the plasma jet, where it is heated and accelerated towards the substrate. The impacting particles form a coating on the substrate surface by deforming and rapidly solidifying. The resulting coating microstructure strongly depends on the dynamics of the particle deformation and solidification. The simulation of the particle impact on the substrate surface allows a better understanding of these phenomena. In this work, a modified momentum source based solidification model, which is suitable for the simulation of multiple particle impacts, is developed. It represents a more robust alternative to the existing momentum source formulation and the viscosity based solidification model commonly used in the literature. The developed solidification model is implemented in a three-dimensional coating build-up simulation to acquire the coating microstructure.

Keywords: Modelling · Particle impact · Solidification · Plasma spraying

1 Introduction

Plasma spraying is a coating process, which is a substantial process variant of thermal spraying. In thermal spraying, the feedstock material is heated and accelerated towards the substrate, thus forming a coating upon impact. In plasma spraying the feedstock material, usually in the form of powder particles, is injected into a plasma jet, the temperatures and velocities of which can exceed $T_p > 12{,}000$ K and $v_p > 1{,}000$ m/s respectively [1]. Due to the high plasma temperatures, nearly any material can be brought to the liquid phase and impacted on the substrate material at relatively high velocities [2]. The liquid particles deform, cool down rapidly and solidify on the substrate surface during their impact, thereby building up a coating (Fig. 1). Measured and calculated cooling rates of the ceramic particles during the impact are reported in the literature to be in the range of $\dot{q} = 10^7$ to $\dot{q} = 10^{10}$ K/s [3–5]. Since the dynamics of particle deformation and solidification control the resulting coating microstructure and hence the properties of the final coating, it is important to model the particle impact to increase the understanding of this process. Furthermore, the experimental observation of the particle deformation process poses challenges due to the short time frames of a few microseconds [6]. Due

U. Reisgen et al. (Eds.): EMPOrIA 2020, LNME, pp. 91–100, 2021.
https://doi.org/10.1007/978-3-030-70332-5_9

to these reasons, several works can be found in the literature that numerically analyze the particle impact and splat formation [7–9]. While the simulation of a single particle impact improves the understanding of the dynamics of particle deformation and solidification, simulating the impact of multiple particles would allow in the future to determine the coating properties such as porosity and effective thermal conductivity.

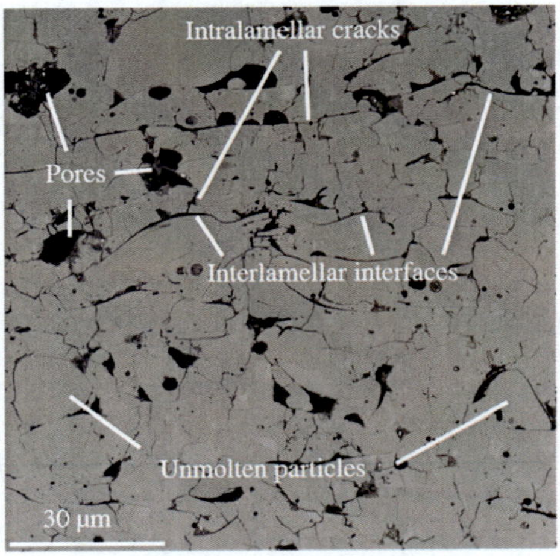

Fig. 1. A cross-section picture of a typical Al_2O_3 coating sprayed with atmospheric plasma spraying

The focus of this work is to present a multiple particle impact model where individual particles can be identified. This gives an unprecedented possibility to track the temperature evolution as well as their cooling rates during and after their impact. To the knowledge of the authors, this has not been demonstrated in the literature for computational fluid dynamics based multiple particle impact simulations. A brief overview of the numerical modelling used to simulate the particle impact and solidification will be given in the next chapter. It will be followed by a previously performed benchmark simulation of an impact of a single nickel particle to verify the model against the reference simulations in the literature [10].

This previous model was extended for this work to simulated the impact of 20 particles and the material was changed to alumina. The results of these simulations will be presented. Furthermore, the temperature evolution as well as the cooling rates will be shown for the individual Al_2O_3 particles. Hereby a better understanding of particle impact in atmospheric plasma spraying will be achieved, which in turn will enable process improvements in the long term.

2 Numerical Modelling

A common way to model the particle impact in thermal spraying is the computational fluid dynamics approach, which allows to resolve the deformation and splashing of liquid particles. To distinguish between the liquid particle phase and the gaseous ambient phase, a multiphase modelling approach Volume of Fluid (VOF) is used. VOF is a region following scheme, which enables the modelling of two or more immiscible fluids. This method stores the volume fraction information of each fluid phase in the cells of computational domain. Since the computational fluid dynamics approach inherently assumes the particles to be fluids, additional numerical modelling is needed to take the particle solidification and solidified splats into account. Commonly used modelling approaches in the literature are the modified viscosity method [11] and enthalpy porosity method [12].

In the modified viscosity method the solidification is modelled by assigning a temperature dependent viscosity profile to the particle material which rapidly reaches large values as the temperature decreases to the solidification temperature. A major disadvantage of this method is the numerical requirement of the time step to decrease inversely proportional to the increasing viscosity in order for the solution to converge. This, in turn, requires much larger numbers of time steps for the simulation to complete, which results in longer and often critical computation times. Contrary to the modified viscosity method, the enthalpy porosity method does not require a decrease of the time steps for the solution to converge when the particle solidification occurs and is therefore computationally much easier. Due to this advantage, the enthalpy porosity method was used to model the particle impact in this work.

Since in the enthalpy porosity method, the material viscosity is not increased artificially to imitate a solidified material, its viscosity at temperatures below solidification temperatures does not affect the kinematics of the particle material. Therefore, the kinematics of the solidified material need to be controlled by an additional momentum source variable, which is included in the conservation of momentum equation in the numerical solver. In this work, the commercial computation fluid dynamics software ANSYS Fluent 19 (ANSYS. Inc. Canonsburg, United States of America) was used to model and simulate the particle impact. This software allows for the user to manipulate the momentum source with the help of a user defined function and thus to artificially accelerate or decelerate the particle depending on a flow condition. To model the solidification, the user defined momentum source was defined in such a way, that below the solidification temperature, the particle material is rapidly decelerated to imitate the inertia of a solidified material. The added momentum source term (S) can be found in Eq. (1).

$$S = -C \cdot VF_p \cdot K(\rho) \cdot v \tag{1}$$

The magnitude of the deceleration force acting on the particle material is proportional to its instantaneous velocity v and the particle volume fraction VF_p. The numerical constant C and the density dependent parameter $K(\rho)$ control this deceleration as explained in [10]. S acts on the particle material starting the moment its temperature drops below the solidification temperature, gradually diminishes as the particle slows down and virtually fades away as the particle comes to a halt in its solidified state.

Although Ansys Fluent already comes with a solidification model based on the enthalpy porosity method which could be directly used out of the box, it turned out to be inapplicable for the simulation of multiple particle impacts due to a numerical artifact. The numerical artifact becomes evident when a liquid particle comes in contact with an already solidified particle. To avoid this artifact while simulating the impact of more than one particle, the user defined momentum source was defined as described above. More detailed information about the enthalpy porosity method of Ansys Fluent, its numerical artifact concerning multiple particle impact and the momentum source formulation defined as a user defined function to remedy this issue can be found in the previous work of authors [10].

2.1 Model Validation

As mentioned earlier, experimental verification on the particle impact model is challenging due to the short time scales at which the deformation and solidification occur in thermal spraying. Due to this, the developed particle impact model was verified against a reference particle impact model found in the literature [13]. Additionally, the authors of the aforementioned study have experimentally validated their model for the impact of nickel particles. Thus, comparing the simulation results of the model developed in this work to the simulation results of an experimentally validated model found in literature serves as an indirect experimental validation. To achieve this, a benchmark simulation was built with particle parameters corresponding to those given in the reference simulation. A more detailed description of the simulation set-up as well as the material properties for the nickel particles can be found in the previous work [10]. Figure 2 gives main characteristics of the simulation set-up while visually comparing the two results. Based on the visual attributes of the formed splats as well as the flattening degrees, a good correspondence between the developed and the reference models can be concluded.

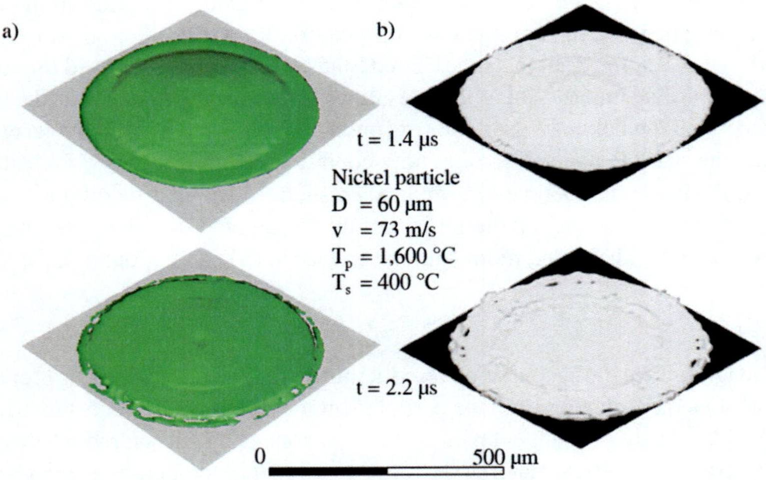

Fig. 2. Simulation results of an impact of a single nickel particle. a) model developed in this work, b) experimentally verified model found in literature [13]. Here, D corresponds to particle diameter, v to particle velocity prior to the impact, T_p to particle temperature and T_s to substrate temperature. The image originates from [10] and is used under CC BY 3.0, slight repositioning of the text has been performed.

2.2 Simulation Setup

After the particle impact model was verified, it was utilized to simulate the impact of 20 Al_2O_3 particles. The parameters defining the particle impact are summarized in Table 1. The particle size is within the range of commonly used particle size distribution for the feedstock material used in atmospheric plasma spraying. The particle velocity and particle temperature correspond to the characteristic particle in-flight parameters occurring during spraying of Al_2O_3 particles at a stand-off distances typical for coating deposition. The time interval between the consecutive particle impacts is defined by the impact time interval. An uncharacteristically short impact time interval of $\Delta t = 5\ \mu s$ was chosen for this set-up to be able to simulate more particle impacts during the total simulation time of $t = 100\ \mu s$ than it would have been possible under realistic coating conditions. The values of the particle size, velocity, temperature as well as the impact time interval correspond to the mean values. The actual values of the aforementioned parameters were randomly varied based on the normal distribution with relatively tight standard deviations, which is given in Table 1 as SD. The particle impact locations on the substrate were generated randomly. Since the substrate temperature was held constant during the simulation without being influenced by the impacting hot particles, the material of substrate was not taken into account and therefore is not given in the table.

Table 1. Simulation parameters for the 3-D multiple particle impacts

Parameter	Value
Particle material	Al_2O_3
Number of particles	20
Mean particle size, SD	30 μm, 2μm
Mean particle velocity, SD	200 m/s, 3 m/s
Mean particle temperature, SD	3,000 °C, 10 °C
Mean impact time interval, SD	5 μs, 0.5 μs
Substrate temperature	20 °C

Al_2O_3 is a ceramic feedstock material commonly used in atmospheric plasma spraying. Its material properties are shown in Table 2, which were taken from Burcat and Ruscic [14]. The solid heat capacity is function of the material temperature and therefore is given as a temperature range.

Table 2. Material properties of Al_2O_3 [14]

Property	Value
Melting temperature	2,327 K
Latent heat of melting	1,060 J g^{-1}
Solid material density	3.95 g cm^{-3}
Liquid material density	2.7 g cm^{-3}
Solid heat capacity	0.6–0.9 J g^{-1} K^{-1}
Liquid heat capacity	1.95 J g^{-1} K^{-1}
Thermal conductivity	6 W m^{-1} K^{-1}

3 Results and Discussion

Due to the nature of fluid dynamics simulation, identifying and tracking the individual particles poses a challenge once they come into contact with each other. As the particles contact each other, they merge into the unified block of the same particle phase with no distinguishing feature that would allow to differentiate one from another and to recognize the borders between the particles. This is characteristic to all particle impact modelling works found in the literature that use the computational fluid dynamics approach. Overcoming this inherent limitation would allow to track the temperature evolution and the cooling rates of individual particles. In the model developed in this work, an additional variable that holds the identity information for each particle was introduced into the solution process. Each particle is assigned a unique identifier as it emerges into the computational domain and the identifier is incremented for the next particle. In this manner, the particles are basically counted in the order of their emergence into the simulation.

Figure 3 illustrates an isometric view of the simulation results after all 20 particles have impacted on the substrate and solidified. When only distinguishing between the gas and particle phases, all of the solidified particles seem to be consisting of a single block of material. While this block contains pores, the individual boundaries between singles particles cannot be distinguished. When visualizing the particles based on their unique identifier, the individual particles can be clearly differentiated.

For the formation of different phases during the solidification of particles, the temperature profile and its corresponding cooling rate are of great interest. The presented model is capable of calculating the temperature evolution of the individual particles as illustrated in Fig. 4. Here it can be seen that each particle emerges into the simulation domain at the temperature of 3,000 K, maintains this temperature for a short period of time during its flight towards the substrate and rapidly cools down upon the impact. The cooling down of the particles continues even after solidification has taken place. It is worth noting that the early particles that make direct contact with the substrate surface seem to cool down faster than the particles impacting on the already solidified particles. Another important aspect is that the particle temperatures are not strictly monotonically decreasing. The temperature of the solidified particles show significant spikes as newly emerged, hot particles impact on them. There temperature spikes can also be observed

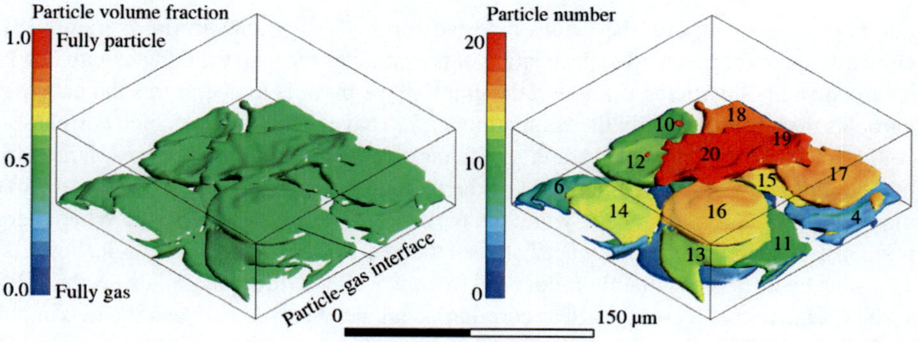

Fig. 3. Result of the impact simulation with 20 Al_2O_3 particles. Comparison of visualizing the solidified particles based on the particle volume fraction versus visualizing them based on their identifier.

for the particles which do not directly contact the new and hot particles. In this case, the spikes arise with a delay, resulting from the heat transfer through the surrounding material.

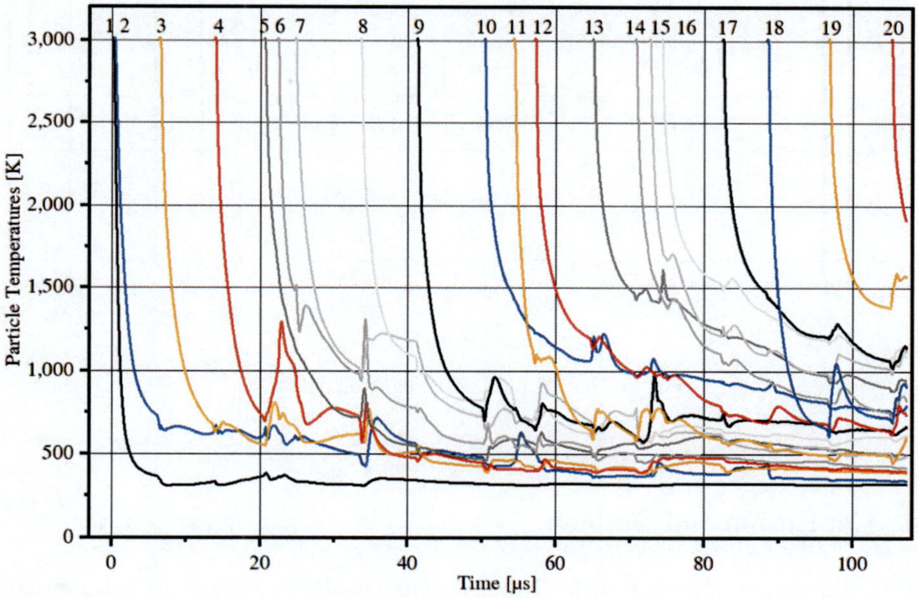

Fig. 4. Temperature evolution of individual particles numbered based on their identifier

A better insight into the cooling rates of the particles can be gained by differentiating the particle temperature with respect to time. Particle heating rates for the individual particles, that were obtained in this manner, are shown in Fig. 5. Here, the pronounced downward spikes in the particle heating correspond to the initial rapid cooling of the particles at the moment of their impact. The overall magnitude of the particle cooling rate is in agreement with the values reported for Al_2O_3 particles in the literature [5]. Upwards spikes correspond to the heating of the particles through the contact with newly emerged and hot particles. Although the graph of the particle temperatures indicate that particles that emerge later in the simulation tend to have gentler slopes, such correlation could not be observed on the graph of the particle heating. This can be explained by the fact that the maximum magnitude of the particle heating rate occurs during the first moments of the particle impact, while the particle temperature is still high. When there is still a high temperature difference between the new and the solidified particle, how fast a particle cools down is mainly influenced by how well the liquid particle wets the solid surface. Due to this reason, no clear correlation can be observed between the maximum cooling rates of the particles and their time of impact.

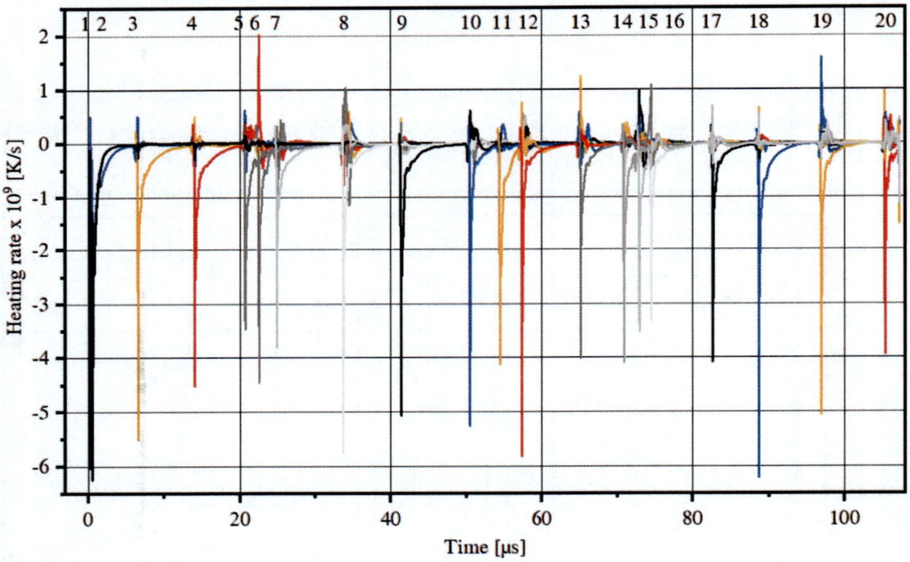

Fig. 5. Heating rates of the individual particles

4 Conclusion and Outlook

This work presented a computational fluid dynamics model for multiple particle impacts. The multiphase formulation to distinguish the gas phase from the particle phase was done using the volume of fluid method. Solidification was modelled by modifying momentum source with the help of user defined routines, which is adapted from the enthalpy porosity method to suit the multiple particle impact. An additional variable was assigned to each particle at the time of its emergence to identify the individual particles after the impact. With the help of this identification method, it was for the first time possible to track the temperature evolution as well as the cooling rates of individual particles separately in a multiple particle impact simulation.

Identifying individual particles opens the possibility to identify the interfaces between the particles, as can be seen in Fig. 6. Interlamellar interfaces in real thermally sprayed coatings are much thinner than the splats, yet they influence the effective thermal conductivity of the coatings. Directly resolving the interlamellar interfaces in the particle impact model is challenging because the grid refinement level would be too high for practical applications. With the help of interface identification it would be possible to model the properties of the interlamellar interfaces instead of directly resolving them with grid refinement. This opens the possibility to simulate the influence of different particle in-flight properties on the resulting coating and its properties.

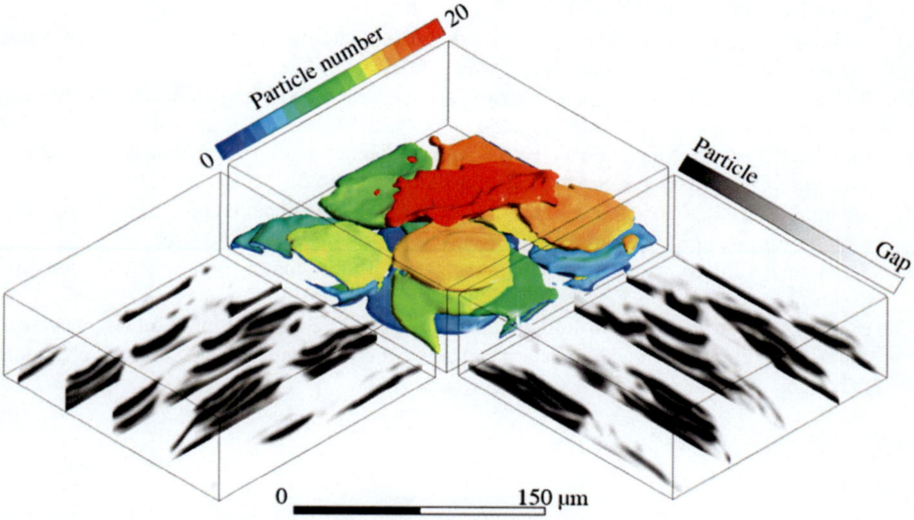

Fig. 6. Virtual cross-sections of the simulated coating build-up, where particles as well as the interlamellar gaps can be identified

Acknowledgments. The presented investigations were carried out at RWTH Aachen University within the framework of the Collaborative Research Centre SFB1120–236616214 "Bauteilpräzision durch Beherrschung von Schmelze und Erstarrung in Produktionsprozessen" and funded by the Deutsche Forschungsgemeinschaft e.V. (DFG, German Research Foundation). The sponsorship and support is gratefully acknowledged.

References

1. Bobzin, K., et al.: A numerical investigation: influence of the operating gas on the flow characteristics of a three-cathode air plasma spraying system. In: Thermal Spray 2013: Proceedings of International Thermal Spray Conference, pp. 400–405 (2013)
2. Heimann, R.B.: Plasma Spray Coating. Principles and Applications, Wiley-VCH, Weinheim (2008)

3. Vardelle, A., et al.: A Perspective on plasma spray technology. Plasma Chem. Plasma Process **35**(3), 491–509 (2015)
4. Bianchi, L., et al.: Microstructural investigation of plasma-sprayed ceramic splats. Thin Solid Films **299**(1–2), 125–135 (1997)
5. Li, L., et al.: Suppression of crystallization during high velocity impact quenching of alumina droplets: observations and characterization. Mater. Sci. Eng. A **456**(1–2), 35–42 (2007)
6. Goutier, S., Vardelle, M., Fauchais, P.: Last developments in diagnostics to follow splats formation during plasma spraying. J. Phys.: Conf. Ser. **275**, 12003 (2011)
7. Vardelle, M., et al.: Influence of particle parameters at impact on splat formation and solidification in plasma spraying processes. J. Therm. Spray Technol. **4**(1), 50–58 (1995)
8. Chandra, S., Fauchais, P.: Formation of solid splats during thermal spray deposition. J. Therm. Spray Tech. **18**(2), 148–180 (2009)
9. Ghafouri-Azar, R., et al.: A stochastic model to simulate the formation of a thermal spray coating. J. Therm. Spray Tech. **12**(1), 53–69 (2003)
10. Bobzin, K., et al.: Modelling of particle impact using modified momentum source method in thermal spraying. J. Phys.: Conf. Ser. **480**, 12003 (2019)
11. Bobzin, K., et al.: Simulation of PYSZ particle impact and solidification in atmospheric plasma spraying coating process. Surf. Coat. Technol. **204**(8), 1211–1215 (2010)
12. Zheng, Y.Z., et al.: Modeling the impact, flattening and solidification of a molten droplet on a solid substrate during plasma spraying. Appl. Surf. Sci. **317**, 526–533 (2014)
13. Pasandideh-Fard, M., et al.: Splat shapes in a thermal spray coating process: simulations and experiments. J. Therm. Spray Technol. **11**(2), 206–217 (2002)
14. Burcat, A., Ruscic, B., Third millennium ideal gas and condensed phase thermochemical database for combustion with updates from active thermochemical tables. Argonne National Laboratory, Lemont (2005)

Simplex Space-Time Meshes for Droplet Impact Dynamics

Violeta Karyofylli(✉) and Marek Behr

Chair for Computational Analysis of Technical Systems, RWTH Aachen University, 52056 Aachen, Germany
{karyofylli,behr}@cats.rwth-aachen.de

Abstract. Droplet impact dynamics is an example of a complex two-phase flow. However, we manage to reduce the complexity of this problem by reducing its dimensionality. This paper highlights an axisymmetric interface-capturing method [1, 2]. The level-set method is used for modeling the evolution of the front, because of its inherent ability to account for large topological changes of the interface [3], and is combined with a continuum surface force (CSF) model. We use a space-time finite element discretization, which is achieved by means of simplex space-time elements and leads to entirely unstructured grids with varying levels of refinement both in space and in time. Therefore, despite the complexity of the contact dynamics, the efficiency of our simulations is improved when using adaptive temporal refinement in areas of interest. Two different benchmark cases from [4] are used for verifying our numerical approach.

Keywords: Space-time · Simplex · Axisymmetric · Two-phase flows · Contact lines

1 Introduction

Problems involving moving contact lines are common in many processes of production engineering, such as continuous liquid film coating, soldering and brazing, flow in porous media, and other critical technological areas [5]. The way that liquid droplets impact and spread is an ongoing research area, since it is encountered in several industrial applications, and the improvement of the manufacturing processes (e.g., ink-jet printing or organic light-emitting diode fabrication) depends on the knowledge of drop impingement [6]. The problem of the droplet impacting belongs to multi-phase flows and is complicated since numerous physical phenomena take place at varying timescales. When a droplet touches a surface, it starts spreading until it reaches a maximum width. Finally, the dominant capillary effects act on the droplet and force it to either stick to or rebound from the surface [7]. According to [6], the most important parameters which influence the impact of droplets are the impact velocity, the surface tension, and the surface roughness. The way a droplet spreads onto a solid can be classified into four phases, namely the kinematic, spreading, relaxation, and wetting phase. In Fig. 1, experimental data is presented, which shows the impact of a drop of water on a solid when different combinations of impact velocity and equilibrium contact angle are defined.

U. Reisgen et al. (Eds.): EMPOrIA 2020, LNME, pp. 101–109, 2021.
https://doi.org/10.1007/978-3-030-70332-5_10

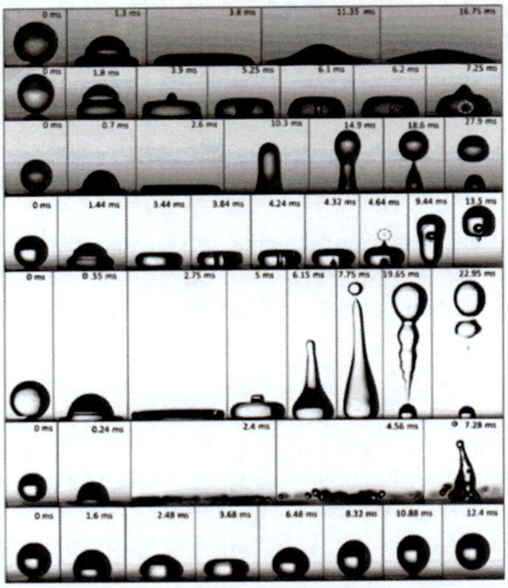

Fig. 1. Water droplet impacting onto either hydrophilic or hydrophobic surfaces at various time instances. Every row represents a different set of impact velocity and equilibrium contact angle. Dashed circles indicate air bubbles entrapment or droplet emission. The figure was adapted from [7].

In the current paper, we use an unstructured space-time finite element method for solving two-phase incompressible flows, including moving contact lines. During the last two decades, the growing interest in 4D simplex space-time discretizations is evident. The fourth dimension is time. In the work of Karyofylli et al. [8], a 4D simplex space-time finite element discretization was exploited in the context of mold filling since simplex space-time meshes allow for temporal refinement close to the evolving front. Therefore, the accuracy of the computations is locally increased at a reduced computational cost. The compressible Navier-Stokes equations were discretized by von Danwitz et al. [9] with unstructured space-time finite element meshes, and the compressible flow was simulated in a valve that fully closes and opens again. Gesenhues [10] computed geophysical flows and their rheology with a simplex space-time finite element solver while employing continuous rheology models. Moreover, Gaburro [11] used a space-time formulation combined with the finite volume method. Zwart [12] initially developed an integrated space-time finite volume method for transient fluid flows. This method satisfies the conservation law in space-time.

2 Governing Equations of Axisymmetric Two-Phase Flows

The computational domain Ω in the case of axisymmetric two-phase flow problems is a subset of $\mathbb{R}^2$. It contains two immiscible Newtonian fluids depicted by the subdomains Ω_1 and Ω_2, where $\Omega_1 \cup \Omega_2 = \Omega$. Here, $\Gamma = \partial\Omega$ represents the boundary of Ω, while $\Gamma_{int} = \partial\Omega_1 \cap \partial\Omega_2$ denotes the interface between the two immiscible phases (Fig. 2).

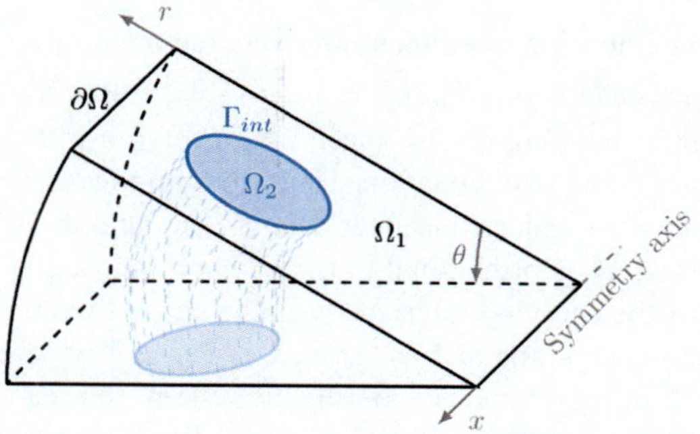

Fig. 2. Computational domain Ω of an axisymmetric two-phase flow problem, containing two fluids in the subdomains Ω_1 (white) and Ω_2 (light blue) (adopted and modified from [13]).

We can express in terms of the coordinate x and the radial coordinate r the velocity as $\boldsymbol{u} = \{u_x, u_r, u_\theta\}$, where $u_\theta = 0$. Moreover, the velocity $\boldsymbol{u}$, the pressure p, and the level-set field ϕ, in each phase i in subdomain $\Omega_i(t)$ are obtained by solving at each instant $t \in (0, t_{total})$ the incompressible Navier-Stokes and the level-set transport equations:

$$\rho_i\left(\frac{\partial u_x}{\partial t} + u_x u_{x,x} + u_r u_{x,r} - g_x\right) + p_{,x} = \mu_i\left\{u_{x,xx} + \frac{1}{r}(r u_{x,r})_{,r}\right\} + f_x^\sigma\,\delta(\phi),$$
$$\rho_i\left(\frac{\partial u_r}{\partial t} + u_x u_{r,x} + u_r u_{r,r}\right) + p_{,r} = \mu_i\left\{u_{r,xx} + \left[\frac{1}{r}(r u_r)_{,r}\right]_{,r}\right\} + f_r^\sigma\,\delta(\phi),$$
$$u_{x,x} + \frac{1}{r}(r u_r)_{,r} = 0,$$
$$\frac{\partial \phi}{\partial t} + u_x \phi_{,x} + u_r \phi_{,r} = 0. \tag{1}$$

where ρ_i denotes the density, μ_i represents the dynamic viscosity, f^σ is the surface tension force and $\delta(\phi)$ is the Dirac delta function. Partial derivatives with respect to the subsequent variable are described by commas and the components of a vector field are symbolized by subscripts. Moreover, the current phase is denoted by the indexing $i = 1, 2$. For the discretization of Eq. (1), we use $P1P1$ finite elements, as described in [8]. The Galerkin/least-squares (GLS) stabilization method is applied [9, 14], because the aforementioned elements are not compliant with the LBB condition.

In [8], we have derived the Galerkin formulation of the Navier-Stokes equation in a Cartesian coordinate system, but we did not take into consideration moving contact lines and the impact mechanism. Hence, we assumed that the equilibrium contact angle is 90°. Here, we demonstrate, following the examples of [4, 15], how the Galerkin formulation of the Navier-Stokes equation should be altered for including more complex wetting effects:

$$\int_{Q_n} \mathbf{w}^h \cdot \gamma\kappa \mathbf{n}^h_{(P_{int})_n} \delta(\phi)\, dQ = \int_{P_n} \gamma \cos\theta_e \mathbf{w}^h \cdot \mathbf{n}^h_L \delta(\phi)\, dP - \int_{Q_n} \gamma\, \underline{\nabla}\mathbf{id}^h_{(P_{int})_n} : \underline{\nabla}\mathbf{w}^h \delta(\phi)\, dQ. \tag{2}$$

where γ is the surface tension coefficient and κ is the curvature. The symbol $\mathbf{n}^h_{(P_{int})_n}$ stands for the normal vector on $(P_{int})_n$, whereas $\underline{\nabla}$expresses the tangential gradient and $\mathbf{id}^h$ is the identity mapping on the space-time evolving interface $(P_{int})_n$, and the superscript h stands for the discretization. Here, θ_e symbolizes the equilibrium/static angle between the moving space-time interface, $(P_{int})_n$, and the space-time boundary P_n. The normal vector, $\mathbf{n}^h_{\mathrm{L}}$ is orthogonal to the contact line $(L_{int})_n$, which is the result of the intersection of the evolving interface with the space-time boundary, and tangent to P_n. Moreover, the space-time slab is defined as Q_n and $\mathbf{w}^h$ are the test functions. For expressing Eq. 2 in axisymmetric coordinate system, the reader should check the formulation in [2] regarding the tangential gradient. Furthermore, the curvature κ can be computed in the axisymmetric case as in [1].

3 Results

3.1 Static Droplet on a Solid Surface

A typical benchmark case of two-phase flows is one of a static bubble inside the computational domain. In the context of contact problems, we assume a static droplet, contacting a solid surface with different equilibrium contact angles. We assume three different scenarios. The general setting is drawn in Fig. 3. In each scenario, the droplet has a radius of $r = 0.1$, the surface tension coefficient is equal to $\gamma = 0.5$, and the density and viscosity for both materials are equal to one. We also make use of the Navier-slip boundary condition, as defined in [8], and the Navier-slip coefficient is uniform and set to $\beta_w = 0.05$, as in [4]. Furthermore, the time slab thickness was 0.01 in all scenarios.

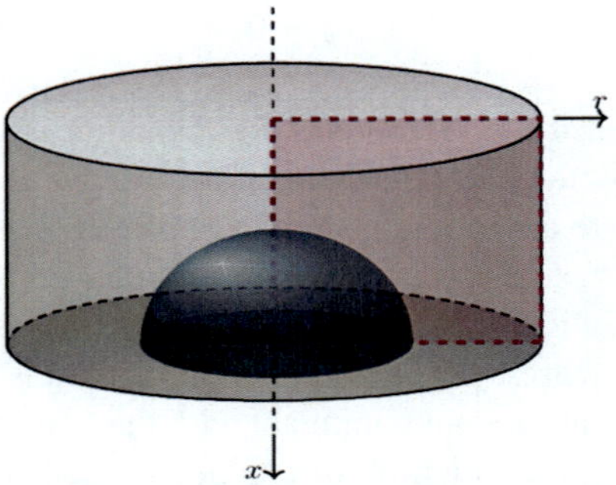

Fig. 3. Axisymmetric static water droplet on a solid surface in 3D: computational domain.

In the first scenario, we used an axisymmetric formulation in combination with a simplex space-time mesh (SST), and the equilibrium contact angle was equal to $\theta_e = 90°$. An SST mesh allows a higher resolution both in space and in time close to the interface, as shown in Fig. 4 (left column). The computational domain was a square of dimensions 0.15×0.15, consisting of 84, 714 elements. For the second scenario, we used an axisymmetric formulation in combination with an SST mesh again. This time, the equilibrium contact angle was equal to $\theta_e = 120°$, as it is shown in Fig. 4 (middle column). The computational domain was a square of dimensions 0.20×0.20, consisting of 69, 529 tetrahedral elements. The third scenario is the 3D equivalent of the second scenario, and a flat space-time mesh (FST) was used, as presented in Fig. 4 (right column). The computational domain was a parallelepiped of dimensions $0.40 \times 0.20 \times 0.40$, consisting of 160, 000 tetrahedral spatial elements.

According to the Laplace-Young equations, the pressure inside the bubble should be equal to $\Delta p = 2\gamma/r$, where Δp is the pressure jump between the two phases, whereas the velocity should be zero everywhere in the domain. As we can observe in Fig. 4, the pressure jump in every scenario is equal to $\Delta p = 10$, which is confirmation that our simulations were in a good agreement with the analytical prediction. Moreover, the results of the second scenario (middle column in Fig. 4) were similar to the 3D equivalent (right column in Fig. 4).

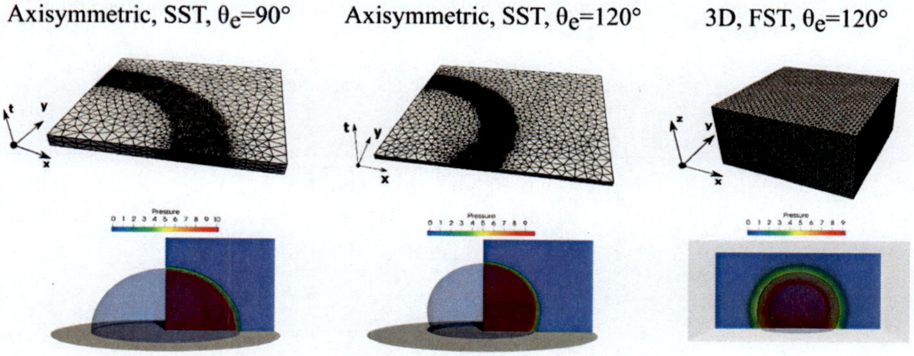

Fig. 4. Grids and pressure distribution of the three different test cases.

3.2 Impact of a Water Droplet on a Solid Surface

In this subsection, we will investigate the impact of an axisymmetric water droplet on a solid surface. The droplet reaches the solid surface and then starts spreading. Moreover, the wettability of the surface can vary. The same example was numerically examined by [4]. Its origin is a physical experiment conducted by [6].

The computational domain Ω and its dimensions are presented in Fig. 5 and it contains air and a water droplet which touches the solid surface with impact velocity: $v_0 = 0.35\,\mathrm{m/s}$. Since the droplet is considered to be spherical and the domain is cylindrical, the example belongs to the category of axisymmetric two-phase flow problems. Therefore, the complexity and dimensionality of our problem can be reduced, and we are allowed to use only the purple-colored slice (see Fig. 5) for simulating this example. The material properties for water and air can be found in Table 1.

In this example, we will present only the case of an equilibrium contact angle, $\theta_e = 179°$. As in [4], we use 179° instead of 180° to avoid any numerical instabilities. The computational domain consists of 80000 triangular elements and the time-slab thickness

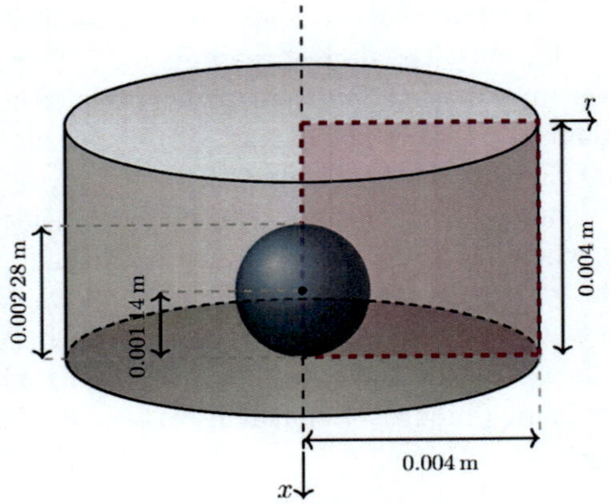

Fig. 5. Axisymmetric water droplet impacting on a substrate in 3D: computational domain.

Table 1. Material properties of water and air.

Properties (units)	Water	Air
$\rho\left\{g/cm^3\right\}$	1	0.0012
$\mu\{g/(cm \cdot s)\}$	0.01	0.000182
$\gamma\left\{g/s^2\right\}$	73	

is equal to $\Delta t = 0.01$ ms. We also make use of the Navier-slip boundary condition, as defined in [8]. The Navier-slip coefficient, β_w, is defined as follows:

$$\beta_w = \beta_\infty \cdot \delta_w(\phi) + \beta_0,\, \delta_w(\phi) = \begin{cases} 0, & |\phi| \leq a \\ \frac{|\phi|}{a} - 1, & a < |\phi| < 2a \\ 1, & |\phi| \geq 2a \end{cases} \tag{3}$$

where β_0 is the wetting coefficient around the contact line, β_∞ is the far-field wetting coefficient, and a denotes the characteristic length scale around the interface. For this test case, we set $\beta_0 = 0.01$, $\beta_\infty = 100000$ and $a = 0.0025$.

In Fig. 6, our results obtained with our in-house space-time solver, XNS, are shown in the bottom row. We compare them with experimental data [6] and other numerical results, computed with the solver DROPS [4]. As we can see, we represent the experimental data well. Hence, the water droplet in our computation is able to bounce back at $t = 12$ ms. The shape of the XNS-simulated droplet is almost identical to the one in the experiment at every time instance. Hence, our computations preserve the droplet characteristics observed during the experiment. For example, our droplet features the same sharp tip at $t = 3.5$ ms, as in the case of [6] at $t = 3.55$ ms. Furthermore, our computation results do not face any issue with the spreading velocity, and the simulation timings are matching quite well the experimental ones.

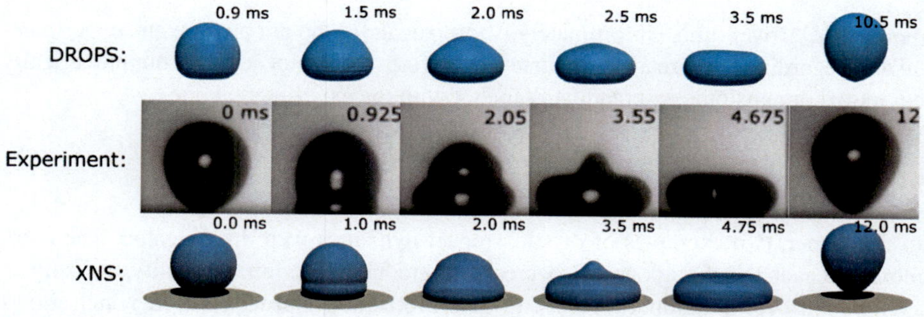

Fig. 6. The impact of a water droplet onto a solid surface with equilibrium contact angle $\theta_e = 180°$. In the top row, the numerical results are obtained with the solver DROPS and are taken and adapted from [4]. In the bottom row, the results of our in-house solver, XNS, are presented. In the middle row, the experimental results are adapted from [6].

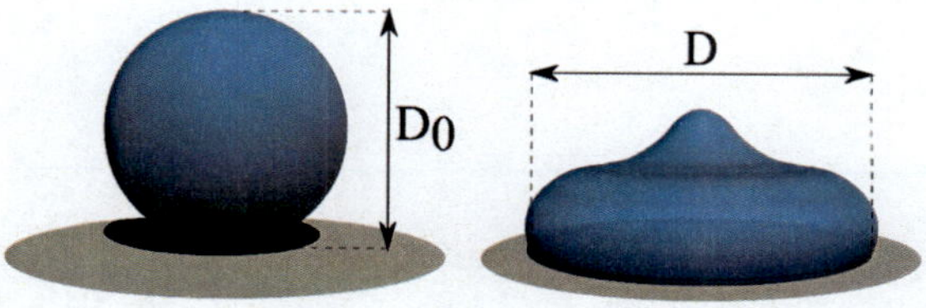

Fig. 7. The droplet's initial diameter (left) and spreading diameter (right).

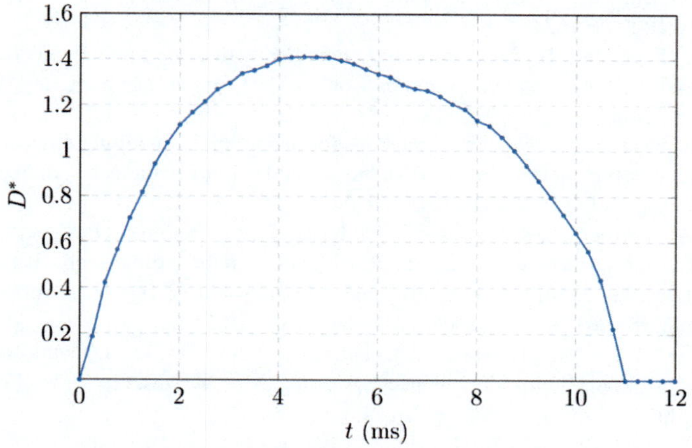

Fig. 8. Dimensionless spreading diameter, D^*, over time.

Following the example of Dong et al. [16] and Klitz [17], we compute the dimensionless droplet spreading diameter $D^* = D/D_0$ over time. Here, the spreading diameter, D, is computed directly at the contact line and D_0 is the droplet's initial diameter, as illustrated in Fig. 7. In Fig. 8, we present the diagram of the dimensionless spreading diameter, D^*, over time. Unfortunately, Vadillo et al. [6] did not provide any experimental data regarding the spreading diameter. Therefore, we cannot compare our numerically computed dimensionless spreading diameter with the experimental one.

4 Conclusion

In this paper, two examples of droplet impact dynamics were investigated. The complexity of each benchmark case is decreased by reducing its dimensionality, utilizing an axisymmetric formulation. Besides, a novel discretization approach is used, which allows arbitrary temporal refinement of the space-time slabs in the vicinity of the interfaces.

Acknowledgment. The authors gratefully acknowledge the support of the German Research Foundation (DFG) under program SFB 1120 "Precision Melt Engineering". The computations were conducted on computing clusters provided by the RWTH Aachen University IT Center and by the Jülich Aachen Research Alliance (JARA).

References

1. Chessa, J., Belytschko, T.: An enriched finite element method and level sets for axisymmetric two-phase flow with surface tension. Int. J. Numer. Meth. Eng. **58**, 2041–2064 (2003). https://doi.org/10.1002/nme.946
2. Ganesan, S., Tobiska, L.: An accurate finite element scheme with moving meshes for computing 3D-axisymmetric interface flows. Int. J. Numer. Meth. Fluids **57**, 119–138 (2008). https://doi.org/10.1002/fld.1624
3. Elgeti, S., Sauerland, H.: Deforming fluid domains within the finite element method: five mesh-based tracking methods in comparison. Arch. Comput. Methods Eng. **23**, 323–361 (2016). https://doi.org/10.1007/s11831-015-9143-2
4. Zhang, L., Reusken, A.: Numerical methods for mass transfer in falling films and two-phase flows with moving contact lines (2017). https://publications.rwth-aachen.de/record/684694. https://doi.org/10.18154/RWTH-2017-02014
5. Baer, T.A., Cairncross, R.A., Schunk, P.R., Rao, R.R., Sackinger, P.A.: A finite element method for free surface flows of incompressible fluids in three dimensions. Part II. Dynamic wetting lines. Int. J. Numer. Methods Fluids. **33**, 405–427 (2000). https://doi.org/10.1002/1097-0363(20000615)33:3%3c405::AID-FLD14%3e3.0.CO;2-4
6. Vadillo, D.C., Soucemarianadin, A., Delattre, C., Roux, D.C.D.: Dynamic contact angle effects onto the maximum drop impact spreading on solid surfaces. Phys. Fluids **21**, 1–8 (2009). https://doi.org/10.1063/1.3276259
7. Lin, S., Zhao, B., Zou, S., Guo, J., Wei, Z., Chen, L.: Impact of viscous droplets on different wettable surfaces: impact phenomena, the maximum spreading factor, spreading time and post-impact oscillation. J. Colloid Interface Sci. **516**, 86–97 (2018). https://doi.org/10.1016/j.jcis.2017.12.086
8. Karyofylli, V., Wendling, L., Make, M., Hosters, N., Behr, M.: Simplex space-time meshes in thermally coupled two-phase flow simulations of mold filling. Comput. Fluids. **192**, 104261 (2019). https://doi.org/10.1016/j.compfluid.2019.104261
9. von Danwitz, M., Karyofylli, V., Hosters, N., Behr, M.: Simplex space-time meshes in compressible flow simulations. Int. J. Numer. Meth. Fluids **91**, 29–48 (2019). https://doi.org/10.1002/fld.4743

10. Gesenhues, L.: Advanced methods for finite element simulation of rheology models for geophysical flows (2020). https://doi.org/10.18154/RWTH-2020-05371
11. Gaburro, E.: Well balanced Arbitrary-Lagrangian-Eulerian finite volume schemes on moving nonconforming meshes for non-conservative Hyperbolic systems (2018)
12. Zwart, P.J.: The integrated space-time finite volume method (1999)
13. Chessa, J., Belytschko, T.: An extended finite element method for two-phase fluids. J. Appl. Mech. **70**, 10 (2003). https://doi.org/10.1115/1.1526599
14. Pauli, L., Behr, M.: On stabilized space-time FEM for anisotropic meshes: incompressible Navier-Stokes equations and applications to blood flow in medical devices. Int. J. Numer. Meth. Fluids **85**, 189–209 (2017). https://doi.org/10.1002/fld.4378
15. Ganesan, S., Tobiska, L.: Modelling and simulation of moving contact line problems with wetting effects. Comput. Vis. Sci. **12**, 329–336 (2009). https://doi.org/10.1007/s00791-008-0111-3
16. Dong, H., Carr, W.W., Bucknall, D.G., Morris, J.F.: Temporally-resolved inkjet drop impaction on surfaces. AIChE J. **53**, 2606–2617 (2007). https://doi.org/10.1002/aic.11283
17. Klitz, M.: Numerical simulation of droplets with dynamic contact angles (2014)

Additive Manufacturing

Melt Pool Formation and Out-of-Equilibrium Solidification During the Laser Metal Deposition Process

Jonas Zielinski[1](✉), Henrik Kruse[1], Marie-Noemi Bold[1], Guillaume Boussinot[2], Markus Apel[2], and Johannes Henrich Schleifenbaum[1,3]

[1] Chair for Digital Additie Production DAP, RWTH Aachen University, Campus Boulevard 73, Aachen, Germany
jonas.zielinski@dap.rwth-aachen.de
[2] Access e.V., Intzestr. 5, Aachen, Germany
[3] Fraunhofer Institute for Laser Technology ILT, Steinbachstr. 15, Aachen, Germany

Abstract. In Laser Metal Deposition (LMD) processes, a complex interplay between the blown particles, the laser beam and the substrate determines the resulting microstructure.

We investigate the melt pool formation during LMD in dependence of the process parameters with particular attention to kinetic effects of the particles on that change the melt pool shape, both experimentally and with means of simulations.

The solidification takes places far from equilibrium due to the large temperature gradients at the phase front and fast solidification velocities. With the help of phase field simulations and microstructure analysis, we will unravel the solidification process during LMD. The dendrite arm spacing is compared with experimental data to prove the validity of the simulation.

In this work, Inconel 718, a material common in LMD turbomachinery applications, is used.

Keywords: Laser Metal Deposition · Microstructure · Solidification conditions

1 Introduction

1.1 Approach

Laser metal deposition (LMD) uses a focused laser beam to melt a substrates surface. With a powder nozzle, a stream of metal particles is blown in the formed melt pool and therefore fuses metallurgical with the substrate (after cooling down). Moving the laser beam and the powder nozzle over substrate generates a weld track that is well bonded to the material. Multiple tracks above each other allow the build op of complex 3d structures. An advantage of LMD in contrast to conventional methods is the near net shape form of the specimen and the small dilution of the added material with the substrate (or lower layers). In the last decade, LMD has become an established processing technique for the manufacturing and repair of metal parts [1].

U. Reisgen et al. (Eds.): EMPOrIA 2020, LNME, pp. 113–122, 2021.
https://doi.org/10.1007/978-3-030-70332-5_11

The resulting microstructure for a specific material depends strongly on the local chemical composition and the local solidification conditions. In well-known and investigated processes, the phase front velocity and the temperature gradients are small compared to the gradients in metal Additive Manufacturing (AM) processes. The first challenge is to determine the solidification conditions at the solidification phase front. From emissivity measurements only the surface temperature can be determined. With thermal elements, the temporal temperature distribution at a fixed point, far away from the phase front (in comparison to the melt pool size) can be measured. Therefore it is advised, to combine those methods with thermal simulations of the corresponding process. In our case the Laser Metal Deposition (LMD) process with the material Inconel 718 is investigated. The dependence of the microstructure (primary and secondary dendrite arm spacing) on the process parameters and therefore indirectly on the solidification conditions is the subject of our research.

The scientific question to be answered in this paper, if a pure simulation based approach is capable of determining the microstructure of an as-built LMD sample. It is also discussed which parts give rise to largest deviation due to model simplifications.

In the first step, the LMD model is explained followed by a validation with single track experiments under the variation of the process parameter by changing the line energy (Laser power divided by scanning velocity) $E_l = P_L/v_S$ up to $\pm 30\%$ from the default parameter. The influence of the process parameter on the solidification condition is investigated there.

In the second section, the model for the calculation of the microstructure, based on the phase field model, is explained. After that, the resulting microstructure from the experiments is compared with the calculated one.

In the last section, the findings are discussed and an outlook for future research is given.

2 From Process Parameters to Solidification Conditions

2.1 Melt Pool Modelling

At the heart of most metal AM processes is the melt pool. Its behavior is strongly influenced by the first order process parameters like Laser Power P_L, the scanning velocity v_S as well as the intensity distribution of the laser in the working plane and newly cladded material in each layer (comparable to a mass rate $\dot{m}_p$) as well as the distance of neighboring melt tracks (hatch distance h_S) [2]. In our work, some of the second order influence factors of the LMD process are included: particle flight path and particle interaction with the laser beam yielding a reduced transmitted laser intensity distribution in the working plane but increased particle temperature, cooling effects by radiation, absorption depending on the melt surface curvature. Higher order terms like for example the influence of gravity on the particle path and the melt behavior are neglected.

The laser source is treated as a surface source with absorption values taken from literature [3]. The track geometry is determined by the solution of the Young-Laplace equation with the constraint of mass conservation from the ingoing powder particles. The powder particles are a volumetric heat source (or sink; depending on their temperature) if their kinetic energy and angle of impact is sufficient; the particles are only deflected if

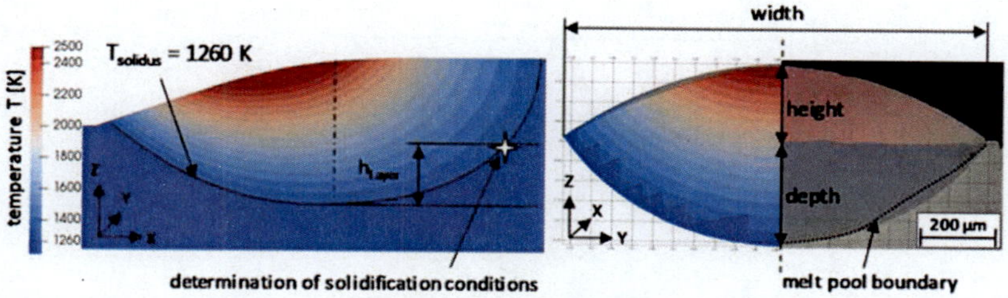

Fig. 1. Temperature distribution of simulated melt track geometry side view and cross section (perpendicular to scanning direction).

they are too slow to overcome the surface tension or when the angle is not steep enough. The thermophysical material properties [4] are assumed to be temperature dependent and the latent heat is included with the method of effective heat capacity. **The full mathematical formulation of the model can be found in our previously published work** [5].

The implementation of Marangoni flow is discussed in literature. The forces that lead to a surface force driven slow are two fold: 1) the temperature dependency of the surface tension and 2) the gradient of the chemical composition on the melt pool surface. For the first influence facto is has been shown in literature that the influence is negligible [6]. Furthermore the capillary length of nickel base alloys ($\approx$4.8 mm [7]) is larger than the melt pool dimension (roughly 1 mm) supports the idea of neglecting Marangoni flow and also gravity. The second factor is unknown and no value can be associated with it.

The temperature distribution for the default parameter set is shown in Fig. 1. As a validation for the model, the melt pool dimensions (depth, height and width) are compared with experimentally obtained single track data where the single tracks were deposited with varied process parameters.

2.2 Melt Pool Model Validation and Influence of Volume Energy

Since the goal is to obtain the solidification conditions, which cannot be observed directly, a method is required which yields reliable results. Our model is tested in the range of 30% variation of the volume energy (or line energy); the energy that is averagely introduced per unit volume (length) in the material. The default parameter is at laser power of 300 W and a scanning velocity of 8.3 mm/s. The same volume energy variation can be obtained by varying either the laser power or scanning velocity (and powder mass rate) to a different (reciprocal) degree. We will investigate later if the same volume energy, in the 30% bounds, yields the same results depending on the first order process parameter that leads to the change in volume energy. The cross sections for each parameter set (take at three different points in the weld track) are compared with the calculated values of our model (2.1). The 13 different parameters with the corresponding line energy are shown in Fig. 2.

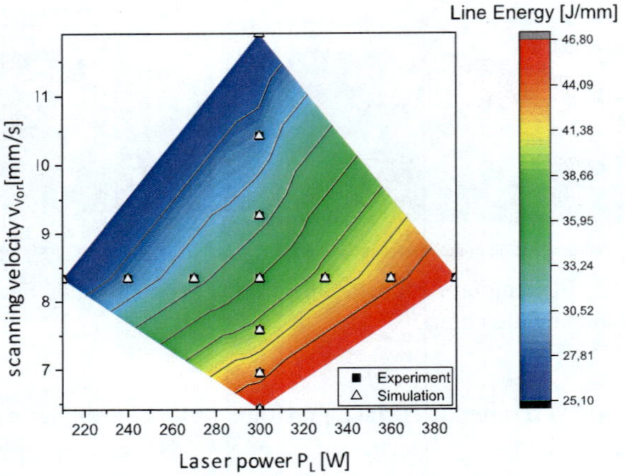

Fig. 2. Parameter variations in laser power and scanning velocity with the according line energy

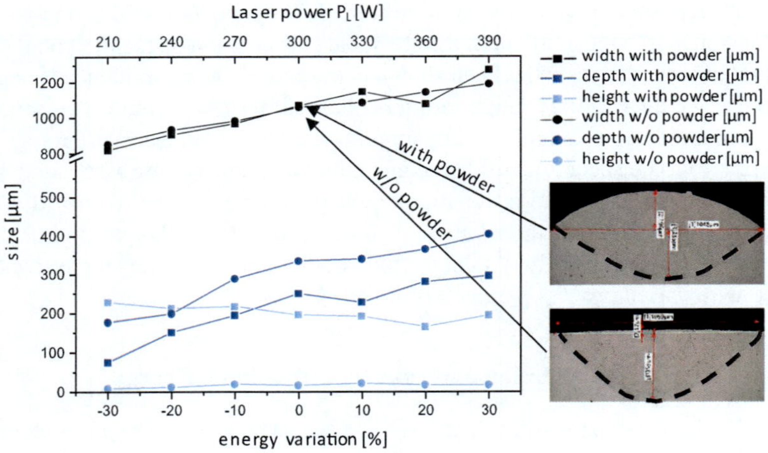

Fig. 3. The effect of the powder mass flow during the LMD process on the depth, width, height and shape (symmetry) of the melt pool.

The data for the depth, height and width is extracted from the isosurface of the solidus temperature for simulations and light microscopically measured from a polished cross section for experimental data (Fig. 3). The height is measured from substrate level to the highest point of the melt track in each cross section. The width is taken on substrate level and the depth is measured from substrate level to the deepest point of the melt pool. For the experimental data always the average of three measurements is used for evaluation. Also the effect of the powder on the melt pool boundary can be seen. It is evident that the first order influence is the added mass to the melt pool, resulting in the typical melt track shape. More interesting is the fact, that an asymmetric boundary between melt and substrate material is formed by the particle melt interaction. The origin of the asymmetry is found in the asymmetry of the powder "beam". Therefore a scanning direction dependency of the melt pool shape (and temperature distribution) also exists in LMD even if coaxial nozzles are used (asymmetries occur more often when nozzles are slightly damaged, but still in use). Even though the asymmetry is much smaller in contrast to off-axis nozzles.

The melt pool widths do not change whether powder is used in the process or not. The width grows proportionally to the volume energy.

The melt pool depth and height depend on the use of powder, because the volume energy can only melt a certain volume of material, thus the depth is decreased if powder is used. Also, the depth scales proportionally with the volume energy E_V. Additionally, a remelting process without powder supply results in a height of the remelted track up to 15 µm. This increase in volume is due to a conversion to the tetragonal γ'' phase, which has a lower atomic packing efficiency than the Inconel fcc γ-matrix.

The simulation data is always taken in a quasi-equilibrium state (where the melt pool - / temperature change within the next time step is almost zero). The comparison of experimentally obtained and simulated data is shown in Fig. 4. It can be noted, that generally the accordance between experiment and simulation is quite well with relative deviations of less than 10%.

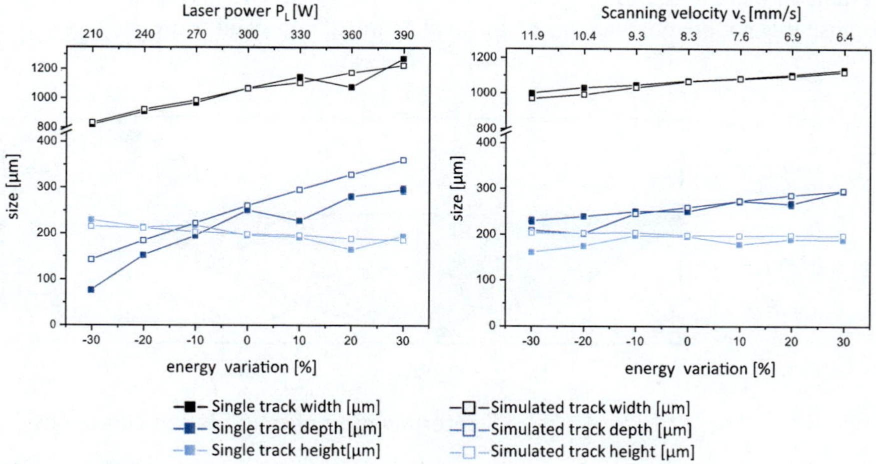

Fig. 4. Comparison of track geometries for experimental samples with simulated results. On the left the volume energy is changed by changing the laser power while on the right side the scanning velocity and powder mass rate are adapted.

For the variation of the scanning velocity, the melt pool depth is calculated very accurately while the variation of the laser power yields more deviating results. A possible origin of the increased deviation could be that effects of the radiation intensity on the melt pool surface are neglected or that the underestimation of particle shadowing effects and neglected evaporation effects (for higher powers) could cancel each other out at the volume energy level of the default parameter.

From the transient three dimensional temperature distribution the solidification conditions (solidification velocity v_{sol}, thermal gradient ∇T and cooling rate $\dot{T}$) can be extracted.

2.3 Determination of Solidification Conditions

As an input for the following microstructure simulation, the solidification conditions (temperature gradient G, cooling rate $\dot{T}$, solidification phase front velocity v_{sol}) are required. In reality, solidification conditions vary all over the solidification phase front; thus picking a representative position is mandatory to obtain comparable 2d microstructures. The assumption is, that if the melt pool geometry (temperature distribution) is calculated correctly, the solidification conditions are also correct (see validation in Sect. 2.2).

Since we are interested in the microstructure inside the build volume and have to take multiple layers into account, it is not advised to pick a point in the track that would be remelted in the consecutive layer To keep the results comparable we determine an effective layer height $h_{Layer} = 4/3 \cdot 0,9 \cdot h_S$ on the dependence of the experimental melt pool height and picked the point that is in the middle of the melt point and layer height above the deepest point on the solidification front. This procedure is shown for the default parameter in Fig. 5. This function for the effective layer height is chosen because it is not be remolten during the application of consecutive layers.

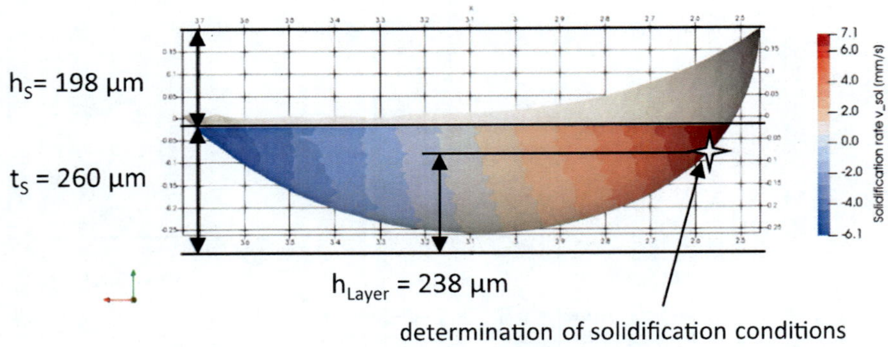

Fig. 5. Point (Star) chosen for the extraction of the solidification conditions, the melt pool depth and height as well as the effective layer height for the default parameter. Scanning direction from right to left.

The solidification conditions and their respective process parameters are listed in Table 1. The solidification velocity strongly depends on the scanning velocity; the effect of the laser power is 20 to 50% smaller. The temperature gradient, on the other hand is influenced more by the laser power than the scanning velocity. The cooling rate depends stronger on the scanning velocity. It can be seen, that the interplay between process parameters and solidification conditions is complex and that there is no easy way to predict them without running simulations.

Table 1. Process parameters and according solidification condition. The default parameter is highlighted in grey.

P_L [W]	d_{Laser} [µm]	v_{Scan} [mm/s]	E_V Var. [%]	G [K/m]	$\dot{T}$ [K/s]	v_{Sol} [mm/s]
300	1000	8.3	0	1,30E+06	9,23E+03	7.1
210	1000	8.3	-30	1,50E+06	1,13E+04	7.5
390	1000	8.3	+30	1,20E+06	7,08E+03	5.9
300	1000	11.9	-30	1,30E+06	1,35E+04	10.4
300	1000	6.4	+30	1,30E+06	6,89E+03	5.3

The extracted conditions are the required input for the microstructure simulation with the MICRESS software.

3 From Solidification Conditions to Microstructure

3.1 Microstructure Modelling

In [8], we have presented phase field simulations of IN718 directional solidification under LPBF conditions including interface kinetics [9]. In this present study we apply the same technique to the conditions in the LMD process. During our investigations, we have varied the solidification speed over a wide range from $v_{sol} = 5,3\,\text{mm/s}$ to $v_{sol} = 10,4\,\text{mm/s}$ (the corresponding temperature gradients and cooling rates are given in Table 1).

The phase field simulations are performed using the MICRESS software, based on a multi-component phase field model [10] that is coupled to the TCNI8 Thermo-Calc database. The composition of the alloy that was chosen for our simulations lies in chemical composition boundary of IN718 and is given in Table 2 (Fe as balance).

Table 2. Chemical composition of Inconel 718 alloy (min-max mass percentage) in comparison to the chemical composition used for the microstructure model (Fe as balance).

Ele.	Ni	Cr	Nb	Mo	Ti	Al	Co	C
Min [w%]	50.00	17.00	4.75	2.80	0.65	0.20	-	-
Max [w%]	55.00	21.00	5.50	3.30	1.15	0.8	1.00	0.08
Modell [w%]	53.20	19.00	5.13	3.05	0.9	0.5	0.5	0.08

In the simulation, the primary gamma phase as well as γ', γ'', Laves and Sigma are implemented. The 2D domain size is 15 μm $\times$ 24 μm and the discretization grid spacing is 12 nm. The thermal gradient ∇T is considered homogeneous in space and constant in time (so-called frozen temperature approximation), and diffusion is neglected in the solid phase. Periodic boundary conditions are applied along the horizontal axis. The thermal gradient points upwards and the preferential growth direction <100> is tilted by 10°.

The results for the default parameter's solidification conditions are shown in Fig. 6.

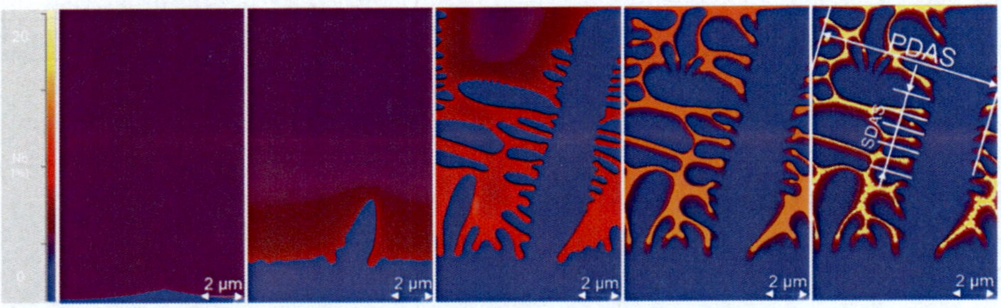

Fig. 6. Dendrite growth during solidification for different time steps with the solidification conditions of $\nabla T = 1, 3E+06\,\mathrm{K/m}$, $\dot{T} = 1E+04\,\mathrm{K/s}$;. The primary dendrite arm spacing is 14,8 μm and the secondary is 2 μm. The Niob content (w.%) is color scaled.

The simulations results show a strong branching behavior, normally expected for slower solidification processes with smaller thermal gradients. Even tertiary dendrites are identified, normally associated with slower solidification. In the interdendritic area Niob precipitates and as the solidification proceeds laves phase (yellow) stabilizes The determined primary/secondary dendrite arm spacings are $14.8 \pm 0.2\,\mu$m and $2.0 \pm 0.2\,\mu$m, respectively.

3.2 Comparison with Experimental Data

The experimentally obtained microstructure for the default parameter is shown in Fig. 7, the position of the sample is in the chosen according to the position given in Fig. 5. Several regions with different micro segregation patterns can be observed ranging from unidirectional growth in the plane of the image to areas of an isotropic 2D dimensional networking, indicating a dendritic growth perpendicular to the plane of the image. The occurrence of secondary dendritic arms can be detected at several points. The primary and secondary dendrite arm spacings are measured to $3.5 \pm 0.2\,\mu$m and $0.9 \pm 0.2\,\mu$m, respectively, this is roughly a factor of 4 smaller than the simulation results.

The deviation between simulation and experiment seems large at first glance, but the simulated tertiary dendrites could be mistaken, in the experiment, for primary dendrites when the branching occurs some 10 μm away from the cross section plane. This would reduce the deviation down to a factor of 2, thus yielding a satisfying accordance between simulation and experiment. Further investigations have to be conducted to test this hypothesis.

Fig. 7. Microstructure sample of LMD manufactured material with the default parameter (Table 1).

4 Discussion and Outlook

The melt pool simulation for LMD yields accurate results. Nevertheless a small systematic deviation can be observed and thus is topic of our ongoing research in the frame of the "Precision Melt Engineering" Collaborative Research Center SFB1120. For the default Inconel 718 parameters, however, the results are accurate enough to consider the calculated solidification conditions as reliable, and to use them as input for a phase field method based microstructure simulation (MICRESS).

The calculated microstructure, primary dendrite arm spacing, is in the order of magnitude of the experimentally measured data. However, the occurrence of tertiary dendrite arms in the simulation is quite a surprise. In our continuous research work, we will try to investigate this phenomena with a sensitivity analysis of the phase field model with respect to the solidification conditions. With this approach we can finally identify whether the solidification conditions chosen (or the position) is representative/good or the LMD process/microstructure model simplifications are eligible.

The point where the solidification conditions are taken is still up for debate. As in this article shown, the solidification conditions change strongly in the meltpool, thus its hard to find a representative volume. Further work could include multiple microstructure simulation domains in the meltpool with different solidification conditions. Those resulting microstructure should then tested against each other to identify the energetically favoured (similar to comparison of growth directions in [8]).

Acknowledgements. All presented investigations were conducted in the context of the Collaborative Research Centre SFB1120 "Precision Melt Engineering" at RWTH Aachen University and funded by the German Research Foundation (DFG). For the sponsorship and the support we wish to express our sincere gratitude.

References

1. Poprawe, R. (ed.): Tailored Light 2: Laser Application Technology. Springer, Heidelberg (2011). (RWTH edition)
2. Pirch, N., Linnenbrink, S., Gasser, A., Wissenbach, K., Poprawe, R.: Analysis of track formation during laser metal deposition. J. Laser Appl. **29**(2), 22506 (2017)
3. Sainte-Catherine, C., Jeandin, M., Kechemair, D., Ricaud, J.-P., Sabatier, L.: Study of dynamic absorptivity at 10.6 μm (CO_2) AND 1.06 μm (Nd-YAG) wavelengths as a function of temperature. Le Journal de Physique IV 01 **C7**, C7-151–C7-157 (1991)
4. Pottlacher, G., Hosaeus, H., Kaschnitz, E., Seifter, A.: Thermophysical properties of solid and liquidInconel 718 Alloy*. Scand. J. Metall. 31. Nr. **3**, 161–168 (2002)
5. Pirch, N., Linnenbrink, S., Gasser, A., Schleifenbaum, H.: Laser-aided directed energy deposition of metal powder along edges. Int. J. Heat Mass Transf. **143**, 118464 (2019)
6. Pirch, N., Kreutz, E.W., Möller, L., Gasser, A., Wissenbach, K.: Melt dynamics in surface processing with laser radiation. In: Proceedings of the 3rd European Conference on Laser Treatment of Materials Coburg 1990, S. 65 (1990)
7. Amore, S., Valenza, F., Giuranno, D., Novakovic, R., Dalla Fontana, G., Battezzati, L., Ricci, E.: Thermophysical properties of some Ni-based superalloys in the liquid state relevant for solidification processing. J. Mater. Sci. 51. Nr. **4**, 1680–1691 (2016)
8. Boussinot, G., Apel, M., Zielinski, J., Hecht, U., Schleifenbaum, J.H.: Strongly out-of-equilibrium columnar solidification during laser powder-bed fusion in additive manufacturing. Phys. Rev. Appl. **11**(1), 567 (2019)
9. Boussinot, G., Brener, E.A.: Interface kinetics in phase-field models: isothermal transformations in binary alloys and step dynamics in molecular-beam epitaxy. Phys. Rev. E Stat. Nonlinear Soft Matter Phys. **88**(2), 22406 (2013)
10. Eiken, J., Böttger, B., Steinbach, I.: Multiphase-field approach for multicomponent alloys with extrapolation scheme for numerical application. Phys. Rev. e Stat. Nonlinear Soft Matter Phys. **73**(6 Pt 2), 66122 (2006)

Understanding Cylinder Temperature Effects in Laser Beam Melting of Polymers

Sandra Greiner[1,2](✉), Andreas Jaksch[1,2], and Dietmar Drummer[1,2]

[1] Collaborative Research Center 814 - Additive Manufacturing (CRC 814), Am Weichselgarten 9, 91058 Erlangen, Germany
sandra.greiner@fau.de

[2] Institute of Polymer Technology (LKT), Am Weichselgarten 9, 91058 Erlangen, Germany

Abstract. Laser beam melting of polymers allows for the fabrication of highly complex parts with excellent properties. For part generation, temperatures close to the softening range of the polymer have to be set on the powder bed surface and in z-direction. These high temperatures are directly linked to high durations of the cooling phase, high delay times for part availability and progressing material aging. Within this work, the impact of the heat control in z-direction is analyzed. Therefore, the cylinder heater temperature is varied and the fabricated parts are characterized extensively. Thermal simulation of the temperature distribution within the build chamber is performed to predict the aging behavior of the material and the maximum cooling time. It could be shown that the majority of part properties is unaffected by reducing the cylinder temperature, whereas cooling time and material aging of the surrounding powder could be reduced. This indicates the possibility of a solidification-optimized processing strategy with targeted temperature control in z-direction.

Keywords: Additive manufacturing · Laser beam melting · Powder bed fusion · Laser sintering

1 Introduction

Powder bed fusion of polymers, which is also called selective laser beam melting or laser sintering (SLS) has developed from a prototype production to a direct manufacturing technology [1]. Compared to conventional production technologies, such as injection molding, SLS offers high geometrical freedom and the possibility to fabricate polymer parts at low quantities at a short design to manufacturing time [2]. The emerging topics concerning advanced SLS polymer processing are the increase of material availability, enhanced reproducibility and the reduction of processing times for faster part availability [3]. All of these topics are somehow linked to the predominant transient temperature fields within the SLS process. According to several authors [4–6] especially the temperature inhomogeneities during the process affect the reproducibility of part quality characteristics and material aging. Multiple energy sources and sinks within the SLS process affect the temperature fields. The temperature distribution on the powder

U. Reisgen et al. (Eds.): EMPOrIA 2020, LNME, pp. 123–141, 2021.
https://doi.org/10.1007/978-3-030-70332-5_12

bed surface is basically determined by the surface heating system. The powder is locally molten by laser. Every layer, surface cooling is induced by recoating a new layer of powder onto the powder bed surface. In z-direction, platform heaters and cylinder heaters (or removal chamber or frame heaters) as well as heat conduction processes affect the temperature fields.

According to the findings of the authors, there is an unused potential for process optimization that can be used, when the impact of the cylinder heater on powder material and part properties is fully understood. For this reason, the basic influence of the cylinder temperature on part properties will be analyzed in the following.

2 State of the Art

In Fig. 1, typical simplified temperature profiles within the three process phases of the SLS process are shown. More precisely, the heating, build and cooling phase, are shown for the surface $T_{Surface}$ or build chamber temperature T_B [7] and the cylinder temperature $T_{Cylinder}$ (T_C). In the heating phase, which lasts between two and three hours depending on the system type and the size of the build chamber, the temperature of the powder bed surface and the build cylinder is gradually increased to their target temperatures.

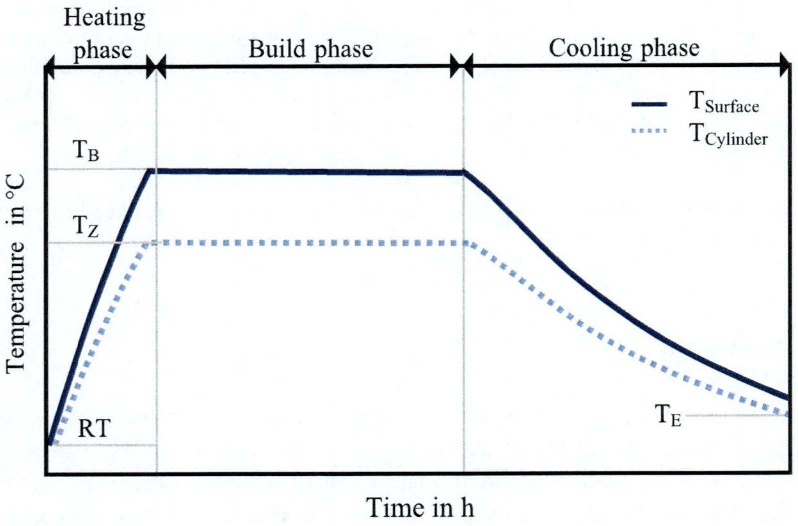

Fig. 1. Schematic illustration of the temperature profiles on the powder bed surface and at cylinder temperature in z-direction during the SLS-process [7].

Within the build phase, the surface and the cylinder temperatures are held at a constant temperature level. Usually, for semi crystalline thermoplastics, the surface temperature is chosen according to the model of isothermal laser sintering [8, 9] slightly below the melting temperature. The temperature of the cylinder heater is set to a significantly lower temperature level, which varies for polyamide 12 (PA12) mostly in the range between 130 °C and 150 °C depending on the system size. The energy to overcome the melting enthalpy of the material is locally introduced by a CO_2-laser, leading to a two-phase mixture area, where the melt is coexisting with unmolten powder. Local laser energy input leads to high heating rates that usually can be neglected within the overall temperature household of the build chamber due to the quick decay time [10]. The powder coating process regularly interrupts this thermodynamically stable state, by adding new and cooler powder onto the powder bed surface. Afterwards, the surface temperature has to be raised to the target temperature, again. Amongst many other parameters, the duration of the build phase is depending on the height of the build job and the layer time and can take more than 24 h. During the final cooling phase, all heating systems are deactivated and the system as well as the powder cake cool down passively. The duration of the cooling phase mainly depends on the thermal conductivity of the polymer powder, the thermal mass and usually corresponds to the duration of the build phase. Part cooling is accepted to happen slowly to avoid warpage by inhomogeneous shrinkage caused by locally differing cooling rates [5, 6, 11]. Due to this slow cooling, the properties of SLS parts are morphologically different from injection molded parts which solidify within a time period of seconds [12].

In Fig. 1, a strongly optimized temperature chart is shown. In reality, the interactions between different temperature fields are more complex and their observation is essential for process understanding and optimization. Regarding the three process phases, most research is performed on understanding the build phase. It is stated, that especially the amount of energy input into the SLS process determines the completeness of the melting within the part, which can for instance be adjusted by increasing the laser power, lowering the scan spacing or by setting higher surface temperatures [13]. In [14], the influence of energy density (E_D) and surface temperature on part density is studied empirically and numerically, indicating a positive correlation between T_B or E_D and part density. Especially the temperature distribution within the molten material affects the quality of the resulting parts. Therefore, the choice of optimum exposure parameters, such as effective laser power or scan speed, has been found to be decisive for the resulting surface and melt temperatures and therefore part properties [15–19].

However, even for optimum surface and exposure parameters, an influence of part positioning within the build chamber on the part characteristics can be detected, even for identical geometries [12, 20]. Wudy [7] and Josupeit [21] showed that every parts thermal history, which is among many other influences mainly depending on the system and the position within the build chamber, is unique. This can be attributed to inhomogeneous temperature fields that are present within the build chamber and especially along the z-axis of most laser sintering systems. [7, 21] These locally varying temperature histories lead to altered part properties due to different material crystallization and solidification. Gibson [22] showed that part density and mechanical properties are affected by part orientation, and building height. Especially parts in the center of the build space tend to have higher tensile strength and part density, which was referred to lower cooling rates and more homogeneous temperature fields. Zharringhalam [13] performed experiments under variation of the cooling period, where parts were extracted from the powder bed at different time spans after fabrication. For comparison, the results are presented in Table 1. It was found that after high times to extraction, higher values for the tensile strength were achieved, which is in accordance to Gibson's findings in [22]. Short times to extraction led to higher values of the elongation at break and Young's Modulus. However, no significant influence on relative crystallinity was observed, indicating that crystallization process was finished earlier [13].

Table 1. Impact of extraction time on mechanical properties and crystallinity [13].

Mechanical properties	Unit	Estimated time to extraction		
		66 h	1 h	0.1 h
Elongation at break	%	13.5	16.8	19.1
Tensile Strength	MPa	42.5	41.0	40.6
Young's Modulus	GPa	1.69	1.65	1.80
Crystallinity	%	27.9	27.5	27.2

The understanding of crystallization pf PA12 within the SLS process has changed significantly over the last decades. In the early 2000s, it was assumed that according to the model of isothermal laser sintering, the molten material would remain liquid until the entire part bed begins to cool down in the cooling phase [23]. Later on, Drummer [24] showed that under processing conditions and depending on the used material system, isothermal crystallization is present and the implementation of time-dependent phase transitions into the model of isothermal laser sintering is necessary to explain different material behavior. Wegner [18] summarized that the validity of the model of isothermal laser sintering was limited to a narrow period of time, which can be correlated to a low part size or build-height. In [25, 26] isothermal crystallization and the degree of crystallization were modelled for PA12 and polypropylene in the first ten layers considering the influence of powder coating cooler material onto the molten material. It could be shown that within 10 layers the majority of crystallization has been completed. In [27], time and temperature dependent material solidification due to isothermal crystallization of PA12 could be observed within the process and within process adapted material characterization. At a build chamber temperature of 168 °C, measurable material consolidation takes place after around 10 min after exposure was found. A revision of the temperature control setup of the build chamber was announced, allowing for an active cooling of the part cake in z-direction after isothermal crystallization and solidification of the material is finished.

The knowledge of material consolidation under isothermal conditions in combination with the knowledge on the inadequacies of the existing technical solution must therefore lead to the temperature system of the build chamber to be reconsidered. Already in 1997, Gibson [22] proposed to improve the temperature control strategy within SLS systems to achieve more homogeneous part properties independent from positioning. In [28], Li examined the cooling of polyamide 12 (PA12) components through simulations and optimized the build chamber geometry and the cooling parameters in terms of maximum temperature homogeneity. It was found that the geometry of the build chamber has a major influence on the temperature homogeneity of the cooling process. For an optimum cooling process a container width small width, long length and high height should be used [28]. Due to more uniform part temperature histories, these adaptions have the potential to increase the reproducibility of part properties and to enhance the efficiency of the SLS process significantly. Furthermore, the cooling phase would no longer determine the production time and part availability [27]. Therefore, a basic understanding on the influence of the cylinder temperature on part properties has to be generated. In the following, experiments are performed aiming for a material dependent temperature control system for an optimized SLS process.

3 Materials and Methods

Within the following section, information on the used material, the experimental setup and the performed series of experiments is shown. The aim is to determine the basic influences of isothermally tempered cylinder heating on different temperature levels on the material and component properties in order to ensure the potential for process acceleration through a reduced cylinder temperature. The long-term goal is to develop a locally and temporally adapted tempering strategy in the z-direction to exploit the potential of consolidation due to isothermal crystallization of PA12 as reported in [27].

3.1 Material System

The experiments were conducted using PA12 powder PA 2200 from EOS GmbH (Krailling, Germany). A refreshed powder mixture of a 50:50 weight ratio of virgin and recycled powder was used. The used powder was taken from overflow containers, and therefore only slightly affected by the high ambient temperature. Therefore, a low thermal impact on the powder and the experiments can be expected. Prior to fabrication, the bulk density and viscosity number (VN) of the powder mixture were determined to 0.44 ± 0.01 g/cm^3 and 56.5 ± 1.4 ml/g.

3.2 Laser Beam Melting System

A freely parameterizable laser beam melting system was applied for the experiments. For studying the impact of cylinder temperature on the resulting part properties, a miniaturized build chamber was used. The schematic illustration of the experimental setup is depicted in Fig. 2 a). The build chamber dimensions in xy-plane are reduced from 350×350 mm^2 to 100×100 mm^2 and the build chamber height is reduced from 500 mm to 150 mm. The miniaturized build chamber was applied, as tempering a small volume of powder can be correlated to a higher control over the present temperature fields, compared to the higher build volume of the standard system. The build chamber is heated by a silicone heater, which is mounted around the build cylinder (see Fig. 2 b). For setting the temperature field on the powder bed surface, eight individually controllable infrared quartz radiator arrays are installed within the system. For laser exposure, a 50 W CO_2-laser with a focus diameter of 0.4 mm was used.

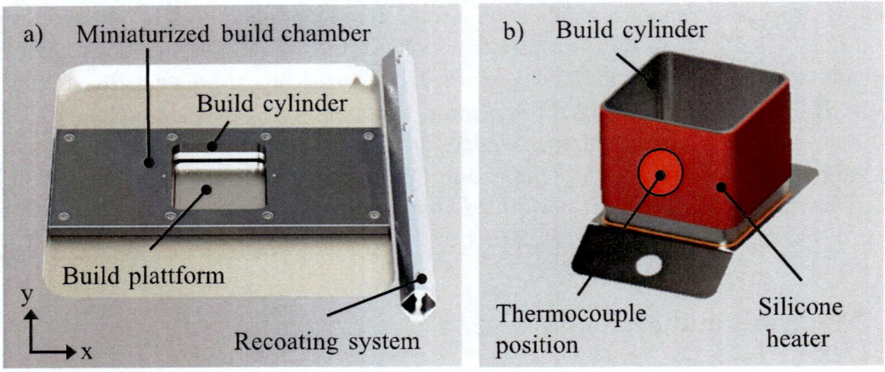

Fig. 2. Schematic illustration of the experimental set-up depicting a) the miniaturized build chamber within the standard build chamber and b) the cylinder heating system.

3.3 Fabrication Experiments

In order to study the influence of the cylinder temperature on the resulting part properties, multipurpose test specimens were fabricated. The dimensions of the sample are sketched in Fig. 3. Within the miniaturized build chamber, six samples can be positioned at equal distancing in one layer. Five layers of test specimens were fabricated and for process stability, all of the samples were oriented in recoating direction. For thermal isolation, 50 layers of powder were recoated before the first and after the last exposed layer, respectively. The build job setup is schematically visualized in Fig. 4. For all experiments, the exposure parameters and the build chamber temperature were kept constant. The parameters are listed in Table 2. In addition, no scaling factors were set. The cylinder temperature was varied between 100 °C and 150 °C with increments of 10 °C. A reference experiment was performed with turned off cylinder heater, which is named "0 °C" experiment in the following sections. After fabrication and cooling to room temperature, the parts were carefully depowdered by using a brush. In addition, powder samples were extracted from material located between the tensile bars. Immediately after extraction, the 30 multipurpose test specimens were stored in a vacuum cabinet at room temperature to avoid water absorption.

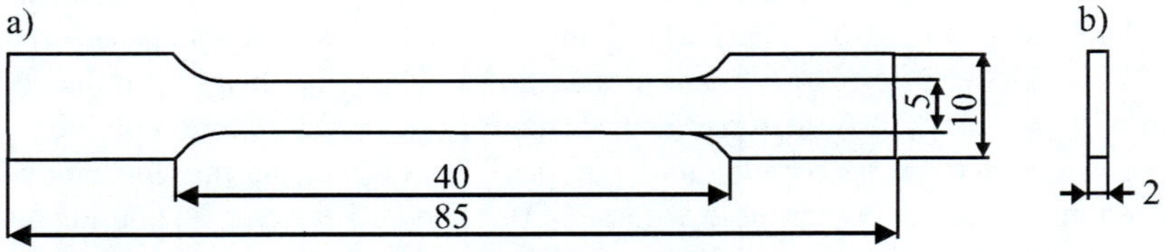

Fig. 3. Geometric description of test samples a) front view and b) side view.

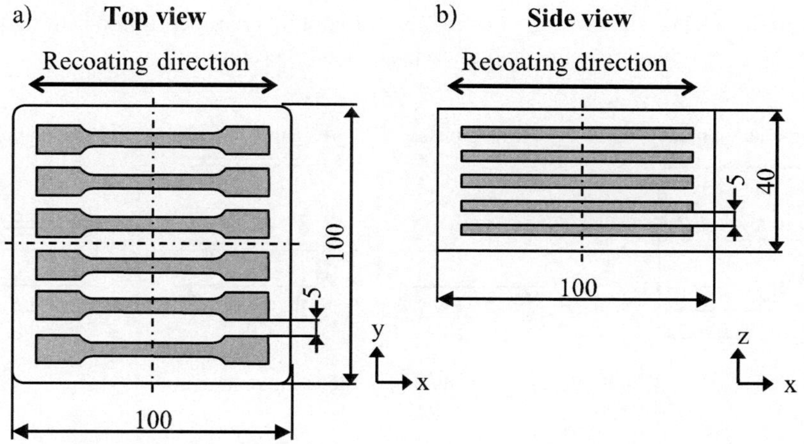

Fig. 4. Schematic of the build job setup in a) top view and b) side view.

Table 2. Constant processing parameters for the fabrication of specimens.

Process parameter	Value	Unit
Laser power P_L	16	W
Scan speed v_s	2,000	mm/s
Hatch distance h_s	0.2	mm
Layer height d_h	0.1	mm
Energy density E_D	0.4	J/mm^3
Build chamber temperature T_B	174	°C

3.4 Temperature Measurements

The thermal boundary conditions were measured at two different sites. The temperature of the powder bed surface was determined by the systems pyrometer in the center of the powder bed. The cylinder temperature input was detected at the site of the silicon heater. For these measurements, the data acquisition rate was 25 Hz, which means that every 40 ms a measuring point was recorded.

3.5 Thermal Simulation

The thermal simulation was performed in the multiphysics software Abaqus CAE 2019 to characterize the local temperature distribution and the chemical aging of the part cake for each time step of the process. To reduce the simulation complexity, a 2D-model was used, and only half of the build chamber was modeled. This simplification was chosen based on symmetrical thermal boundary conditions. Furthermore, ten layers were combined to one simulation layer resulting in a layer height of 1 mm and an average layer time of 405 s. No parts were considered in the simulation. Figure 5 shows the position of the 2D xz-plane located on the top view (xy-plane) of the build chamber. In the heating phase, the initial powder layers get heated up to process temperature.

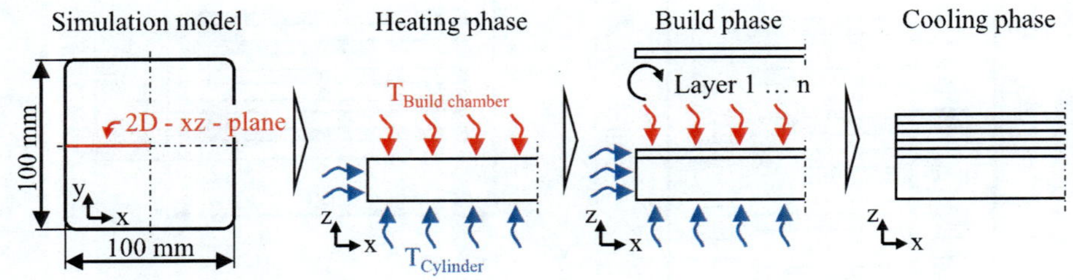

Fig. 5. Schematic depiction of the thermal simulation.

The following build phase adds layers with each simulation step. In the cooling phase, the part cake is cooled down. The thermal boundary conditions were modeled based on the temperature measurements of the experiments introduced in Sect. 3.4 (see Fig. 8). The local calculation of the VN is performed for each step based on the local temperatures and the model in [29]. The simulation parameters are shown in Table 3.

Table 3. Boundary conditions of the thermal simulation.

Simulation parameter	Value	Unit
Heating Phase time/step time	7,200/400	s
Layer time/step time	405/40.5	s
Cooling phase time/step time	36,000/360	s
Starting height	20	mm
Build height	40	mm
Layer thickness	1	mm

3.6 Material and Part Characterization

The VN value was determined for the initial powder mixture by Ubbelohde viscometer using concentrated sulphuric acid as solvent. In order to study the thermal impact of different cylinder heating levels the VN of the powder samples and the test specimens was measured, respectively. The part cake powder was extracted from spaces between the fabricated samples and homogenized. For part characterization, especially the material in the center of the specimen was analyzed in order to minimize the influence of fused particles on the surface.

For basic part characterization, the part dimensions, width and height respectively, were determined by outside micrometer (measurement accuracy: ± 0.01 mm) at three different sites for every sample. The measuring positions are highlighted in Fig. 6.

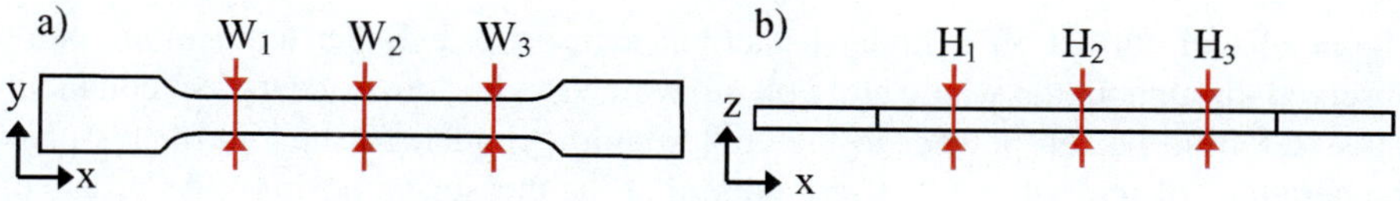

Fig. 6. Schematic description of the measuring positions for the determination of the a) part width and b) part height.

The morphology of the parts was characterized in terms of layer connectivity, porosity and crystallinity by transmitted light microscopy under polarized light at 25-x magnification using the microscope Axio Imager.M2 from Carl Zeiss AG (Germany). Exemplarily, samples fabricated with turned off cylinder heater and parts fabricated with cylinder temperatures of 110 °C, 130 °C and 150 °C were analyzed. Therefore, thin sections of approximately 10 μm were extracted from the center of the tensile bars, which were located in the center of the first, third and fifth row of parts. The extraction site is depicted in Fig. 7. Subsequently, the part porosity was determined by gray value analysis in the center of the samples with a constant threshold value for binarisation of 55.

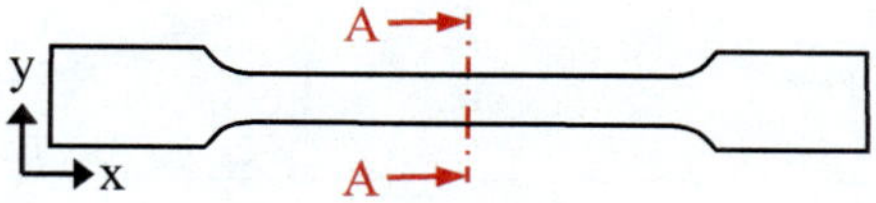

Fig. 7. Sampling position for the extraction of the thin cuts of 10 μm.

Part mechanics were determined by tensile tests following DIN EN ISO 527–2 using the universal testing machine Zwick 1465 from ZwickRoell GmbH & Co. KG (Germany). For the determination of the Young's modulus, the testing speed was set to 0.5 mm/min. For failure analysis and to characterize the tensile strength and the elongation of break, the testing speed was adjusted to 25 mm/min.

4 Results and Discussion

The results of the surface and cylinder temperature measurements are shown in Fig. 8. During the initial heating phase, the different heating temperature levels and plateau phases can be detected for the surface of the powder bed. The surface temperature was measured by pyrometer and the temperature of the cylinder heater was acquired by thermocouple measurements. After the end of the heating phase, the average surface temperature was determined to 174 °C, which is the build chamber temperature T_B. Furthermore, rapid temperature rises to temperature levels of more than 200 °C are visible that can be correlated to a random detection of the exposure process. In addition, temperature reductions to temperatures levels of around 150 °C are present, which can be traced back to the heating phase after powder coating of new and cool powder layers onto the surface of the powder bed. In the build phase, the target temperatures of the cylinder temperature are reached immediately. For the untempered experiment

($T_C = 0$) with turned off cylinder heater, an increase in cylinder temperature can be observed throughout the whole build phase, which can be correlated to heat conduction processes from heated powder to the cooler build cylinder. At the end of this phase, temperatures of more than 90 °C are present at the measuring position. As no plateau phase was reached yet, it can be expected, that an increase in build duration would lead to a further increase of cylinder temperature. In contrast to that, the cylinder temperature is constant for the tempered experiments, although there is no active heat extraction from the system. For that reason, for this system and build setup, a thermal balance at measuring position can be expected at a cylinder temperature of around 100 °C.

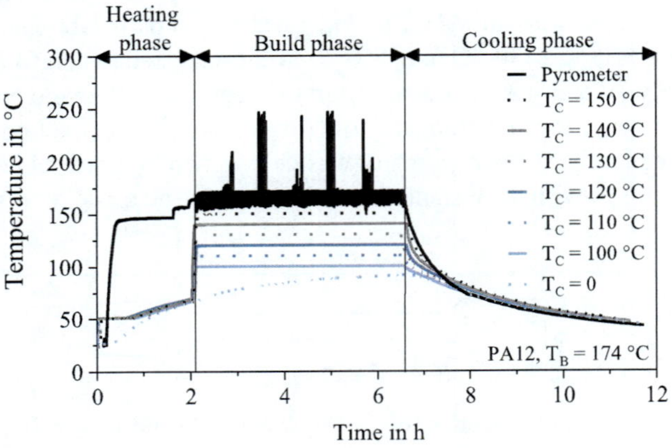

Fig. 8. Results of the surface and cylinder temperature measurements.

Within the cooling phase, an almost exponential decrease in surface and cylinder temperature is present. It can be seen, that higher cooling rates are present at the surface compared to the cylinder heaters in z-direction. Concerning cylinder heaters, the cooling rate is depending on the temperature level. The higher the initial cylinder temperature, the higher the cooling rate, especially in the beginning of the cooling phase. It can be seen, that the reduction of cylinder temperature leads to an almost identical cooling profile at the site of the build cylinder. This can be explained by the temperature gradient between the build cylinder and the surrounding air.

Figure 9 shows the calculated temperature distribution of the xz-plane for different cylinder temperatures and cooling times. The thermal boundary conditions for this simulation are chosen based on the temperature measurements shown Fig. 8. Comparing the different cylinder temperatures at the end of the building phase (cooling phase: t = 0 h), a strong influence of the local temperature distribution in z-direction can be detected. With decreasing cylinder temperature, a three-dimensional temperature gradient develops from the surface temperature towards the colder cylinder walls. This results in a thermal history of the powder and parts, depending on the xy- and z-position in the part cake. Hence, a part build in the SLS process is not in an isothermal state over the processing time. In fact, it is cooling down to a temperature between build chamber and cylinder temperature shortly after the exposure. Furthermore, the calculated results show that the temperature of the upper layers is mainly controlled by the build chamber heating system. The upper layers are not influenced by the temperature fields induced by the cylinder heaters. After turning off the heater system, the temperature of the powder bed surface decreases rapidly (cooling phase t = 0.5 h). The highest calculated temperature moves from the upper layers to the center of the part cake. This leads to a core-shell temperature distribution. With ongoing cooling time, the influence of the cylinder temperature reduces, and the temperatures start to converge.

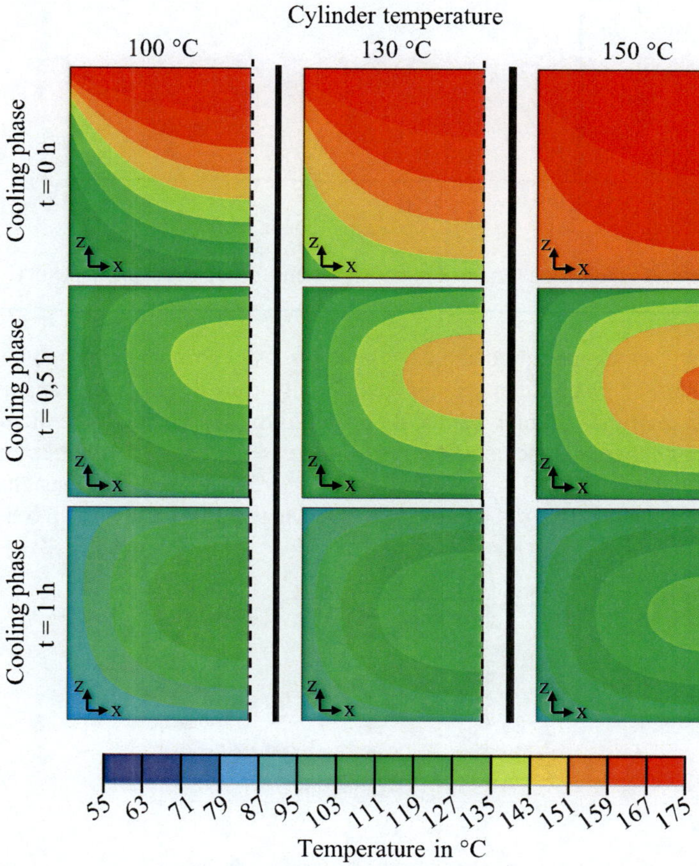

Fig. 9. Thermal simulation of the local temperature distribution of the cooling phase.

Figure 10 shows the calculated maximum local temperatures over xz-plane against the cooling time. At the beginning of the cooling phase, the maximum temperature is 174 °C for all three cylinder temperatures, which is the build chamber temperature. With continuous cooling and turned off heating systems, the maximum temperature decreases rather quickly and a temperature difference of around 10 °C can be detected between the 150 °C and 100 °C cylinder temperature. While the cylinder temperature affects the temperature distribution significantly, the effect on the maximum temperature is rather small. The maximum temperature is located in the center of the powder bed and mainly dominated by the specific heat coefficient and the heat conductivity of the powder. A higher heat conductivity, e.g. by adding fillers, could lead to higher cooling speed.

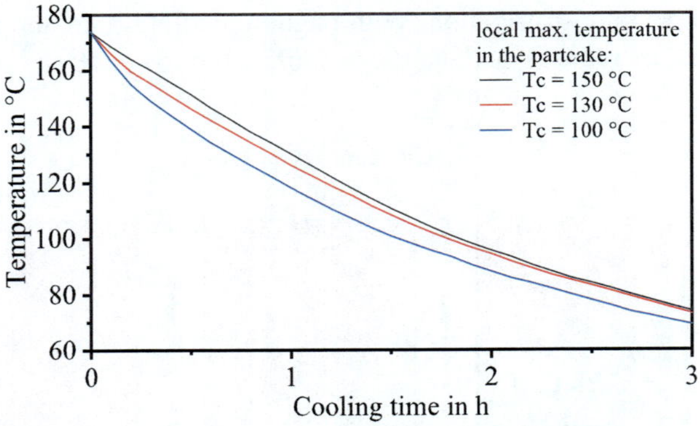

Fig. 10. Thermal simulation of the maximum local temperature in the part cake during cooling.

The results of the measured and calculated viscosity numbers are shown in Fig. 11. For PA12, thermal post condensation is a commonly known aging mechanism that leads to an extension of the polymer chains. By a simplifying assumption, the increase of the polymer chain length can be correlated to an increase in viscosity number. In addition, thermal degradation from laser exposure leading to a reduced polymer chain length should be considered. The VN value of the initial powder mixture was 56.5 ± 1.4 ml/g.

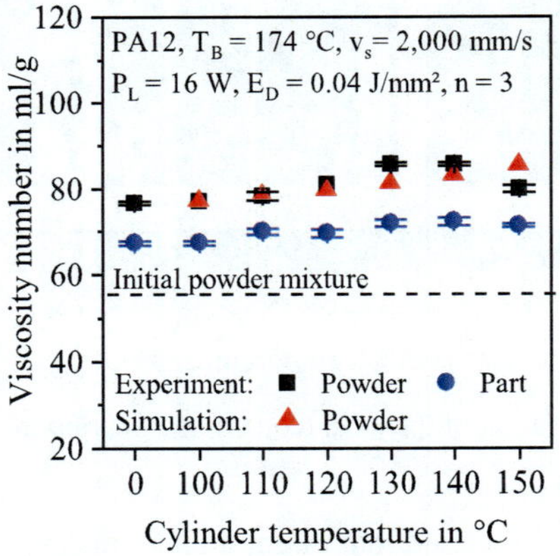

Fig. 11. Results of the measurements of the viscosity number.

It can be seen that compared to the initial powder mixture, an increase of the VN value is present for the resulting part cake powder and the resulting part, respectively. Furthermore, the VN of the part cake powder is approximately 10 ml/g higher than the

VN of the part, which can be referred to differing aging mechanism within the melt and within solid material and to the higher surface area of the small particles compared to that of the melt pool. In addition, an increase in cylinder temperature results in an increasing VN value for the part and the part cake. The calculated VN, based on the thermal simulation, fit with the measured experimental VN. However, at a cylinder temperature of 150 °C a decrease in VN of the part cake powder is visible while the VN of the simulation increases, which cannot be clearly explained. The reduction of the VN could be explained either by increasing thermal degradation at higher cylinder temperature or by insufficient definition of the powder extraction site. Both effects are not represented in the calculated value and cannot be addressed in the used model. In [5] the dependency of material aging on the position within the part cake was shown. A stronger ageing effect was measured in the center compared to outer regions of the part cake. However, to gain more reliable information on material aging, gel permeation chromatography measurements could be performed in future to determine the molecular weight of the powders and parts.

The results of the geometrical characterization are shown in Fig. 12. The nominal dimension of the part width is 5 mm. It is clearly visible, that the widths of the parts are geometrically oversized for all parameter combinations. The mean values are almost at the same level of around 5.2 mm, with a slight decrease for cylinder temperatures of more than 110 °C. However, the standard deviations are too high for a clear assessment of the influence and can be correlated to the part positioning. Furthermore, the standard deviations are decreasing with increasing cylinder temperature. This can be correlated to a theoretically more homogeneous temperature distribution along the build chamber cross-section in z-direction and local effects.

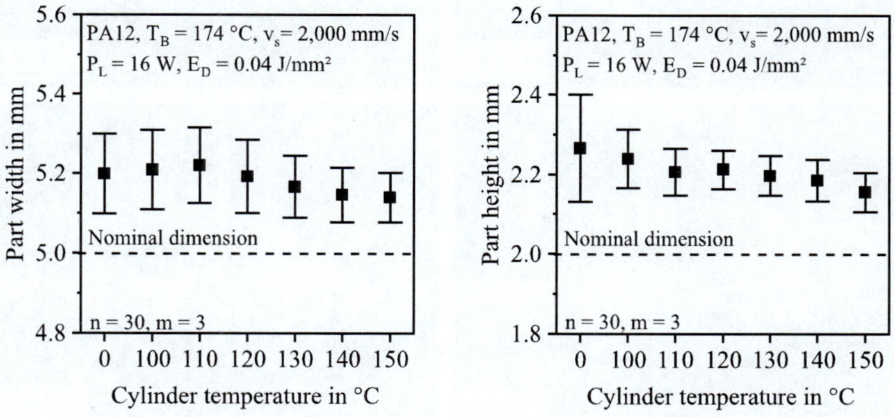

Fig. 12. Results of the measurements of the part width (left) and the part height (right).

Considering part height, an almost similar behavior can be observed. The nominal dimension of the part height is defined to 2 mm. In general, the samples are oversized in their height dimension, as values of around 2.2 mm are present for most of the parameters. Again, a slight decrease in height dimension and in standard deviation is caused by an increasing cylinder temperature. This can be referred to more homogeneous temperature fields at higher cylinder temperature. In summary, it can be stated that with increasing cylinder temperature, a tendentially smaller part cross section can be expected, influencing the part mechanics.

In Fig. 13, the microscope images of the thin cuts under polarized light are shown. For all cross sections, low porosity is visible within the parts, indicating sufficient energy input. For all components, adhering, unmolten particles are visible at the edges of the cross section. No clear statement can be made about the quantity of adhering particles. This result can be explained by the thermal simulations in Fig. 9. During the construction phase, almost identical thermal boundary conditions prevail for all components due to the strong influence of surface heaters. Within deeper layers, the components are influenced by the lower cylinder temperatures. Taking into account the results in [27], it can be concluded that the formation of part properties such as component geometry is mainly dependent on the surface temperature fields and is only slightly influenced by the cylinder temperature due to already performed isothermal crystallization and solidification processes.

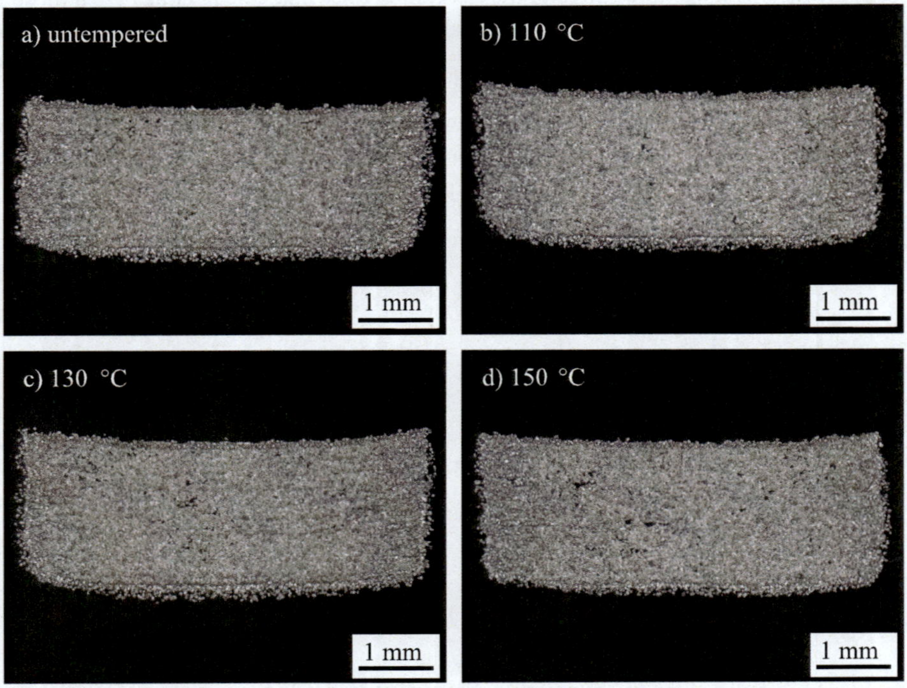

Fig. 13. Thin sections of samples fabricated with different cylinder temperatures a) untempered, b) 110 °C, c) 130 °C and 150 °C (exemplarily shown for the third row of parts).

Considering the morphology of the samples, visible pores are preferably oriented in the building direction along the build plane, which can be usually assigned to inhomogeneous particle arrangements and too low energy input. The results of the geometric characterization of the components can be verified by the optical measurement of the microscopic images, since here too an oversize of the component height and the component width can be proven for all components. In addition, the typical SLS formation of a convex component bottom side and a concave component top side is visible for all components. This is due to friction and flow effects of the melt on the powder bed, respectively. Clamping this component geometry in the measuring device increases the notch effect during mechanical testing and the probability for early part failure, which would be noticeable by low values for the elongation at break. Regarding the semi crystalline structure of the parts, no significant differences can be found. In the centers of the parts fine and homogenous-looking spherulitic structures can be found. At the side areas of the parts, the layer structures are visible as layers of unmolten particles are incorporated into the melt. This can be explained by the locally differing energy input and cooling in the center and in the side regions of the parts.

The results of the porosity evaluation are shown in Fig. 14. Comparatively low mean values of below 2% were measured by optical analysis of microscopy images from three different parts per parameter set. Since the surface roughness is not taken into account when calculating the porosity optically, comparably low porosity values are achieved that are subject to high standard deviations. The highest porosities are found for the untempered experiment whereas no significant difference can be observed for the tempered experiments. The inhomogeneous tempering for the untempered experiment could have led to this slightly elevated value of porosity.

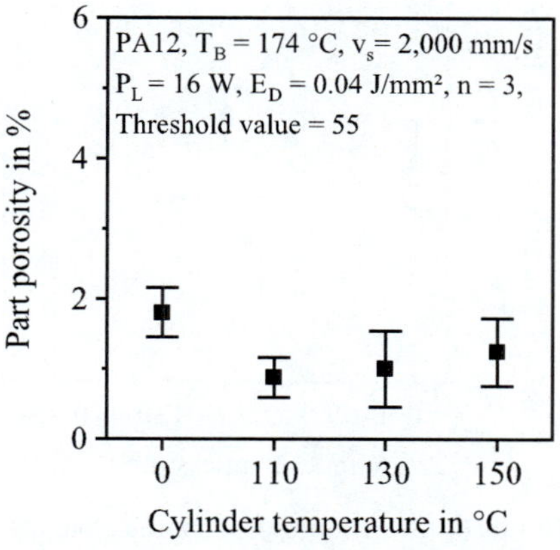

Fig. 14. Results of the optical measurement of the part porosity.

The results of the tensile tests are depicted in Fig. 15 and 16. The Young's modulus is determined to values of around 1,850 MPa and comparatively high standard deviations of mostly around ±50 MPa are present. The variation of the cylinder temperature has no significant influence on the Young's modulus or the standard deviation. The mean values of the tensile strength are determined to values of around 46 MPa, which is comparable to freshly sintered or dry PA12 samples. Again, no significant influence of cylinder temperature on the determined characteristic value was found.

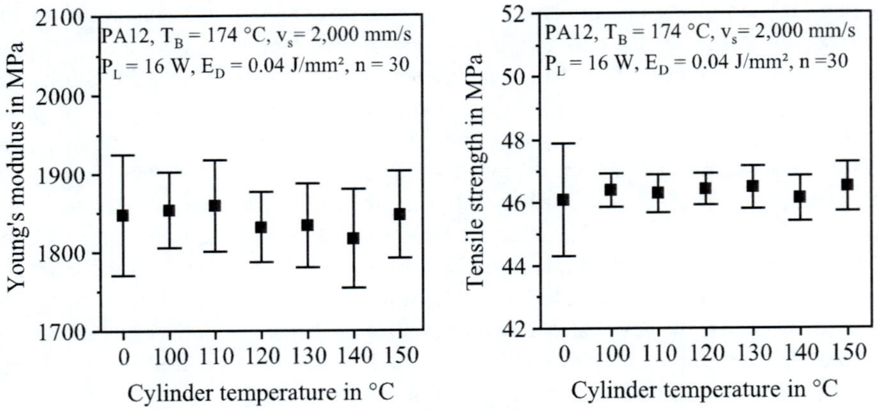

Fig. 15. Results of the measurements of the Young's modulus (left) and the tensile strength (right).

In Fig. 16, the results of the measurement of the elongation of break are depicted in dependence on the cylinder temperature. It must be emphasized that the characteristic value of elongation at break in particular is sensitive to material and process influences and therefore should be used as a significant mechanical characteristic value.

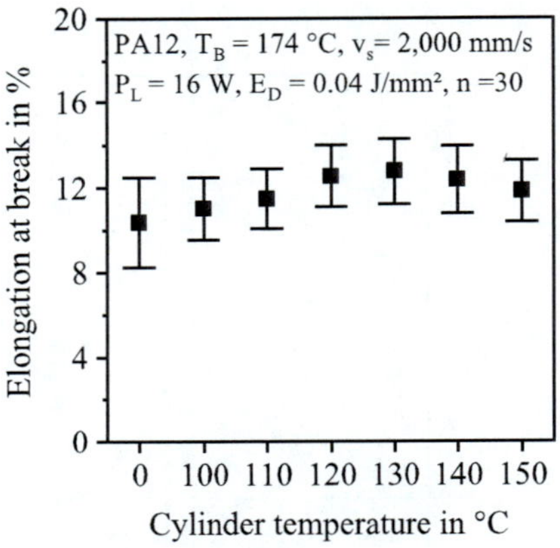

Fig. 16. Results of the measurements of the elongation of break.

In some cases, very low values of less than 10% of elongation at break were achieved, which is due to the choice of test specimens scaled to the half of their original size and conditioning of the parts, which can be described as freshly sintered and dry. In addition, the shape of the cross-section promotes the notch effect during tensile testing. However, maximum values of 12.8% were measured for a cylinder temperature of 130 °C, which can be described as optimum parameter for the size of the miniaturized build chamber. At cylinder temperatures between 0 °C and 130 °C, a slight increase in the mean values of the elongation at break can be observed, while at higher temperatures there is a tendential decrease in elongation at break. Especially the latter influence cannot be clearly be characterized due to the comparatively high standard deviations. Nevertheless, an influence of the cylinder temperature on the elongation of break is clearly visible, which indicates that a simple cylinder temperature reduction with unchanged part quality is not possible for process optimization.

5 Summary and Outlook

Within the shown investigations, the influence of an isothermal cylinder heating system on the resulting part properties was shown in order to evaluate the possibility to reduce the cylinder temperature for an acceleration of the cooling phase. Thermal simulations indicate that within the upper layers, the samples were subjected to almost identical thermal boundary conditions, due to the strong influence of surface heaters. Within deeper layers, the components are influenced by the lower cylinder temperatures. Hence, for the used experimental setup, a minor influence of cylinder temperature on part geometry, part morphology and part mechanics, such as Young's modulus and tensile strength was observed. However, the part porosity and the elongation at break benefit from elevated temperature levels of the build cylinder.

It can be concluded that the formation of most of the observed part properties is mainly dependent on the surface temperature fields and is only slightly influenced by the cylinder temperature due to already performed material solidification in the upper layers. According to these and earlier findings, a time dependent temperature profile in z-direction will be applied to the cylinder heating system. Close to the surface of the build chamber, high temperatures close to T_B or the optimum cylinder temperature depending on build chamber size will be the target temperature. The temperature decrease in z-direction is realized by an active cooling system according to the progressing isothermal crystallization kinetics and the linked material consolidation. This will lead to an accelerated and more efficient SLS process in future.

Acknowledgements. Funded by the Deutsche Forschungsgemeinschaft (DFG, German Research Foundation) - Project-ID 61375930 - SFB 814 "Additive Manufacturing" TP B03 and T01.

References

1. Gibson, I., Rosen, D., Stucker, B.: Additive Manufacturing Technologies-3D Printing, Rapid Prototyping, and Direct Digital Manufacturing, 2nd edn. Springer, Heidelberg (2015)
2. Van Hooreweder, B., Kruth, J.-P.: High cycle fatigue properties of selective laser sintered parts in polyamide 12. CIRP Ann. **63**(1), 241–244 (2014)
3. Goodridge, R.D., Tuck, C.J., Hague, R.J.M.: Laser sintering of polyamides and other polymers. Prog. Mater Sci. **57**, 229–267 (2012)
4. Wegner, A., Witt, G.: Ursachen für eine mangelnde Reproduzierbarkeit beim Laser-Sintern von Kunststoffbauteilen. RTejournal **10** (2013)
5. Josupeit, S., Schmid, H.-J.: Temperature history within laser sintered part cakes and its influence on process quality. Rapid Prototyp. J. **22**(5), 788–793 (2016)
6. Shen, J., Steinberger, J., Göpfert, J., Gerner, R., Daiber, F., Manetsberger, K., Ferstl, S.: Inhomogeneous shrinkage of polymer materials in selective laser sintering. In: International Solid Freeform Fabrication Symposium, pp. 298–305 (2000)

7. Wudy, K.: Alterungsverhalten von Polyamid 12 beim selektiven Lasersintern. Dissertation Friedrich-Alexander-Universität Erlangen-Nürnberg (2017)
8. Alscher, G.: Das Verhalten teilkristalliner Thermoplaste beim Lasersintern. Dissertation, Berichte aus der Kunststofftechnik, Shaker Verlag, Aachen (2000)
9. Schmachtenberg, E., Seul, T.: Model of isothermic laser sintering, In: Society of Plastics Engineers, ANTEC 2002 conference proceedings, 3030–3035 (2002)
10. Greiner, S., Wudy, K., Wörz, A., Drummer, D.: Thermographic investigation of laser-induced temperature fields in selective laser beam melting of polymers. Opt. Laser Technol. **109**, 569–576 (2019)
11. Yuan, M., Bourell, D., Diller, T.: Thermal conductivity measurements of polyamide 12. In: Proceedings of the Solid Freeform Fabrication Symposium, pp. 427–437 (2011)
12. Bourell, D. L., Watt, T. J., Leigh, D. K., Fulcher, B.: Performance limitations in polymer laser sintering. In: International Conference on Photonic Technologies, Physics Procedia, vol. 56, pp. 147–156 (2014)
13. Zarringhalam, H.: Investigation into crystallinity and degree of particle melt in selective laser sintering. Dissertation, Loughborough University (2007)
14. Tontowi, A.E., Childs, T.H.C.: Density prediction of crystalline polymer sintered parts at various powder bed temperatures. Rapid Prototyp. J. **7**(3), 180–184 (2001)
15. Drummer, D., Drexler, M., Wudy, K.: Impact of heating rate during exposure of laser molten p arts on the processing window of PA12 powder. Phys. Procedia **56**, 184–192 (2014)
16. Drexler, M.: Zum Laserstrahlschmelzen von Polyamid 12 – Analyse zeitabhängiger Einflüsse in der Prozessführung. Dissertation, Friedrich-Alexander-Universität Erlangen-Nürnberg (2016)
17. Wegner, A., Witt, G.: Understanding the decisive thermal processes in laser sintering of polyamide 12. In: AIP Conference Proceedings, vol. 1664, p. 160004 (2015)
18. Wegner, A.: Theorie über die Fortführung von Aufschmelzvorgängen als Grundvoraussetzung für eine robuste Prozessführung beim Laser-Sintern von Thermoplasten. Dissertation, Universität Duisburg-Essen (2015)
19. Kiani, A., Khazaee, S., Badrossamay, M., Foroozmehr, E., Karevan, M.: An investigation into thermal history and its correlation with mechanical properties of PA 12 parts produced by selective laser sintering process. J. Mater. Eng. Perform. **29**, 832–840 (2020)
20. Rietzel, D., Drexler, M., Kühnlein, F., Drummer, D.: Influence of temperature fields on the processing of polymer powders by means of laser and mask sintering technology. In: Solid Freeform Fabrication Symposium, pp. 252–262 (2011)
21. Josupeit, S.: On the influence of thermal histories within part cakes on the polymer laser sintering process. Dissertation, Universität Paderborn, Shaker (2019)
22. Gibson, I., Shi, D.: Material properties and fabrication parameters in selective laser sintering process. Rapid Prototyp. J. **3**(8), 129–136 (1997)
23. Kruth, J.-P., Levy, G., Klocke, F., Childs, T.H.C.: Consolidation phenomena in laser and powder-bed based layered manufacturing. Annals CIRP **56**(2), 730–759 (2007)
24. Drummer, D., Rietzel, D., Kühnlein, F.: Development of a characterization approach for the sintering behavior of new thermoplastics for selective laser sintering. In: LANE 2010, Physics Procedia, vol. 5, pp. 533–542 (2010)
25. Amado, A., Wegener, K., Schmid, M., Levy, G.: Characterization and modeling of non-isothermal crystallization of Polyamide 12 and co-Polypropylene during the SLS process. In: Polymers & Moulds Innovations Conference, pp. 207–216 (2014)
26. Amado Becker, A. F.: Characterization and prediction of SLS processability of polymer powders with respect to powder flow and part warpage. Dissertation, ETH Zürich (2016)
27. Drummer, D., Greiner, S., Zhao, M., Wudy, K.: A novel approach for understanding laser sintering of polymers. Additive Manuf. **27**, 379–388 (2019)

28. Li, X., Van Hooreweder, B., Lauwers, W., Follon, B., Witvrouw, A., Geebelen, K., Kruth, J.-P.: Thermal simulation of the cooling down of selective laser sintered parts in PA12. Rapid Prototyp. J. **24**(7), 1117–1123 (2018)
29. Drummer, D., Wudy, K., Drexler, M.: Modelling of the aging behavior of polyamide 12 powder during laser melting process. In: AIP Conference Proceedings, vol. 1664, p. 160004 (2015)

Additive Manufacturing of Multi-material Polymer Parts Within the Collaborative Research Center 814

Robert Setter[1,3], Thomas Stichel[1,2], Thomas Schuffenhauer[1,2,4], Sebastian-Paul Kopp[1,2,4], Stephan Roth[1,2,4], and Katrin Wudy[1,3](✉)

[1] Collaborative Research Center 814 – Additive Manufacturing, Am Weichselgarten 9, 91058 Erlangen, Germany
katrin.wudy@tum.de

[2] Bayerisches Laserzentrum GmbH (blz), Konrad-Zuse-Straße 2-6, 91502 Erlangen, Germany

[3] Department of Mechanical Engineering, Professorship of Laser-Based Additive Manufacturing, Technical University of Munich, Munich, Germany

[4] Erlangen Graduate School in Advanced Optical Technologies (SAOT), 91052 Erlangen, Germany

Abstract. During the past years additive manufacturing (AM) has revolutionized the manufacturing world by enabling rapid generation of geometrically-intricate designs. However, up to now in laser and beam-based AM of polymers only single powder materials whether filled or pure plastics can be processed. One aim of the Collaborative Research Center 814 (CRC 814) – Additive Manufacturing is to establish new process technologies to produce multi-material polymer parts in AM. Therefore, two different strategies basing on Laser Sintering (LS) will be explored: On the one side, selective powder deposition technologies like vibrating nozzles or electrophotography are investigated, which enable to lay down different powders beside each other in one process. On the other side, a liquid reactive UV-curing thermoset is implemented into the LS process chamber. After curing of the UV-curing thermoset the powder beside the cured thermoset is molten by the use of the CO_2 laser. Both strategies allow the generation of multi-material parts consisting of material regions with different functional properties.

Keywords: Additive Manufacturing · Laser Sintering · Binder Jetting · Multi-material parts

1 Introduction

Laser Sintering (LS) of polymer powders is a powerful additive manufacturing (AM) technology, which can be used to generate fully functional prototypes or complex plastic design models by performing a layer-by-layer 3D printing process [1]. Although this method provides a lot of benefits and possibilities, it also brings along a tremendous restriction: the usage of a single material within the building process. Thereby, the realization of multi-material components with combined functionalities is unfeasible.

U. Reisgen et al. (Eds.): EMPOrIA 2020, LNME, pp. 142–152, 2021.
https://doi.org/10.1007/978-3-030-70332-5_13

To diminish this constraint, the Collaborative Research Center 814 (CRC 814) – Additive Manufacturing investigates different approaches in combination with LS which base on the one hand on the selective deposition of multiple powder materials (multi-material powder layers) with new coating technologies and on the other hand on the implementation of a liquid reactive UV-curing thermoset in the LS process chamber in order to selectively adapt material properties.

Here, the current state of the art of AM of multi-material parts is presented. After that, specific results of nozzle-based multi-material deposition, electrophotographic patterned powder layer preparation, as well as new LS routes for the melting of multi-material polymer powder layers are discussed. At last, the dynamic mechanical properties of resin infused polymer powder specimens are analyzed.

2 State of the Art – Additive Manufacturing of Multi-material-Parts

According to Anstaett et al. [2] there are two ways of differentiating multi-materials parts: Discrete or graded changes of materials. A discrete change of materials is defined as a material combination, where two or more materials adjoin each other, having a defined border. A graded change of material is a transition area with a smooth change from one material to another. At this moment, there are multiple concepts in the field of AM, which attempt the combination of different types of materials as well as the local adjustment of certain properties. Fused Deposition Modeling (FDM) is an extrusion-based processing technique that is capable of building multi-material parts. An application example of this technology are additively manufactured multi-material antennas as showcased by Mirotznik et al. [3] at the University of Delaware (Newark, Delaware). Polycarbonate thermoplastic stock with different fillers was used to print dielectric and magnetic components. Poly-Jet-Modeling (PJM) and Multi-Jet Modeling (MJM) print photo reactive polymers in liquid form, which are cured by being exposed to a UV-light source implemented within the printing head. [4]. Components created by PJM/ MJM tend to negatively alter over time through exposure to daylight and oxygen, which deteriorates the mechanical and optical properties [5, 6]. Sugavaneswaran et al. [7] demonstrated that randomly oriented multi-material AM components can be created using PJM. The showcased reinforced components are specimen for tensile testing (ASTM: D412-C), which combine Tango Plus™, an elastomer like matrix material and Vero white™ as the hard-plastic reinforcement. Further, Tibbits et al.[8] were able to implement shape-memory behavior into parts through multi-material PJM (a.k.a. 4D-printing). Through the additional functional implementation, they were able to create multi-material single strands, which are self-folding through energy input or change of environmental conditions. Choi et al. [9] from the University of Texas demonstrated multi-material Stereolithography with rotating vats. As in PJM, the different materials are limited to liquid photopolymers only. LS of powders is currently predominantly commercially available for the creation of mono-material parts[2]. However, laser sintered components show porous integrity, which decreases the functional utilization of the final parts [4]. Promising concepts are currently being worked on to combine different materials within LS. As described by Anstaett et al. [2], there are three different ways to allocate different powder in the building chamber: Vibrating nozzles, coater and electrophotography. Especially the use of electrophotography shows to be a highly promising approach. For the accurate preparation of arbitrary multi-material powder layers for LS two techniques – the vibrating powder nozzles and the electrophotography – were explored thoroughly. Vibrating nozzles enable the highly selective discharge of small amounts of powder within the LS building chamber, which is why they are predestined for the targeted functionalization of sintered parts [9]. Electrophotography, also known as xerography, relies on a photoconductive substance, whose electrical resistance decreases when it is exposed to light. This complicated process is not only essential in the common printing industry because of its ability to handle powders very precisely and fast, but also can lead combined with AM to a powerful 3D printing technology for creating multi-material functional components [10]. However, the stacking of powder layers by electrostatic forces is challenging which was firstly shown by Kumar et al. in [10] in their pioneer work of electrophotographic 3D printing. Thus, also different approaches basing on pressure and heat are investigated to produce 3D objects [11].

3 Multi-material Concepts Within the CRC 814

3.1 Multi-material powder deposition

There are several approaches to pursue the goal of generating multi-material parts in the LS context, all of which are still subject of ongoing research. Vibrational powder nozzles can be employed for the deposition of fine geometric features (e. g. dots, lines) of powder materials, allowing the generation of arbitrary multi-material powder patterns. The nozzle-based approach exploits the reversible bridging mechanism of the powders within the conical nozzle body: Without external stimulus, stable powder bridges are formed in the nozzle orifice. Application of stimuli such as vibration to the nozzle causes the break-up of these bridges which initiates powder mass flow [12, 13]. Due to the elevated processing temperature during LS and its effect on flowability, employment of a nozzle system in a LS machine requires exact control of the nozzle temperature (and hence powder temperature) to ensure constant flowability. For this reason, the used metal nozzles are equipped with internal channels allowing the continuous flow of cooling/heating liquids [14].

The electrophotographic principle applied in common printing technology uses standard toners, which mostly are made from styrene acrylic co-polymer / polyester resin with particle sizes in the range of 6 μm to 20 μm. They are furthermore functionalized with charge control agents [15]. However, in the context of AM typically powders made of polyamide 12 (PA12), polyethylene (PE) and polypropylene (PP) are utilized [16]. The particles contained have a diameter of around 50 μm or more. Since electrophotography is applied to produce two-dimensional prints on papers, the technology transfer to LS requires the solution of certain challenges[15]. Next to the different materials already mentioned, the stacking of layers in order to release a three-dimensional object is crucial. Since the part thickness is increasing with subsequent printing of layers, a transfer driven by an electric field between the photoconductor and the initial substrate plate is not useful. To meet this challenge, a new technology was developed, which allows the contactless deposition of powder picked up by a photoconductor on any substrate. The method uses a transfer metal grid with defined electric potential which is placed in close distance above the substrate or the already printed layers [15] Since the distance between the grid the photoconductor is hold constant, the transfer electric field is independent from the height of the already generated part.

The processing of two or more powder materials with different melting temperatures requires the adaption of the energy deposition method. Preheating and melting can be achieved by using three sources of energy: (1) IR emitters for global preheating of the build chamber to the preheating temperature of the low-melting material, (2) a CO_2 laser for local, selective preheating of the high-melting polymer, and (3) a thulium laser for simultaneous melting of the multi-material part geometry [17]. However, this process requires a quick (1 to 5 s) powder coating process to avoid shadowing of those powder areas selectively preheated by the CO_2 laser radiation which leads to early crystallization and curling of the top layer. Since the production of powder layers using the vibration nozzle system already presented is slow (several minutes per layer with a size of 50x50 mm^2),, an alternate process route for the combination with vibrating nozzles was recently introduced, which just uses two energy sources [18]. While the building chamber is heated to melting temperature of the high-melting material which leads to the melting of the low-melting polymer by the IR emitters directly after its deposition by the nozzle, the high-melting material is selectively melted by a moving CO_2 laser beam comparable to the classical LS process.

3.2 Implementation of Reactive Liquids in the Laser Sintering Process

As can be seen in Fig. 1, a concept of integrating reactive resins within a laser-based AM technique is based on the combination of two processes: LS and Binder Jetting. A micro dispenser is used to selectively inject the reactive liquid into the powder bed. An infrared lamp preheats the building chamber near the melting point of the powder. Depending on the process strategy, the curing of the reactive liquid and the melting of the LS powder are performed either simultaneously or sequentially. The schematic concept in Fig. 1 is depicted for a photo reactive resin that is cured by UV radiation. Accordingly, thermoplastic elastomeric powder has to be used as the "soft" component.

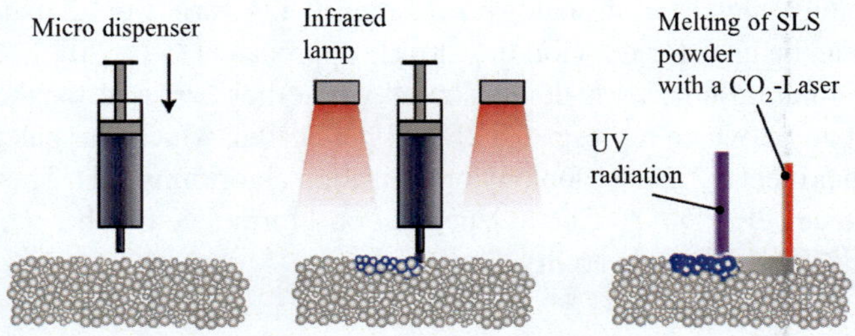

Fig. 1. Schematic concept and set up of a combined LS/ binder jetting process for the implementation and curing of reactive liquids within the building chamber (adapted from [19])

The overall target for the depicted process is the homogenous distribution of the reactive liquid at specific locations of the powder bed without separation of the individual components. As mentioned in Sect. 2, the requirements towards the powder and the reactive liquid are dominantly dependent on the process parameter of the LS process. Besides thermal stability, injection speed and surface tension between powder and liquid component other factors like building chamber atmosphere and part shrinkage are highly influential on the successful implementation of reactive liquids in the powder bed. The approach of implementing reactive resin in LS process is highly promising towards the creation of functional multi-material parts with discrete or graded changes of materials and mechanical properties. However, there are multiple requirements towards the powder and the reactive liquid that have to be taken in account:

- Due to the preheated building chamber, the liquid component must be thermally stable near the melting temperature of the thermoplastic.
- The infiltration speed must be kept at an ideal state, which is mainly influenced by the viscosity of the liquid component and the specific characteristics of the powder.
- For sufficient wetting of the surface, a characteristic ratio in surface tension between powder and liquid component is required.

In the following chapters, experimental setups and results of the above described approaches for creating multi-material parts are discussed and compared to each other.

4 Experimental Methodology

4.1 Nozzle-Based Multi-material Deposition

The setup for deposition of two powder materials with nozzles consists of a high-temperature piezo actuator providing the vibration stimulus and two metal nozzles. This construct is mounted to the standard coater of a LS machine (P380, EOS GmbH, Germany) in such a manner that nozzle movement perpendicular to that of the blade coater is enabled. The nozzles' internal fluid channels can be connected to either an oil heating unit (30 °C–150 °C) or a water cooler (20 °C) for precise control of the nozzle temperatures. Vibration of the actuator is controlled via a function generator creating a sinusoidal voltage signal and an analog power amplifier. A variety of LS materials (PA12, PP, TPE-A) can be used for the vibrating nozzle approach [13, 14, 20]. In Sect. 5.1, the potential of vibrating nozzles is demonstrated while they are used for the preparation of a powder pattern which represents a 2D gripper design, which was calculated by the CRC 814 sub project C2 using topology optimization algorithms [21]. The powder used is polypropylene (PP, PD0580 Coathylene, DuPont) formulated with 1 wt. % Aerosil® (R106, Evonik) for better flowability.

4.2 Electrophotographic Patterned Powder Layer Preparation

Figure 2 shows the individual steps for the powder development of the electrophotographic powder application method modified for powders which are used in AM [15]. First, the photoconductive plate coated with a 100 μm thick layer of the positively-charging photoconductor As_2Se_3 is charged by means of corona wire. The second step is the charging of a layer of powder particles within the powder bed with a scorotron unit. The scorotron consists of a corona wire, to which high voltage is applied, and a metallic scorotron grid with a much lower voltage, which is placed between the corona wire and the powder surface. In this way the powder particles can be homogeneously charged. Simultaneously, a latent charge image is generated by selectively illuminating the photoconductive plate employing a DLP projector. Within the Discharged-Area-Development (DAD) the previously positively charged particles are hereinafter transferred to the discharged areas of the photoconductive plate. The final step (not included in Fig. 2), an electric transfer field is applied in order to deposit the powder particles from the photoconductive plate onto the substrate plate. A variety of LS materials (PA12, PA11, PP, TPU) can be used for the electrophotographic development process depicted in Fig. 2 [15]. Here, the development process is demonstrated for the preparation of a powder pattern which represents the same 2D gripper like in the Sect. 4.1. The powder used is polypropylene (PP, PD0580 Coathylene, DuPont) formulated with 1 wt. % Aerosil® (R106, Evonik).

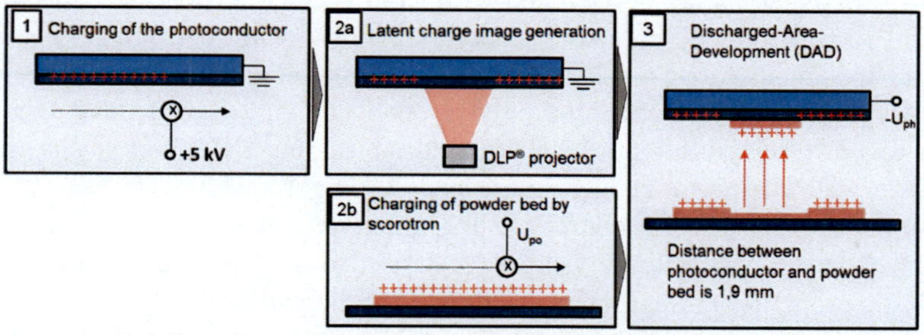

Fig. 2. Individual process steps of the electrophotographic powder development method. Discharged-Area-Development: Step 1: Charging of the photoconductor; step 2a: Latent charge image generation; step 2b: Charging of powder bed; step 2: Powder development

4.3 Generation and Analysis of Multi-material Parts

Using vibrating nozzles in combination with the alternate process route (see Sect. 3.1), simple multi-layer multi-material samples can be produced. The process route used was discussed in detail in [18], so the study here focuses on the morphology of the bonding zone between the different material regions after processing. For the investigation, PP (PD0580 Coathylene, Axalta, Switzerland) formulated with 1 wt. % Aerosil® and urethane-based TPE (TPE-U; Rolaserit PB, AM Polymer Research, Germany) were selected. 5 wt% carbon black (Lamp Black 101, Orion Engineered Carbons, Luxembourg) was admixed to the TPE-U powder to allow the distinction between both materials and the microscopic analysis of the boundary zone. Thin sections were extracted from five-layer thick rectangular TPE-U/PP samples which possess a vertical, flat bonding zone. By means of transmitted light microscopy the boundary zone was evaluated.

4.4 Dynamic Mechanical Properties of Resin Infused Polymer Powder Specimens

The fundamental experimental methodology for the creation of multi-material parts in a combined LS/binder jetting process concentrates on analyzing three influential categories: Material combination, process strategies/energy input and material distribution. First investigations of material combinations and the process strategies have already been documented by Wudy et al. [19, 22].

This investigation concentrates on the material distribution and more specific the macroscopic dynamic properties of the transition area between reactive liquid and thermoplastic powder. Since the actual combination of reactive liquid and powder are not yet possible within a LS-based AM technique, specimens for this investigation are created within a UV oven. The specimens have dimensions of 80 mm × 10 mm × 2 mm. The reactive liquid is an UV-Acrylate UV DLP Hard purchased from PhotoCentric3D, United Kingdom. It consists of acrylate and methacrylate monomers and a photoinitiator. For the powder, an unmodified PrimePart ST PEBA 2301 polyether block polyamide powder (PEBA) from the supplier EOS GmbH, Germany, is used. For the further investigations, the materials are abbreviated with TPE-A and UV-Acrylate. Specimens of 40/60, 30/70 and 20/80 TPE-A to UV-Acrylate ratios (weight percentage) are created, as well as specimens solely made of TPE-A. Furthermore, specimens exclusively made of UV-Acrylate are created with different degrees of cure of 97% and 100%. The degree of cure was determined through UV-Differential Scanning Calorimetry. The curing of the UV-Acrylate was performed by exposition to UV light for 3 min for every specimen, except for specimens of 100% degree of cure, which were exposed to UV-light twice for 3 min. Dynamic Mechanical Analysis (DMA) is used to analyze the dynamic mechanical viscoelastic properties of the specimens. For this investigation a DMA ARES G2 of the company TA Instruments, USA is used. Frequency sweeps are performed to determine the storage modulus G', loss modulus G" and the loss factor tan (δ) at constant room temperature with increasing excitation frequencies between 0,01 Hz and 10 Hz. For this, a torsion vibration DMA apparatus set up is used.

The main goal of this investigation is to show, that through the implementation of a reactive resin within a thermoplastic powder the mechanical properties of the multi-material specimen can be adjusted selectively and over a broad range.

5 Results and Discussion

5.1 Nozzle-Based Multi-material Deposition

These mass flows through a nozzle are highly dependent on the applied voltage amplitude for the vibration excitation The PP powder pattern created by a vibrating nozzle at room temperature is depicted in Fig. 3 (left). Therefore, a nozzle movement speed of 20 mm/s was used which led to a preparation time of about 3 min. The powder pattern was evaluated concerning spatial characteristics (e. g. resolution, line width, layer height, homogeneity) by means of laser scanning microscopy. It has been shown that a minimum line width of about 1.25 mm with a height of about 0.7 mm was achieved; however, strong fluctuations for the line width were detected. The latter are mainly caused by uncontrolled and unspecified parasitic vibrations induced by the moving axes. The results show that the application of vibration nozzles for the preparation of powder pattern can be carried out, but is slow and therefore not suitable for large-scale use.

5.2 Electrophotographic Powder Pattern Development

If using Discharged-Area-Development for powder pattern generation, the charge image of the photoconductor shall hinder the development of the powder in certain areas. However, if the applied electric field outperforms the repulsion due to the charge image, false development occurs, which means that powder is picked up also by the charged areas of the photoconductors. In order to prevent this, the powder charge, which depends on U_{po}, and the development voltage U_{ph} have to be adjusted. Experiments at a development distance of 1.9 mm showed that best results with smallest amount of false development are achieved for PP powder at an electric field of about 780 kV/m and a grid voltage U_{po} of 2.0 kV. These parameters were used to create the powder pattern depicted in Fig. 3 right. The graphical analysis of Fig. 3 right shows that only minor false development occurs, while the powder pattern itself is quite homogenous with a degree of coverage close to 100%. Moreover, the minimal structure details (minimal line thicknesses) are between 0.2 and 0.5 mm which shows that electrophotography enables much more delicate patterns than vibrating nozzles (see Fig. 3 left).

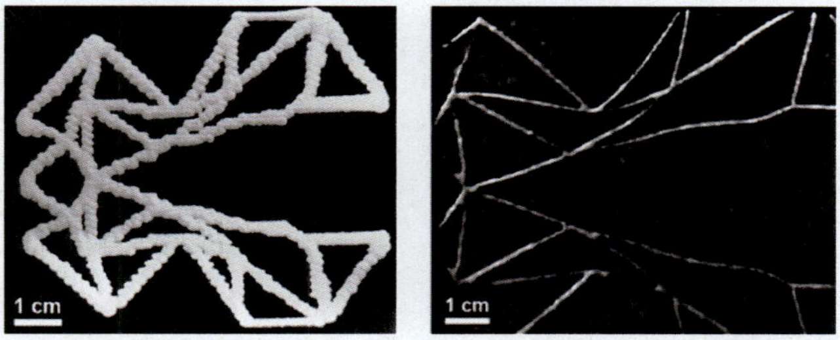

Fig. 3. Photographic images of single-layer powder pattern made out of PP powder created by nozzle discharge with vibration excitation (300 Hz, 30 V) (left) and by electrophotographic Discharged-Area-Development deposited onto an As_2Se_3 photoconductor (right).

5.3 Consolidation of Multi-material Powder Layers Prepared by Vibrating Nozzles

Multi-material samples were successfully sintered via the aforementioned alternate process route and analyzed according to Sect. 4.3. A typical microscopic sectional view of the boundary zone between TPE-U and PP is shown in Fig. 4. It is remarkable that the boundary zone includes a mixed zone which extends over several millimeters. This is caused by the serial powder deposition process using vibrating nozzle, whereas the powder lines are deposited next to each other with a defined overlap.

Since the graded boundary zone enables a quite effective mechanical adhesion[23], combining materials, which are known to be incompatible for conventional 2K injection molding processes due to the lack of intramolecular diffusion processes, is possible.

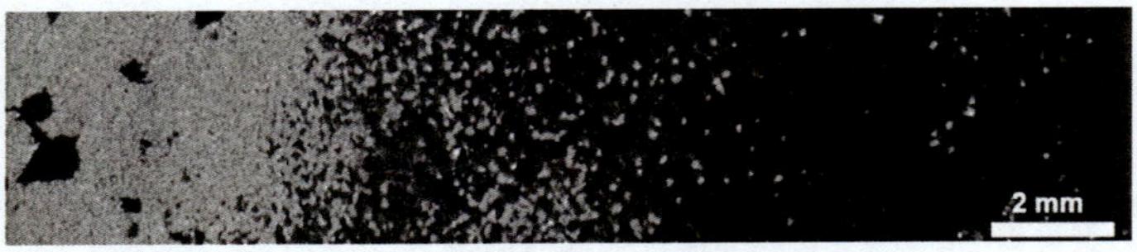

Fig. 4. Microscopic image of a thin section of a multi-material sample (left: PP; right: TPU) generated by means of vibrating nozzles in combination with the alternate process, which is based on two sources (IR heater and CO_2 laser) for energy deposition.

5.4 Dynamic Mechanical Properties of Resin Infused Polymer Powder Specimens

To analyze the dynamic mechanical properties oft resin infused polymer powder specimens, two parameters are taken in to account: Storage modulus and the loss modulus. As can be seen in Fig. 5, the storage modulus increases strongly with increasing frequency for all specimens. This is a sign for increasing stiffness of the materials for higher frequencies. Compared to specimens of plain UV-Acrylate with 97% degree of cure, a significant increase of the storage modulus for UV-Acrylate specimens with 100% degree of cure is visible. Thus, with increasing curing degree the stiffness rises. TPE-A shows the lowest storage modulus. As expected the increase of the TPE-A content within TPE-A and UV-Acrylate blends lowers the storage modulus drastically.

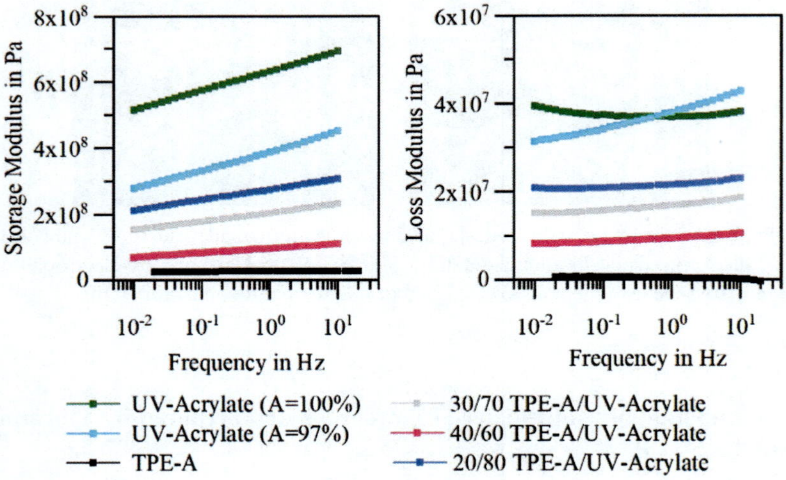

Fig. 5. Storage and Loss modulus in Pa for different TPE-A to UV-Acrylate ratios under constant room temperature during a DMA frequency sweep

The aim of the investigation was to adjust a high variety of dynamic mechanical properties with the combination of the two materials TPE-A and UV-Acrylate, which is proven with these results. The loss modulus shows an atypical decrease for increasing frequencies for 100% cured UV-Acrylate specimens. This is potentially caused by premature specimen failure. Furthermore, it can be seen on a global scale, that the loss modulus of all other specimens is nearly stable for the depicted frequency range. This is beneficial towards the target of a constant amount of irreversible deformation energy for increasing load frequencies. Furthermore, it can be noticed that UV-Acrylate specimens with 97% Degree of cure show a stronger increase of the loss modulus compared to TPE-A and UV-Acrylate blends. As before, a general decrease of the loss modulus for increasing TPE-A content is visible, with plain TPE-A as the global minimum.

6 Conclusion

The results of the deposition of single layer test pattern made out of polypropylene using vibrating nozzles as well as electrophotography indicated that both technologies differ tremendously concerning their application potential. The powder deposition by vibrating nozzles showed itself to be relatively slow with a layer preparation time of several minutes and minimal line widths in the mm range. This makes vibration nozzles particularly useful for the selective functionalization of sintered components, but not for the generation of complex components from multiple materials. In contrast, electrophotography obtains high deposition speeds of a few seconds since whole layers can be prepared at once. In addition, a minimum line width down to ca. 0.2 mm could be achieved showing electrophotography being predestined for the generation of complex multi-material components with high precision. The microscopic analysis of the boundary zone in a multi-material sample produced by the combination of vibrating nozzles and an adapted LS melting strategy revealed a boundary zone of TPE-U and PP which includes a zone of several millimeters in which both materials are gradually mixed. Since the graded boundary zone enables a quite effective mechanical adhesion, the combination of thermodynamically immiscible materials can be realized. In the future, it is planned to combine electrophotography with a LS melting strategy in order to generate defined graded boundary zones with optimized mechanical properties.

The possibility to adjust mechanical properties using multiple materials was demonstrated by a new approach basing on liquid reactive UV-curing thermoset which is implemented into the LS process chamber. After curing of the UV-curing thermoset the powder beside the cured thermoset is molten by the use of the CO2 laser. The measurement results of the storage and loss modulus for resin infused powder specimens lead to the overall conclusion, that the mechanical properties of the multi-material parts can be adjusted selectively and over a broad range. Through increase of the UV-Acrylate content within TPE-A and UV-Acrylate blends, as well as increase of the degree of cure, the storage modulus is drastically raised. However, the final process might not be able to manufacture parts in the depicted range with a fluent transition of the mechanical properties, but instead create variations consisting of two extremes like highly stiff to highly elastic. Furthermore, not all depicted ratios of reactive liquid to powder are realistically applicable for the creation of functional parts. The degree of cure has a more significant influence in terms of adjustability and improvement of the mechanical properties.

Acknowledgement. Funded by the Deutsche Forschungsgemeinschaft (DFG, German Research Foundation) – Project-ID 61375930 – SFB 814, sub-project B7.

References

1. Kruth, J.P., et al.: Consolidation phenomena in laser and powder-bed based layered manufacturing. CIRP Ann. **56**(2), 730–759 (2007)
2. Anstaett, C., Seidel, C.: Multi-Material Processing. Laser Tech. J. **13**(4), 28–31 (2016)
3. Mirotznik, M.S., et al.: Multi-material additive manufacturing of antennas. In: 2016 International Workshop on Antenna Technology (iWAT) (2016)
4. Gebhardt, A.: Generative Fertigungsverfahren 2016: Carl Hanser Verlag, München (2016)
5. Ravve, A.: Light-Associated Reactions of Synthetic Polymers. Springer, Heidelberg (2006)
6. Kaiser, W., Kunststoffchemie für Ingenieure. Carl Hanser Verlag München (2016)
7. Sugavaneswaran, M., Arumaikkannu, G.: Modelling for randomly oriented multi material additive manufacturing component and its fabrication. Mat. Des. (1980–2015) **54**, 779–785 (2014)
8. Tibbits, S.: 4D printing: multi-material shape change. Archit. Des. **84**(1), 116–121 (2014)
9. Choi, J.-W., Kim, H.-C., Wicker, R.: Multi-material stereolithography. J. Mater. Process. Technol. **211**(3), 318–328 (2011)
10. Kumar, A., Dutta, A.: Investigation of an electro-photography based rapid prototyping technology. Rapid Prot. J. **9**, 95–103 (2003)
11. Rojas Arciniegas, A., Esterman, M.: Characterization and modeling of surface defects in EP3D printing. Rapid Prot. J. **21**, 402–411 (2015)
12. Stichel, T., et al.: Powder layer preparation using vibration-controlled capillary steel nozzles for additive manufacturing. Phys. Procedia **56**, 157–166 (2014)
13. Stichel, T., et al.: Mass flow characterization of selective deposition of polymer powders with vibrating nozzles for laser beam melting of multi-material components. Phys. Procedia **83**, 947–953 (2016)
14. Stichel, T., et al.: Multi-material deposition of polymer powders with vibrating nozzles for a new approach of laser sintering. J. Laser Micro Nanoeng. **13**, 55–62 (2018)
15. Stichel, T., et al.: Electrophotographic multi-material powder deposition for additive manufacturing. J. Laser Appl. **30**(3), 032306 (2018)
16. Kruth, J.-P., et al., Consolidation of polymer powders by selective laser sintering. In: Proceedings of the 3rd International Conference on Polymers and Moulds Innovations, pp. 15–30 (2008)
17. Laumer, T., et al.: Simultaneous laser beam melting of multimaterial polymer parts. J. Laser Appl. **27**(S2), S29204 (2015)
18. Schuffenhauer, T., et al.: Process route adaption to generate multi-layered compounds using vibration-controlled powder nozzles in selective laser melting of polymers (2019)
19. Wudy, K., Budde, T.: Reaction kinetics and curing behavior of epoxies for use in a combined selective laser beam melting process of polymers. J. Appl. Polym. Sci. **136**(7), 46850 (2019)
20. Stichel, T., et al.: Polymer Powder Deposition using Vibrating Capillary Nozzles for Additive Manufacturing (2014)
21. Stingl, M.H.D.: On a combined geometry and multimaterial optimization approach for the design of frame structures in the context of additive manufacturing. In: 7th International Conference on Additive Technologies - iCAT 2018, pp. 112–119 (2018)
22. Wudy, K., Drummer, D.: Infiltration behavior of thermosets for use in a combined selective laser sintering process of polymers. JOM **71**(3), 920–927 (2019)
23. Laumer, T., et al.: Realization of multi-material polymer parts by simultaneous laser beam melting. J. Laser Micro Nanoeng. (2015)

Extreme High-Speed Laser Material Deposition (EHLA) as High-Potential Coating Method for Tribological Contacts in Hydraulic Applications

Achill Holzer[1](✉), Stephan Koß[2], Stephan Ziegler[2], Johannes Henrich Schleifenbaum[2], and Katharina Schmitz[1]

[1] Institute for Fluid Power Drives and Systems, RWTH Aachen University, Aachen, Germany
achill.holzer@ifas.rwth-aachen.de

[2] Digital Additive Production, RWTH Aachen University, Aachen, Germany

Abstract. Additive manufacturing enables the use of a variety of material compositions, especially for near-surface layers and coatings, which allows the optimization of tribological systems regarding their properties like chemical resistance. Nevertheless, it is also cost saving for various hydraulic components. By using the innovative Extreme High-Speed Laser Material Deposition (EHLA) process, a large number of different material combinations can be produced on almost any rotationally symmetrical components. In a previous publication, the authors investigated different process techniques in terms of their tribological properties and used a commonly used stainless 316L material "as printed". Since stainless steel generally has very critical tribological properties, measurements were performed with relatively low loads and velocities. In these tests, EHLA-coated surfaces achieved convincing results. Based on these findings, further measurements with more realistic loads and specialized tribological investigations were performed. Therefore, this paper examines the frictional behavior of 316L surfaces produced by laser-based EHLA treatment against different tribological pairings such as standard quenched steel as well as brass. Allowing a broad comparability, all surfaces have been post processed by lapping as most of the planar parts of hydraulic components are lapped.

Keywords: Tribology · Additive manufacturing · Hydraulic components

1 Introduction

1.1 Motivation

Hydraulic applications are widely used in construction equipment, airplanes and even cars. While being used in construction equipment such as excavators and cranes, a long service life is needed. The hydraulic system needs to work for thousands of hours operating without replacing single components. Aircraft systems must of course also work reliable for many years. Passenger cars need to be designed cost efficiently and

U. Reisgen et al. (Eds.): EMPOrIA 2020, LNME, pp. 153–167, 2021.
https://doi.org/10.1007/978-3-030-70332-5_14

environmentally friendly as they are produced in high quantity. These two points are the reason why research is being conducted on new materials for hydraulic components. In hydraulic applications, a wide variety of different materials is used. Three categories can be identified: Nitriding steel, cast iron and brass. Brass is often used as counter face of the steel or cast iron surface [13]. Due to varying pressures and rotational speed levels, all states of friction can occur within tribological contacts in hydraulic units [2, 12]. Starting with boundary lubrication in the very first moment after the speed-up, and following by a phase of mixed lubrication, about 95% hydrodynamic lubrication is strived for in the steady state of operation as it promises low friction and wear. A high-pressure oil supply often allows for the use of hydrostatic bearings or pockets. However, the use of lead-free materials leads to new challenges in choosing materials. According to Czichos and van Bebber, planar contact pressures can be between $1 \cdot 10^{-3}$ and $4 \cdot 10^{3}$MPa [2, 3]. Typically, hydraulic systems consist of various tribological contacts. These can be divided into two categories: Those with movement in longitudinal direction, like pistons, and those with rotational movement, like thrust bearings. In this work, the focus is the rotational movement. As it is very complicated to investigate in components themselves, a model test bench is used for the first measurements (see Sect. 2.4). The model test bench allows to perform comparative measurements akin to those in real hydraulic components. Especially the tribological contact cylinder block – valve plate - one of the most complicated contact in hydraulics. The contact serves as hydrostatic bearing as well as hydraulic valve and can be abstracted very well [14].

Therefore, both the geometric dimensions and the choice of materials were kept close to those of real components. The basis material used in this paper is a standard quenched and tempered steel, 42CrMoV4 (1.7225). The same material has been used as base for the EHLA coating, as well as reference hard-hard counterpart.

Aeterna 3038, which can be considered a standard leaded brass material, was used to represent a hard-soft tribological pairing typical for hydraulics [14]. All discs have been lathe-turned and lapped. Table 2 shows the geometric parameters of the Tribo test discs.

A highly productive and innovative process is available with the EHLA process, with which the surfaces can be coated individually adapted to the areas of application. Compared to thermal spraying and galvanic coating processes, EHLA produces metallurgically bonded and dense coatings. Due to the characteristics of the process, unconventional material combinations can be created and almost any coating material can be used. However, because the main energy input is into the powder materials, the process is also suitable for coating heat-sensitive substrates, such as hardened materials, aluminum and grey cast iron. In initial accompanying tests, the generated coatings have shown promising results.

1.2 Objective

Hydraulic power components are an important field in drive technology, as they provide high forces and easy longitudinal movement. Mainly those systems use combustion engines as they do often work in remote areas. To increase the power density, hydraulic pumps have to operate at higher rotational speed, which also increases the efficiency of the diesel engine. This is a challenge for the design of the tribological joints. Almost all traditional tribology pairings consist of either expensive special threatened steel, with a hardening step or containing environmentally critical lead. This work studies the possibility of replacing costly or hazardous materials by very basic steel materials equipped with a very thin but tribological active coating. It is expected, that the coated parts behave in the same way, as the coated material itself would do. There should not be any influence by the base material and no peel of the coating. Three possible outcomes are expected: Lower material use having a positive impact on the production costs, a renunciation of lead-containing materials, having a positive impact on the environment and an improved power density due to higher rotational speeds.

2 Materials and Methods

2.1 Extreme High-Speed Material Deposition (EHLA)

Extreme High-Speed Laser Material Deposition (EHLA) is a process based on the in-situ supply of powders via a nozzle and the melting of the powder particles in flight before impinging on a substrate, forming a metallurgical bond with the substrate. In contrast to conventional LMD, the powder focus is usually set above the substrate which leads to a higher interaction time of the powder particles and the laser radiation. Simulations show that this increases the absorption of the laser beam by the powder particles [6]. Due to high superficial velocities of 500 m/min, the process is limited to depositions on rotationally symmetric parts (e.g. tubes and discs) and is often used to apply protective coatings. Since most of the laser power is absorbed by the powder in flight; the dilution

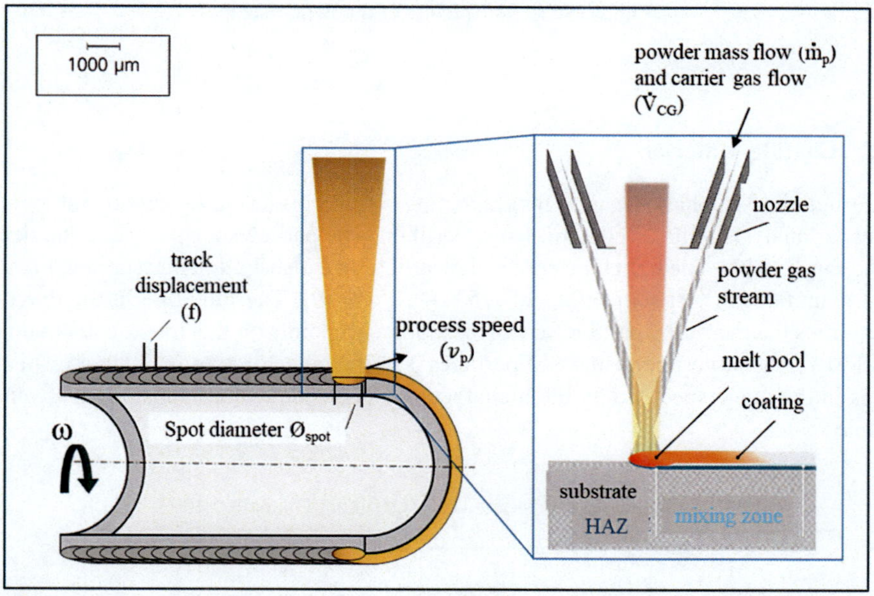

Fig. 1. Sketch of Extreme High Speed Laser Material Deposition (EHLA) including all relevant process parameters

of coating and base material is low which enables a wide range of applicable material combinations [7, 8]. Typical layer thicknesses are in the range of 20 to 350 μm.

Figure 1 is a sketch of the EHLA process and its most important parameters concerning Energy input into powder particles and substrate. The powder focus is positioned above the substrate, so that the intermediation time of the particle through the laser beam suffice to melt the particles. The most important parameters that influence the energy input into the particles are the particle velocity $v_{particle}$ and the powder mass flow $\dot{m}_p$ in the powder gas stream. The energy input into the substrate is mostly influenced by process speed v_p and track displacement f which is the lateral distance between one coating pass and the following. With increasing v_p, the interaction time between the laser beam and the substrate decreases, reducing energy input into the substrate. This leads to a reduction of melt pool size and the size of the heat affected zone. For the process development and the coating of the Tribo discs, a 4-axis machining system (Fig. 2) from Hornet with the laser source TruDisc 8001 is used.

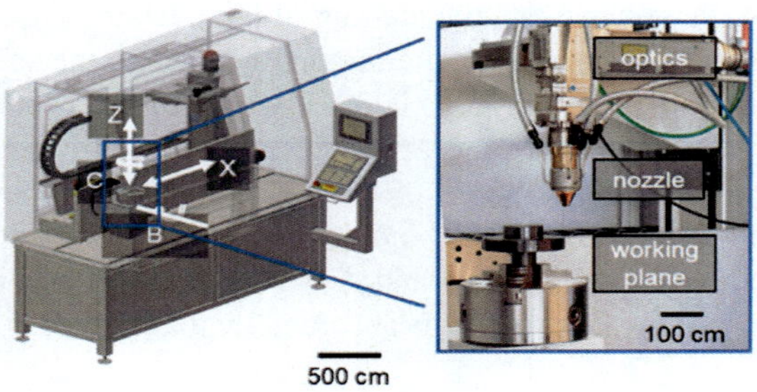

Fig. 2. Schematic picture of the EHLA setup

2.2 Coating Material

The material 316L has been used for the coating of the Tribo test disks. Figure 3 shows the powder analysis of the used 316L from Oerlikon. The particles are spherical and dense (see Fig. 3 a),b)). The light blue area in the particle size distribution measurement marks the manufacturer's specifications of −53 + 20 μm [9]. The blue line in the diagram describes the measured particle size distribution. All points on this line together add up to 100%. The largest percentage of particles is in the range of approx. 30 to 45 μm and thus in the range specified by the manufacturer. The chemical composition is given in Table 1.

Table 1. Chemical composition 316L (Oerlikon, Diamalloy 1003–1) [18]

Fe	Cr	Ni	Mo	Si	C
Balanced	17 wt.%	12 wt.%	2.5 wt.%	2.3 wt.%	0.03 wt.%

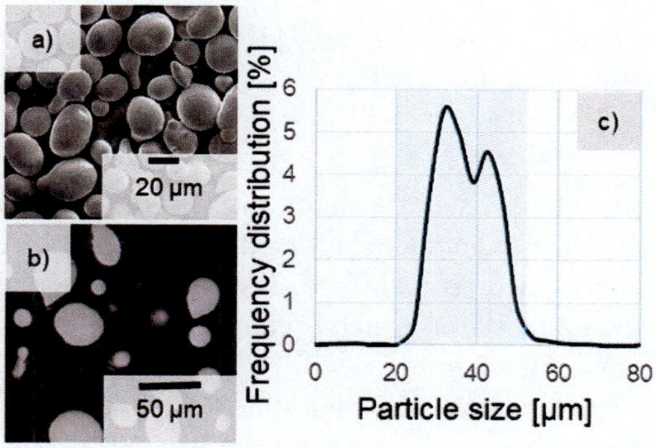

Fig. 3. a) REM picture of 316L paricles (Oerlikon, Diamalloy 1003–1) with the magnification of 400x, b) Incident light microscope image of 316L paricle cross sections with a magnification of 200x, c) Measured particle size distribution of used 316L powder

2.3 Surface Finish

Contrary to the first investigations, in which the coated Tribo test discs were examined in their original state "as printed" [4], real surface treatments as usually applied in hydraulics have been examined here. In hydraulics, the geometry is typically created first by turning to an approximate shape. Then, depending on the material, the gas nitriding process takes place. Finally, the parts are getting lapped or honed.

Lapping is a mainly room-bound process with geometrically undefined cutting edges. It is a production and finishing process, defined as chipping with loose grains distributed in a lapping paste which are usually guided with a shape-transferring counterpart (lapping tool) featuring ideally undirected cutting paths of the individual grains. Surfaces to be lapped are normally flat. If the work pieces have different geometric shapes, correspondingly modified method variations must be applied, in part manually and with auxiliary tools [5].

Control plates from axial piston machines, which are to be abstracted here with the Tribo test disks, can be either flat (flat control plates) or spherical (spherical control plates) [15].

One of the most interesting benefits of lapped surfaces are the undirected processing traces, which means that a pressure gradient can form without dependence on the spatial directions. Fewer connected oil pockets compared to the surface formed by the turning process. Connected grooves lead to high contact pressure peaks at the surface mountains, resulting in severe adhesive wear. The lapping process can be compared to the later lubrication situation in the tribological contact especially in the area of fluid lubrication. Lapping took place on a FLP 900 lapping machine, with a disc diameter of 900 mm by FLP Microprecision GmbH, using a SiC400 lapping paste with a grain size of 12 to 15 μm.

2.4 Tribological Approach and Parameters

A wide range of tribological contacts in hydraulic systems can be abstracted as planar contacts. These are often investigated using a disc - disc tribometer. Ifas conducts its measurements with a device consisting of (see [1]) by turning two discs at a speed up to 15 m/s and contact pressure up to 40 MPa.

Test Bench

The disc – disc tribometer consists of a stator and a rotor, which are pressed together using a hydraulic cylinder. For geometrical details see Table 2. A normal force sensor constantly regulates the applied normal force. The rotor is driven by a hydraulic motor and transmits its torque to the stator, which is connected to a force sensor using a lever. The schematic drawing is shown in Fig. 4.

Both discs are completely covered with a specific hydraulic fluid, which is temperature closed loop controlled. In this examinations a BECHEM Hydrostar HEP 46 EEL biodegradable high-performance hydraulic fluid has been used. Over the different tests, the fluid temperature in the basin was kept at 40 °C.

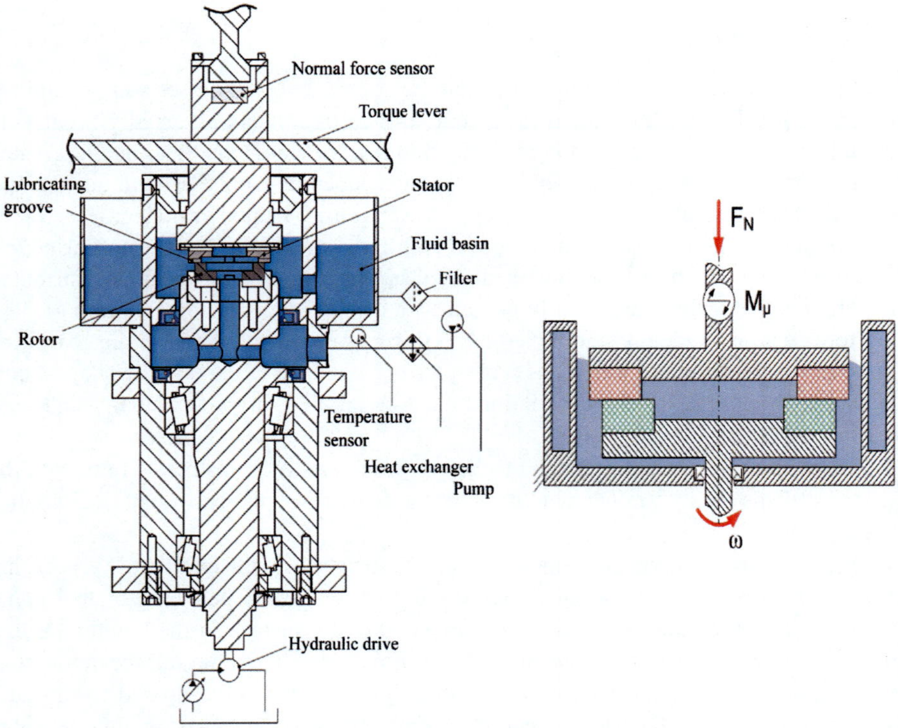

Fig. 4. Cross section of the disc - disc tribometer with rotor and stator (right) [1]

Based on frictional force F_R and normal force F_N, a dimensionless friction coefficient μ is calculated:

$$\mu = \frac{F_R}{F_N} \tag{1}$$

The x-axis, which shows the rotational speed in a Stribeck diagram, is also corrected for the influence of temperature, as the border between mixed lubrication and hydrodynamic lubrication depend mainly on the viscosity η, the pressure p and the angular frequency ω. Therefore, the Gümbel-Hersey number u is used.

$$u = \frac{\eta\omega}{\overline{p}} \tag{2}$$

As - due to the measuring principle - heating always occurs in tribological contact, the influence of temperature must be corrected. This can be done by using the Arrhenius or Andrade Modell [10], approximating the actual viscosity η by a known base viscosity η_0, the activation energy E_A, the gas constant R and the temperature$T$.

$$\eta = \eta_0 \cdot exp\left(\frac{E_A}{R \cdot T}\right) \tag{3}$$

This procedure makes it possible to record Stribeck curves, whereby the parameters temperature, speed and surface pressure must only be recorded accurately, but not precisely controlled, as the Gümbel-Hersey number is calculated by the measurement results of the parameters. The geometric properties of the test discs are shown in Table 2.

Table 2. Geometric parameters of the Tribo test discs

Geometric Parameter	Value
Contact area A_c	946 mm^2
Outer diameter d_a	70 mm
Inner diameter d_i	59 mm
Roughness R_z	0.4 µm

Stribeck Curves

Stribeck curves are a basic but powerful type of diagram to examine all stages of friction of lubricated tribological contacts. The examination consists of an almost load free speed up of the contact. Then, when thermal equilibrium is reached, the load is applied. Because of the high speed, fluid lubrication is usually accessed. Then the speed is reduced by a slow ramp. The tribological contact passes all stated of friction until it reaches boundary lubrication. As shown in Fig. 5, the coefficient of friction is plotted on the y-axis, while the Gümbel-Hersey number is shown on the x-axis.

In the hydraulic application, the contact should mostly run with about 5% of solid state friction. This operation point usually meets the lowest frictional forces and leads to high efficiencies.

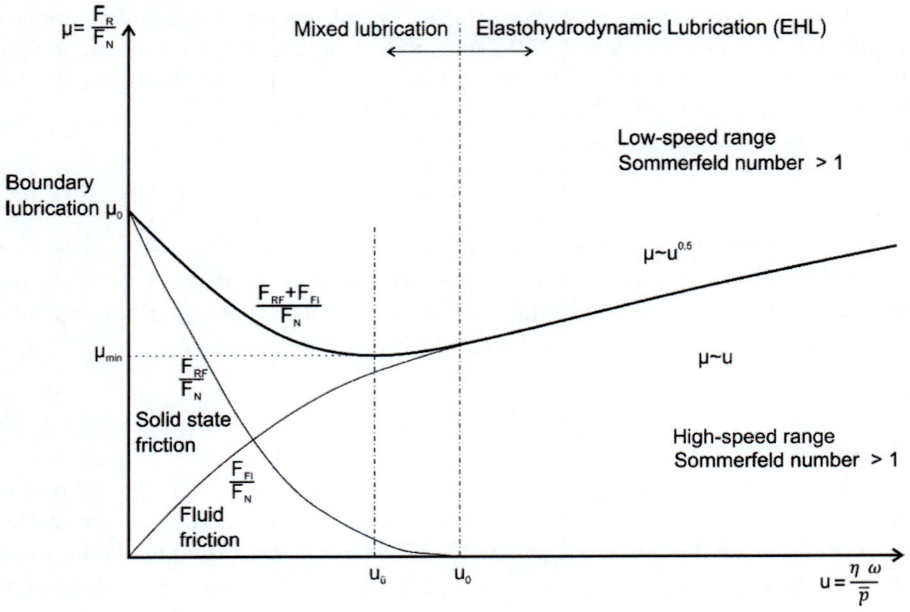

Fig. 5. Lubrication states shown in Stribeck curve

3 Results

In the following chapter the results of two different material parings, EHLA coated 316L against 42CrMoV4 and EHLA coated 316L against an Aeterna 3038 will be shown. Due to the heavy adhesive wear of the 42CrMoV4, it was not possible to show a complete Stribeck curve. The coefficient of friction exceed 0.5, what produced to much heat to be lead away.

3.1 Coating Process

Through parameter variation during process development, a dense metallurgical bonded coating could be achieved (Fig. 7a)). Therefore, the parameters laser power P_L, process speed v_P, track displacement f, carrier gas flow $\dot{V}_{CG}$ and powder mass flow $\dot{m}_p$ have been varied. For the used coating the following parameters could be developed (Table 3).

Figure 6 depicts the measurement results of the hardness measurement of the coating. On the left hand side in the diagram are the measurement results of the substrate material with an average value of 327 HV0.3, and on the right are the results of the 316L coating with an average value of 200 HV0.3. The measured hardness of the coating is within the range of lead-containing brass materials, typically used for hydraulic contacts. Therefore it is expected, that against the steel (hard-hard paring) the coating is soft enough and will behave like the brass material would do. We are expecting the wear to take place on the coating and not on the counterpart. The run against the brass material with the

Table 3. Coating parameters

Laser power (P_l)	2400 W
Process speed (v_P)	50 m/min
Track displacement (f)	0.25 mm/rot
Carrier gas flow ($\dot{V}_{CG}$)	6 l/min
Powder Mass Flow ($\dot{m}_P$)	17.5 g/min

same hardness is more unclear. As both materials have different bases and the crystal structure is not the same, very few wear is expected.

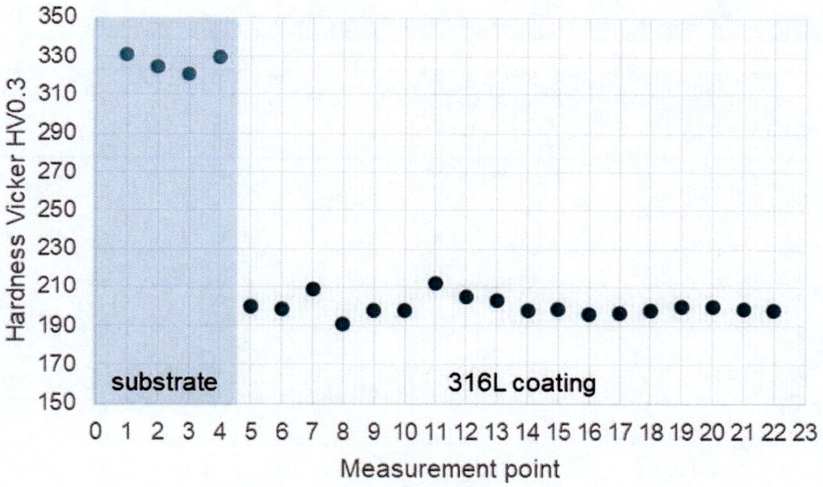

Fig. 6. Result of the hardness measurement (hardness Vickers HV0.3)

The deposited coating has been lapped and examined to determine its tribological properties. As a result of the test, the tribological active surface has been characterized by metallographic analysis (Fig. 7 b)). The cross section of the tested coating shows no cracks or other coating deviations. The contact surface has no erosion, scoring or other indications of uneven wear.

The cross sections of the disc after the Stribeck curve were etched for 2 min at 60 °C with V2A etching solution (distilled water, hydrochloric acid and nitric acid) for microstructure analysis (Fig. 8). Furthermore, the small mixing zone can be seen by the clear bonding zone between substrate and coating.

Through the etching, the individual layers and the melt bath geometry can be identified. However, the microstructure exhibits dendritic solidification.

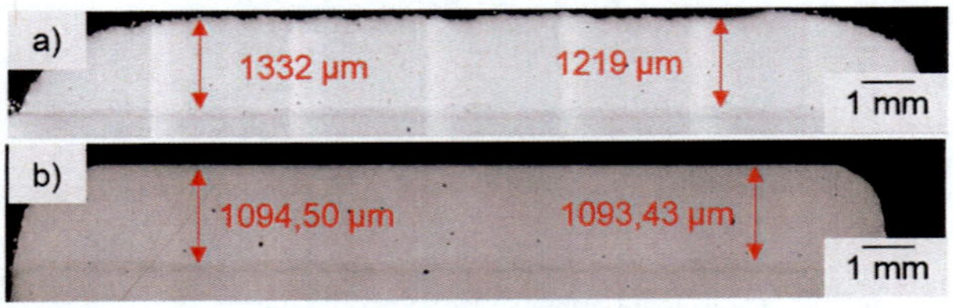

Fig. 7. Cross sections of tribological discs: a) Cross section after EHLA coating with 50x magnification, b) Cross section after tribological test coating with 50x magnification

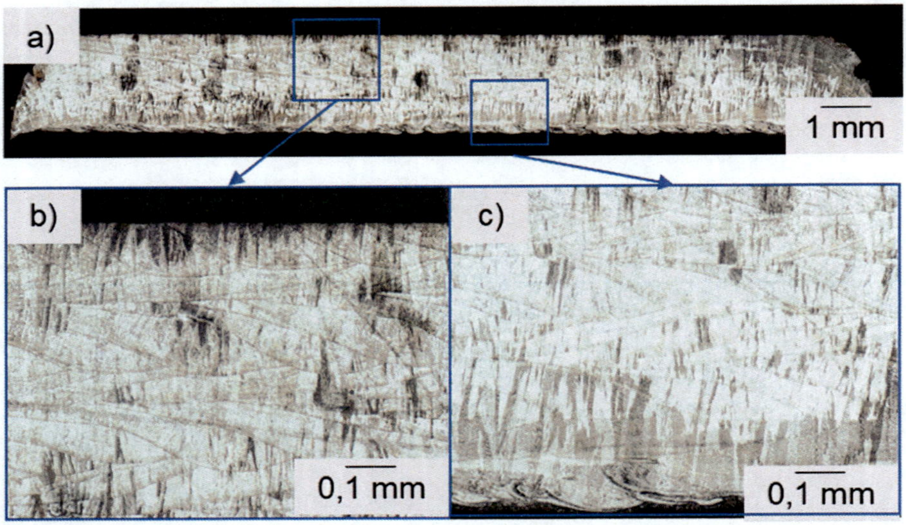

Fig. 8. Etched cross sections of tribological discs: a) Etched cross section after tribological test coating with 50x magnification, b) Etched cross section of the coating with 100x magnification, c) Etched cross section of the bonding zone with 100x magnification

3.2 Wear

Testing the EHLA coated disc against the reference disc (steel) shows the known mechanism of fretting, as shown in the previous publication [4]. Contrary to the former results, the wear for this pairing is very low. This can be explained with the lapped surface which allows the load to be much better distributed over the contact area. The wear profile is shown in Fig. 9. Geometrically the counterpart is in contact at about 2.5 mm of the measured length.

Running the EHLA coated disc against the Aeterna 3038, shows a significant difference. The EHLA coated disc shows almost no visible wear and there is no distinct edge of wear. Instead, there is a smooth transition. The wear profile is shown in Fig. 10. Geometrically the counterpart is in contact at about 2.5 mm of the measure length.

The wear track can be seen by the bare eye on both sides of the tribological pairing (Fig. 11). As there is a clear dark track, it can be assumed that the frictional energy was high enough to activate the extreme pressure additives in the hydraulic fluid [11].

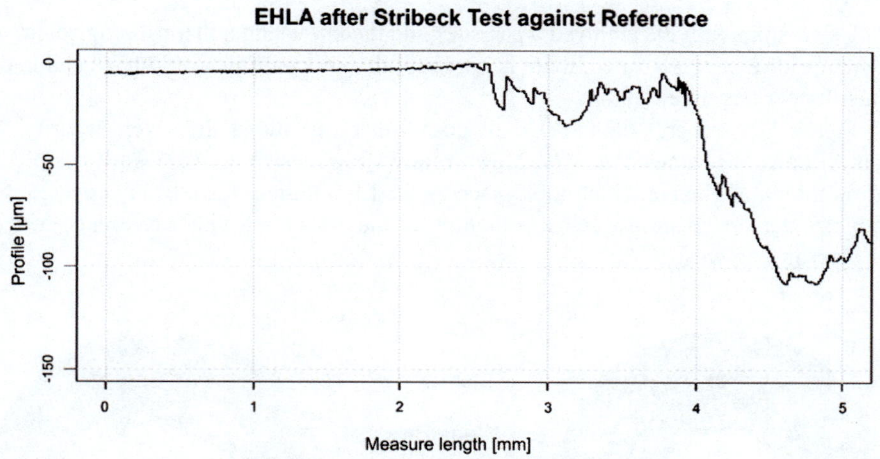

Fig. 9. Wear profile of the EHLA coated disc after Stribeck test against reference

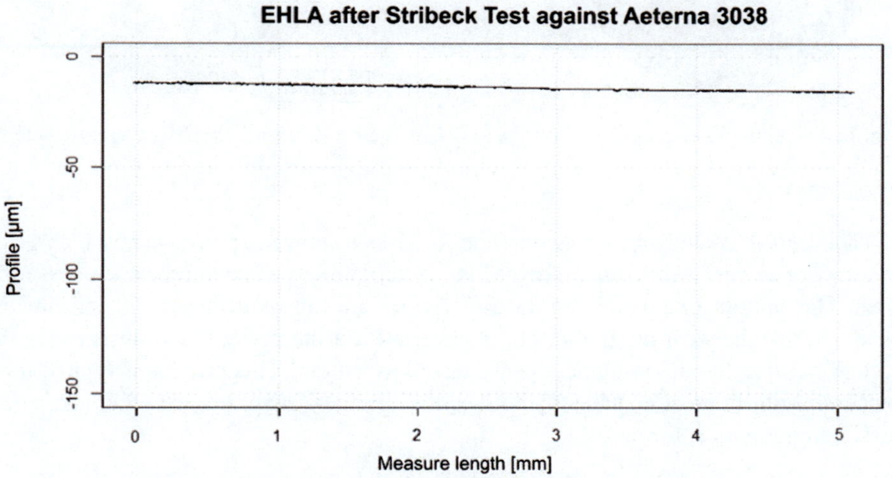

Fig. 10. Wear profile of EHLA after Stribeck test against Aeterna 3038

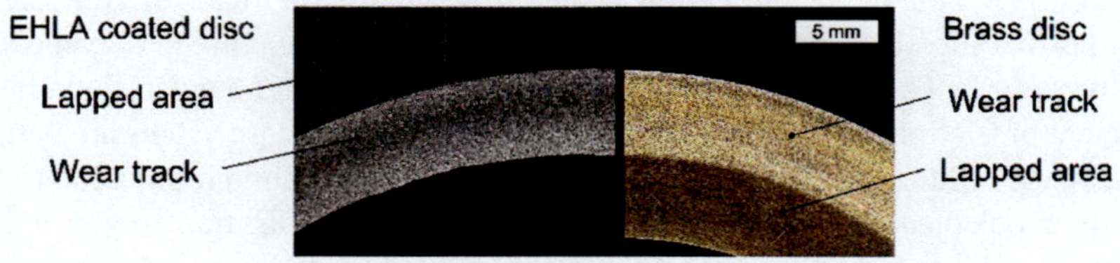

Fig. 11. Optical analysis of the two Tribo test discs

Calculating with the archived wear rates and the knowledge, that tribological active additives have been activated, there is no doubt that this pairing would have endured a much longer test time.

Figure 12 compares the two EHLA coated discs. a) shows the severe wear of the EHLA coated disc against the 42CrMoV4 reference, adhesive wear is clearly visible. b) shows the color change which took place as the EP additives have been activated. On both disc the lapped surface remains on parts of the discs to provide a reference surface for determining wear.

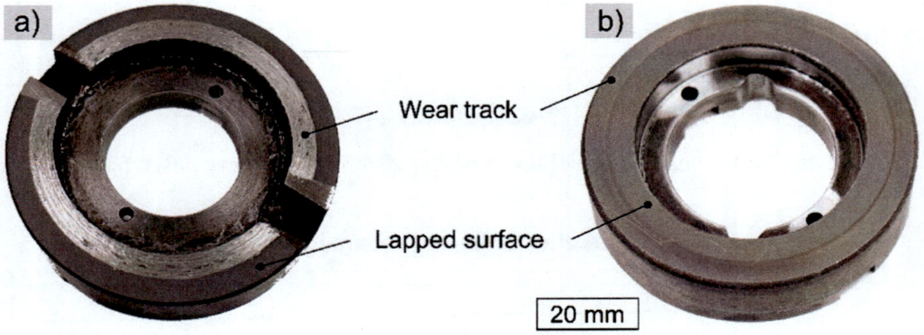

Fig. 12. Wear tracks of the Tribo test discs, a) EHLA against Reference, b) EHLA against Aeterna 3830

The EHLA coated surface against the Reference showed severe wear, which leads to a failure in very short time, whereas the Aeterna 3830 as counterpart showed mild wear. The authors assume that no state of equilibrium can be achieved with such heavy wear and that the wear progresses at the same rate. Furthermore, it was observed in the second pairing that the roughness peaks have leveled out. This process is typical for a successful run-in process, whereby both contact partners adapt to each other and a new surface roughness is formed.

3.3 Stribeck Curves

Several Stribeck curves of the EHLA coated disc against the (leaded) Aeterna 3830, with a normal pressure of 4 MPa have been recorded. The maximum average sliding speed was 6 m/s. The lowest coefficient of friction is 0.012, in boundary lubrication; a coefficient of friction of 0.062 has been measured. However, these values are within the normal range of lubricated contacts [16]. The transition from fluid lubrication to mixed lubrication took place a Gümbel-Hersey number of around 10, the transition from mixed lubrication to boundary lubrication took place at a Gümbel-Hersey number of 0.5. A representative Stribeck curve for the tests performed is shown in Fig. 13.

Due to the heavy adhesive wear, it was not possible to measure a Stribeck curve of the EHLA coated disc against the Reference disc.

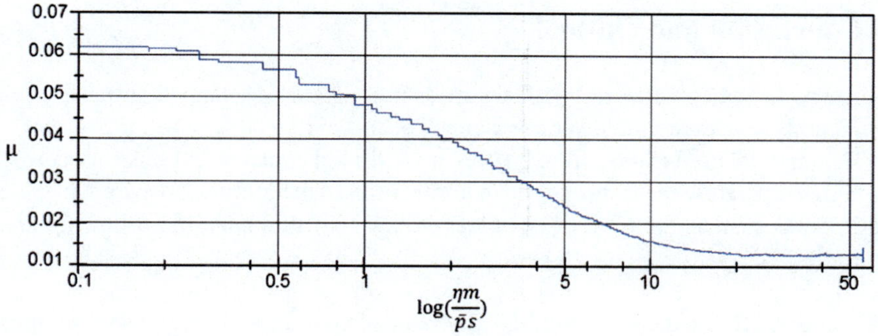

Fig. 13. Stribeck curve of EHLA against Aeterna 3038

4 Discussion

The austenitic steel 316L is corrosion as well as acid resistant and can therefore also be used with highly corrosive lubricants. The coatings generated by using EHLA are dense and do not show any coating defects even after tribological testing. Therefore, this material is suitable for use in tribological contacts with corrosive lubricants and/or in a corrosive environment [18]. Furthermore, no remaining defects or changes in the microstructure could be detected by etching to reveal microdefects and the microstructure. This demonstrates the high temperature resistance of the material as well as the fact that no critically high temperatures occurred in the test [18]. However, further metallographic investigations, such as EDX, are necessary to provide a conclusive analysis of the microstructure.

Against the reference part of 42CrMoV4, the coating shows a wear of approx. 110 μm after the test according to the wear curve in Fig. 9. However, after the metallographic cross sections, a coating with a thickness of averaged approx. 1275 μm was generated, of which approx. 1090 μm residual layer thickness is still present after post processing and tribological examination. As this only corresponds to a reduction of the layer thickness of approx. 185 μm, the initial layer thickness can be significantly reduced to less than 500 μm. In addition, 316L is a comparably soft material with a hardness of approx. 200 HV, so that wear and thus the layer thickness can be further reduced by using an anti-wear coating, such as a hard metal alloy or a Metal Matrix Composite (MMC). For example, MMCs have shown almost no wear in the field of brake rotors in load tests for the automotive sector (DYNO) [17].

The result showed a behavior, as expected of the coating material. The steel – EHLA approved once again, that steel on steel contacts need to be further improved as they suffer from irreconcilable as well as friction coefficients higher than 0.1 [2, 4, 16].

When used against the brass material, the EHLA coated disc showed a behavior, which seems to be similar to the typically used quenched or nitrated steels [1, 2, 13]. There was no grooving nor annealing colors visible, which is another indication for a steady contact.

The coating did not peel off in any circumstances at the test bench. Even in severe wear conditions, where adhesive wear destroyed the surface of the coating to a depth of more than 0.1 mm. This behavior has been observed earlier by the authors [4].

5 Conclusion and Outlook

Compared to the “as printed” surfaces measured earlier [4], lapped surfaces showed significantly less wear and lower coefficients of friction.

The material used is too soft to be used in hard-hard contacts, but it is very corrosion and high temperature resistant and can therefore be used in highly corrosive environments and in areas such as the food or chemical industry. In summary, the following results were achieved.

- Through EHLA a dense metallurgically bonded coating with a thickness of approx. 1275 μm of the material 316L could be produced. After post-processing and testing, the coating has a residual layer thickness of approx. 1090 μm. The binding between substrate and coating was well enough to maintain the link of the layer eve in heavy mixed lubrication and with almost purely adhesive wear. The subsequent metallographic analysis did not present any layer defects after the load in the tribological test.
- The highest wear occurred in the test against the reference demonstrator made of the material 42CrMoV4 with approx. 110 μm. In the test against brass there was almost no wear
- To improve the friction coefficient, which would definitely lower the wear, measurements with different materials need to be done. It has been shown, that EHLA coatings used in hard-soft parings reach good friction coefficients.

EHLA coated surfaces are a promising way of binding together several material properties. Over all tribology tests, even in fully adhesive wear, no signs of a peel-off of the welded layer showed. Especially the use against brass material, which can be seen as state of the art of tribological pairings in hydraulic applications showed very good results. On a tribological point of view, 316L stainless steel coatings produced by EHLA can be recommended when using as hard – soft pairing against a leaded brass material. This enables EHLA coatings to be applied on base steel layers, where it can replace the more expensive nitrated or quenched steel, which can make the production of large components in particular more cost-effective. The approach of using EHLA coated stainless steel as hard counterpart against reference steel, could not fully satisfy. For the future, other materials need to be tested, promising better tribological properties. The next step will be to improve the tribological properties by using more resistive materials as it could be done by filling in a certain amount of carbides. This would increase the strength against wear, allowing to increase the lifetime of the surface.

References

1. Otto, N.: Experimental analysis of sustainable ester- and water-based hydraulic fluids. Ph.D. thesis, RWTH Aachen University, Germany (2018)
2. Czichos, H., Habig, K-H. (Hg.): Tribologie-Handbuch. Tribometrie, Tribomaterialien, Tribotechnik. 4. vollst. überarb. und erw. Aufl. Wiesbaden: Springer Vieweg, Germany (2015)

3. Van Bebber, D.: PVD-Schichten in Verdr¨angereinheiten zur Verschleiß- und Reibungsminimierung bei Betrieb mit synthetischen Estern. Ph.D. thesis, RWTH Aachen University, Germany (2003)
4. Holzer, A., et al.: Tribological investigations on additively manufactured surfaces using Extreme High-Speed Laser Material Deposition (EHLA) and Laser Powder Bed Fusion (LPBF)". In: 12th International Fluid Power Conference, Dresden, March 9–11, Dresden, Germany (2020)
5. Klocke, F., Kuchie, A.: Lapping and polishing. In: Manufacturing Processes 2. RWTHedition. Springer, Heidelberg (2009)
6. Pirch, N., Linnenbrink, S., Gasser, A., Wissenbach, K., Poprawe, R.: Analysis of Track Formation During Laser Metal Deposition. Fraunhofer ILT, RWTH-Aachen Lehrstuhl für Lasertechnik (LLT) (2017)
7. Schopphoven, T., Gasser, A., Backes, G.: EHLA: extreme high-speed laser material deposition. LTJ **14**(4), S. 26–29 (2017). https://doi.org/10.1002/latj.201700020
8. Schopphoven, T., Gasser, A., Wissenbach, K., Poprawe, R.: Investigations on ultra-high-speed laser material deposition as alternative for hard chrome plating and thermal spraying". In: Journal of Laser Applications 28 (2), S. 22501 (2016). https://doi.org/10.2351/1.4943910
9. Oerlikon metco: Material Product Data Sheet, Austenitic Stainless Steel Powder for Thermal Spray. Oerlikon metco (2017)
10. Zerbe, C., et al.: Schmierstoffe2. In: Zerbe, C. (ed.) Mineralöle und verwandte Produkte. Springer, Heidelberg (1952)
11. Burghardt, G., Wächter, F., Jacobs, G., Hentschke, C.: Influence of run-in procedures and thermal surface treatment on the anti-wear performance of additive-free lubricant oils in rolling bearings. Wear 328–329, S. 309–317 (2015). https://doi.org/10.1016/j.wear.2015.02.008
12. N.N. DIN 50281: Friction in bearings; definitions; types; conditions; physical quantities. Withdrawn, Beuth Verlag (1977)
13. Paulus, A.: Reaktionsschichtbildung auf bleifreien Bronze- und Messingwerkstoffen im Kontakt von Zylinder und Steuerscheibe einer Axialkolbenpumpe. Ph.D. thesis, RWTH Aachen University, Germany (2017)
14. Oberem, R.: Untersuchung der Tribosysteme von Axialkolben-Schrägscheibenmaschinen der HFA- Hydraulik. Ph.D. thesis, RWTH Aachen University, Germany (2002)
15. Murrenhoff, H.: Fundamentals of Fluid Power Part 1: Hydraulics. Shaker Verlag, Aachen (2018)
16. Blau, P.J. Friction science and technology. From concepts to applications, 2nd edn. CRC Press/Taylor & Francis, Boca Raton (2009)
17. Rettig, M., Grochowicz, J., Käsgen, K., Eaton, R., Wank, A., Hitzek, A., Schmengler, C., Koß, S., Voshage, M., Schleifenbaum, J.H., Verpoort, C., Weber, T.: Carbidic Brake Rotor Surface Coating Applied by High-Performance Laser-Cladding (2020)
18. Oerlikon: Material Product Data Sheet - Type 316L Austenitic Stainless Steel Powders (2020)

In-Situ Alloying in Gas Metal Arc Welding for Wire and Arc Additive Manufacturing

Uwe Reisgen, Rahul Sharma, and Lukas Oster(✉)

Welding and Joining Institute, RWTH Aachen University, Pontstraße 49, 52062 Aachen, Germany
oster@isf.rwth-aachen.de

Abstract. The use of low transformation temperature (LTT) alloys seems to be a promising way for reducing the residual stress level of fusion welded components. Wire and arc additive manufacturing (WAAM) is a high performance additive manufacturing process for generating large metallic components, which is based on common arc welding processes. The following article describes the investigations regarding generating LTT alloys in WAAM through in-situ alloying. Therefor a multi wire gas metal arc process in spray transfer mode is being used to generate the target LTT alloy. By using two high alloyed cold wires, it was possible to reach a chemical composition for LTT alloys, proposed by Steven and Haynes. The process showed stable behavior and it was possible to build up test specimen in form of wall shaped structures of 15 mm height. By establishing in-situ alloying for the additive manufacturing of LTTs a new field of investigations regarding the structural behavior of LTT-injected components is being opened.

Keywords: Low Transformation Temperature (LTT) alloys · Wire and Arc Additive Manufacturing · Multi wire gas metal arc welding · In-situ alloying

1 Introduction

Wire and arc additive manufacturing (WAAM) is a promising technology for the additive fabrication of large scaled metallic components and is of high interest of current research activities. In WAAM, arc welding processes are being used in the way of classical shape welding to build up components layer wise [1, 2]. The most common processes that are being used for WAAM are gas metal arc welding and the cold/hot wire assisted plasma process. In both cases, an electric arc is being used as heat source to melt the filler wire. High deposition rates of over 5 kg/h, the wide spectrum of usable materials and the low limitations regarding the build volume make WAAM a key technology for large scale additive manufacturing. Two current fields of applications for WAAM processes are the aerospace industry, to reduce the amount of subtractive machining for expensive materials [3], as well as the manufacturing of spare parts with reduced delivery time for the maritime sector [4].

As a result of the high deposition rates, WAAM processes are characterized by a high heat input and therefore the manufactured components are prone to thermal distortion,

U. Reisgen et al. (Eds.): EMPOrIA 2020, LNME, pp. 168–177, 2021.
https://doi.org/10.1007/978-3-030-70332-5_15

the formation of residual stress and a lower geometric accuracy. The underestimation of residual stress fields can lead to fatal component failure during component finishing [5]. This is, why besides of the precise knowledge of the residual stresses in the components, compensation strategies are of high interest for a safe industrial application of WAAM.

Especially in welding applications, the use of Low Transformation Temperature (LTT) alloys has been a promising strategy for reducing the thermal deformation and keeping the amount of residuals stress low at the same time. The key concept of LTT alloys is to achieve a martensitic phase transformation, and therefore an expansion of volume, at reduced temperatures. As the transformation temperature is reduced, the volume increase cannot be compensated by plastic deformation, which leads to the formation of compressive stress. This compressive stress is being utilized to compensate the thermally induced tensile stress. Two important key parameters for estimating the LTT effect are the matensite start temperature (M_s), as well as the martensite finish temperature (M_f). To gain a maximum of stress compensation, the martensite start temperature should be as low as possible, while the martensite finish temperature needs to be over room temperature to enable full martensitic phase transformation. Different formulae exist for estimating the martensite start temperature, based on the amount of alloying elements in mass percent. Regarding LTT effect, the formula of Steven and Haynes is supposed to be most suitable, formula 1 [6].

$$Ms(^{\circ}C) = 561 - 474 * C - 33 * Mn - 17 * Ni - 17 * Cr - 21 * Mo \quad (1)$$

Regarding the current state of the art, a chemical composition of 10% Chromium and 8–12% Nickel, leading to a M_s-temperature near 200 °C, is supposed to be most suitable for achieving a stress compensating LTT effect.

Different research groups have been working on using LTT materials for residual stress reduction in weld seams [7–10]. Kannegiesser et al. characterized different LTT alloys regarding their transformation behaviour using high energy synchrotron diffraction as well as cold crack affinity by performing Tekken tests [8]. The experiments show, that the martensitic transformation is highly dependent on Chromium and Nickel content. Experiments on pure weld metal samples show, that also the dilution coming from the base material needs to be considered. Furthermore a certain amount of retained austenite prevents cold cracks. Additionally it is stated, that the cooling rate is of minor importance for the transformation behaviour, as the M_s-temperature is mainly dependant on the chemical composition. This makes the use of LTT interesting for WAAM, as the interlayer temperatures and cooing rates normally are higher compared to fusion welding.

Reisgen et al. were investigating the effect of LTT wire inlays on the residual stress formation in laser beam welding [9]. Here the final alloy was generated in-situ by melting up the wire inlay and mixing it with the base material. The results show, that a reduction of thermally induced deformation of the base plate can be achieved by this approach. Other experimental work dealt with using a plasma-multi-wire process with alternating wire feed speeds to generate chemically graded fusion welds [11]. This in-situ alloying approach shows that it is possible to locally vary the chemical composition and at the same time establish a stable fusion welding process.

Mochizuki et al. implemented a simulation model to estimate the LTT effect for GMA welding of a 10% Cr 10% Ni LTT filler wire. Here fillet welds were performed, using a GMA process. The simulation results were referenced to physical experiments by comparing the calculated and measured strain rates of the welded component [12].

Kromm et al. investigated the LTT effect in multi layer welding [13]. As the stress compensation effects are reduced by the heat treatment of following welding layers, the investigations focussed on the transformation behaviour in dependence on the inter layer temperature. It was shown, that the stress compensation LTT effect can be used effectively as long, as the inter layer temperature is kept above the martensite start temperature M_S.

For the fabrication of LTTs in WAAM, initial investigations have been performed by Houichi et al.. Here LTT filler wire was processed, using a GMA welding process to generate wall shaped structures. The results show, that a reduction of the longitudinal thermal distortion is possible [14]. However, the effect on the angular distortion was low, compared to the longitudinal distortion. The effect of the inter layer temperature was not kept into account.

In summary it was shown, that a reduction of residual stress can be achieved by the use of LTT alloys. For wire and arc additive manufacturing though, a couple of constrains need to be considered. Especially for the manufacturing of large scaled components the use of high alloyed filler materials should be limited to cut production cost. At the same time, there is a strong need for residual stress control to minimize unexpected component failure. This is why local LTT infusion through in-situ alloying during the printing process for stress control could be a promising process development.

The following works shows first results on generating LTT alloys by multi wire GMA welding for stress control in additive manufacturing.

2 Experimental Setup

The experiments have been performed using a standard gas metal arc process in spray transfer mode. The welding torch was manipulated, using a linear movement. Table 1 gives an overview on the most important constant welding parameters. The electrode wire feed speed was constantly set to 8 m/min, the welding speed 80 cm/min which

Table 1. Welding parameters for in situ welding experiments.

Parameter	Value
Shielding gas	DIN EN ISO 14175: M20-ArC-8
Shielding gas flow rate	15 l/min
Electrode wire feed speed	8 m/min
Electrode material	EN ISO 14341-A: G 3Si1
Cold wire material 1	EN ISO 14343-A: G 19 9 LSi
Cold wire material 2	EN ISO 14343-A: G 25 20
Welding speed	80 cm /min
Welding torch position	PA
Substrate material	DIN EN 10025–2 S235JR

resulted in an constant energy per unit length of 6,09 kJ/cm. As shielding gas, Argon plus 8% CO_2 was used. The contact tip to work piece distance was set to 15 mm. The inter layer temperature was kept constant at 100 °C for all experiments.

Three different material combinations were investigated, as shown in Table 2.

Table 2. Welding parameters for in situ welding experiments.

	Electrode wire	Cold wire 1	Cold wire 2
Combination 1	EN ISO 14341-A: G 3Si1	EN ISO 14343-A: G 19 9 LSi	—
Combination 2	EN ISO 14341-A: G 3Si1	—	EN ISO 14343-A: G 25 20
Combination 3	EN ISO 14341-A: G 3Si1	EN ISO 14343-A: G 19 9 LSi	EN ISO 14343-A: G 25 20

The chemical composition of the welding consumables is shown in Table 3.

Table 3. Chemical composition of the wires, determined by OES analysis.

	Fe	C	Cr	Ni	Mn	Si
Electrode	97.4	0.08	0.413	0.041	1.01	0.056
Cold wire 1	67.3	0.015	19.610	9.680	1.610	0.756
Cold wire 2	51.7	0.11	24.96	20.89	1.32	0.55

The welding torch was extended by two external cold wire feeding systems. The cold wires were fed symmetrically to the welding direction into the molten pool with an angle of 20° relative to the substrate surface and 15° to the movement axes of the welding torch, Fig. 1. The welding experiments were performed in order to evaluate a stable process window, as well as the maximum possible cold wire feed speeds without generating process instabilities.

3 Experiments

In the beginning, the maximum possible wire feed speed was investigated using one of the wires, followed by experiments with both wires. Wall shaped specimen were produced by depositing 5–7 layers. For compensating geometric deviations in the process start and end area, the welding direction was reversed every layer. The specimen height varied between 8 mm and 12 mm, depending on the cold wire feed speed. Cross sections were extracted for microscopic analysis, as well as chemical analysis via optical emission spectroscopy (OES).

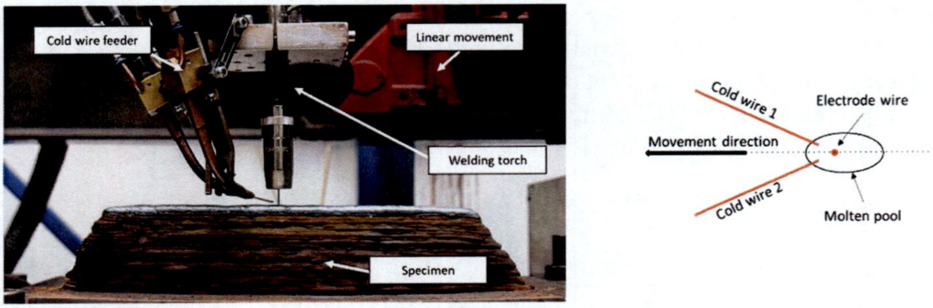

Fig. 1. Left: Welding torch with cold wire feeding system. Right: schematic top view on the wire orientation.

4 Results and Discussion

In total, 23 stable working points were investigated, with a summarized cold wire feeding speed, ranging from 5 m/min to 15 m/min. For feeding the wires from the backside into the molten pool, a minimum wire feed speed was required to enable a stable droplet transfer. Table 4 shows the investigated parameters, as well as the chemical composition, measured by the OES analysis.

4.1 Chemical Composition

The chemical composition was estimated by calculating the average value out of three OES measurements per sample. Figure 2 shows the Chromium and Nickel equivalents that were calculated from the OES analysis, drawn into the Schaeffler diagram. The Chromium and Nickel equivalents were calculated using formulae 2 and 3:

$$Cr_{\{eq\}} = Cr + Mo + 1.5 * Si + 0.5 * Ta + 0.5 * Nb + Ti \quad (2)$$

$$Ni_{\{eq\}} = Ni + 30 * C + 0.5 * Mn + 7.5 N \quad (3)$$

It can be seen, that the estimated Chromium and Nickel equivalents are following the linear interpolation between the welding consumables as expected. Furthermore, it is visible, that the use of single wires barely reaches the Chromium and Nickel contents, that are required to reach the martensitic area. The combination of wire one and wire two enables to reach the part of the martensitic-austenitic zone, which also lets expect an LTT microstructure. Especially samples number 18 to 23 have been considered for a deeper microscopic analysis.

4.2 Morphology

Further investigations to characterize the material were done by microscopic analysis. Figure 3 shows the macroscopic images of the multi wire samples. The images show a typical structure of wall shaped WAAM samples. A strong directional solidification can be identified for the first six layers, while the top layer partly shows a

Table 4. OES results and calculated Chromium and Nickel equivalents.

Num	Cold wire feed speed 1 [m/min]	Cold wire feed speed 2 [m/min]	Ni-Eq	Cr-Eq	C [%]	Mn [%]	Si [%]	Cr [%]	Ni [%]
1	5	–	3.83	4.18	0.06	1.39	0.72	3.07	1.46
2	6	–	4.02	4.84	0.05	1.36	0.69	3.78	1.73
3	7	–	4.19	4.89	0.06	1.41	0.72	3.79	1.78
4	8	–	4.33	5.08	0.06	1.40	0.68	4.03	1.90
5	9	–	5.02	6.80	0.05	1.44	0.70	5.70	2.67
6	10	–	5.17	7.09	0.05	1.44	0.68	6.01	2.83
7	–	5	6.01	4.97	0.08	1.34	0.66	3.96	2.98
8	–	6	7.14	6.41	0.08	1.38	0.66	5.39	3.98
9	–	7	7.61	6.96	0.08	1.38	0.63	5.98	4.45
10	–	8	8.56	7.79	0.09	1.38	0.63	6.81	5.12
11	–	9	8.84	8.15	0.09	1.36	0.61	7.20	5.51
12		10	9.16	8.65	0.08	1.42	0.62	7.68	5.95
18	5	5	9.71	11.61	0.08	1.49	0.63	10.59	6.95
19	5	6	11.04	12.14	0.09	1.49	0.63	11.12	7.50
20	5	7	11.13	12.90	0.07	1.49	0.61	11.91	8.18
21	5	8	11.57	13.39	0.07	1.52	0.62	12.39	8.66
22	5	9	12.13	13.75	0.07	1.51	0.61	12.75	9.13
23	5	10	12.67	14.38	0.08	1.54	0.60	13.39	9.45

non-directional solidification. Bright areas can be identified along the fusion lines of the weld seams, which hints to a macroscopic segregation of Chromium and Nickel. For a further evaluation of the materials, microscopic images have been prepared, Fig. 4.

At higher magnitude, a strong effect of the changing chemical composition on the microstructure is visible. From sample 18 to sample 23 the Nickel content is changing from 6.95% to 9.45%, while the amount of Chromium raises from 10.59% to 13.39%. Especially the microstructure, which sets in from sample 21 looks very similar to the LTT microstructure, described in the literature.

Kromm describes LTT cross sections, taken from weld metal in [7] as cellular martensitic microstructure, characterized by a Chromium and Nickel segregation along the martensite spots. An increasing amount of Nickel leads to larger austenitic segregation zones beside the martensitic spots. A very similar microstructure can be seen in Fig. 4, pictures 21, 22 and 23. According to this interpretation, the brighter areas show segregated austenitic phases with higher concentration of Chromium and Nickel while the darker spots represent the martensitic areas. The Chromium to Nickel relation lies below 1.5, which is why no presence of delta ferrite is being expected. The microstructure stays

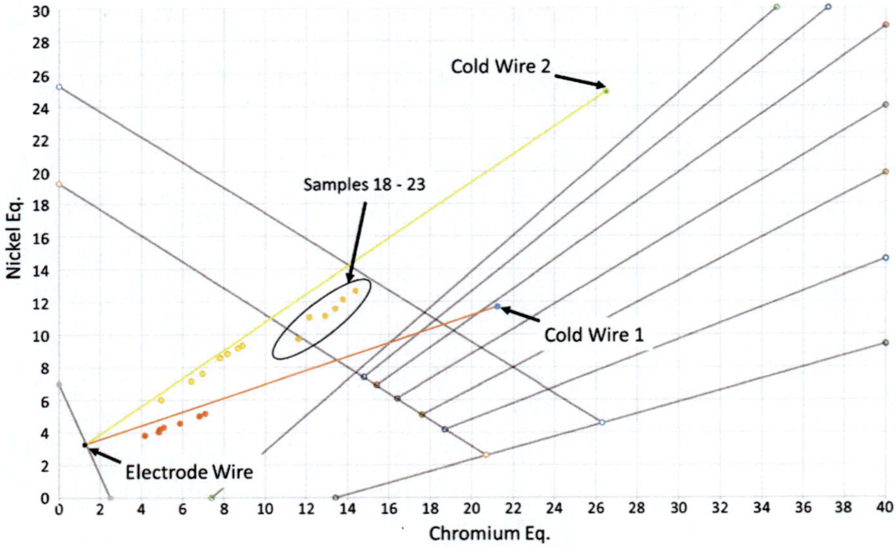

Fig. 2. Schaeffler diagram, showing the results of Chromium and Nickel equivalents, calculated from the OES meauserments.

similar over the longitudinal axis of the samples, except near the fusion line of the first layer, as well as the last layer. Due to base material dilution, the first welding layer shows a fine martensitic structure without bright segregation zones. The last weld seam shows a characteristic non-directional solidification as to be seen in the macro images.

The microscopic analysis in combination with the results from the OES measurements show a high chance, that a martensitic LTT alloy was generated by the double wire approach of material combination 3. The calculated M_S-temperatures lie between 180 °C (sample 18) and 81 °C (sample 23), which makes the in-situ-alloyed samples feasible to enable an LTT effect. As the inter layer temperature was kept at 100 °C, a positive effect of the martensitic phase transformation on the stress level of the base plates can not be expected. The geometrical stability of the component in wire and arc additive manufacturing highly depends on the size, viscosity and surface temperature of the molten pool. One key parameter, that has an impact on those aspects, is the inter layer temperature. In terms of processing LTT alloys in WAAM, this means that a sweet spot needs to be found between a suitable M_S-temperature for maximum stress compensation effect, and inter layer temperature for keeping the manufacturing process stable. The in-situ-alloying approach seems to be a highly promising way, to achieve a tailor made, locally changing chemical composition, which is adopted to the heat dissipation properties of the components geometry. In contrast to the direct use of an LTT wire as electrode, the in-situ approach is much more flexible in terms of multi material additive manufacturing.

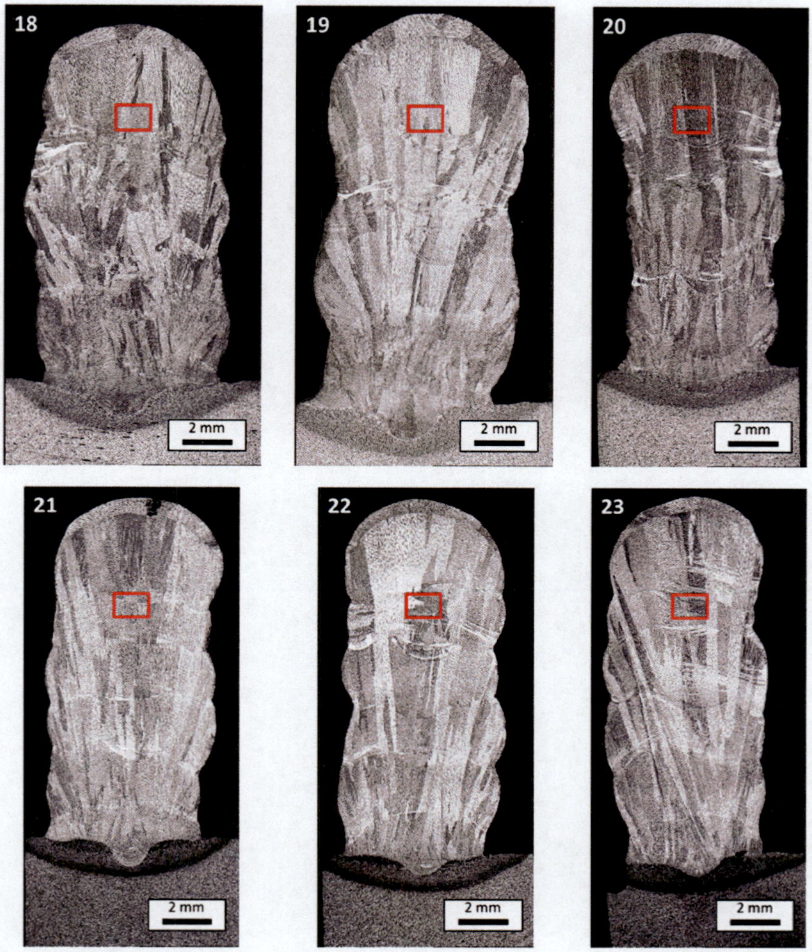

Fig. 3. Macroscopic images of the multi wire samples. The red rectangles mark the positions of microscopic images.

5 Summary

The presented work shows, that it is possible, to establish a stable gas metal arc multi wire process for wire and arc additive manufacturing, that is capable of reaching a chemical composition, suitable for LTT alloys. By varying the wire feed speed, it is possible to fluently set a Chromium content between 3.07% and 13.39%, as well as a Nickel content between 1.46% and 9.45%. This opens a wide variety of possibilities in terms of residual stress control in WAAM. As additive manufacturing processes are characterized by an inhomogeneous material distribution, which leads to a varying thermal distortion and stress distribution. The local design of the chemical composition enables to adapt the martensite start temperature to the thermal cooling behaviour of the component. This gives the possibility to achieve a local LTT effect tailored to the components

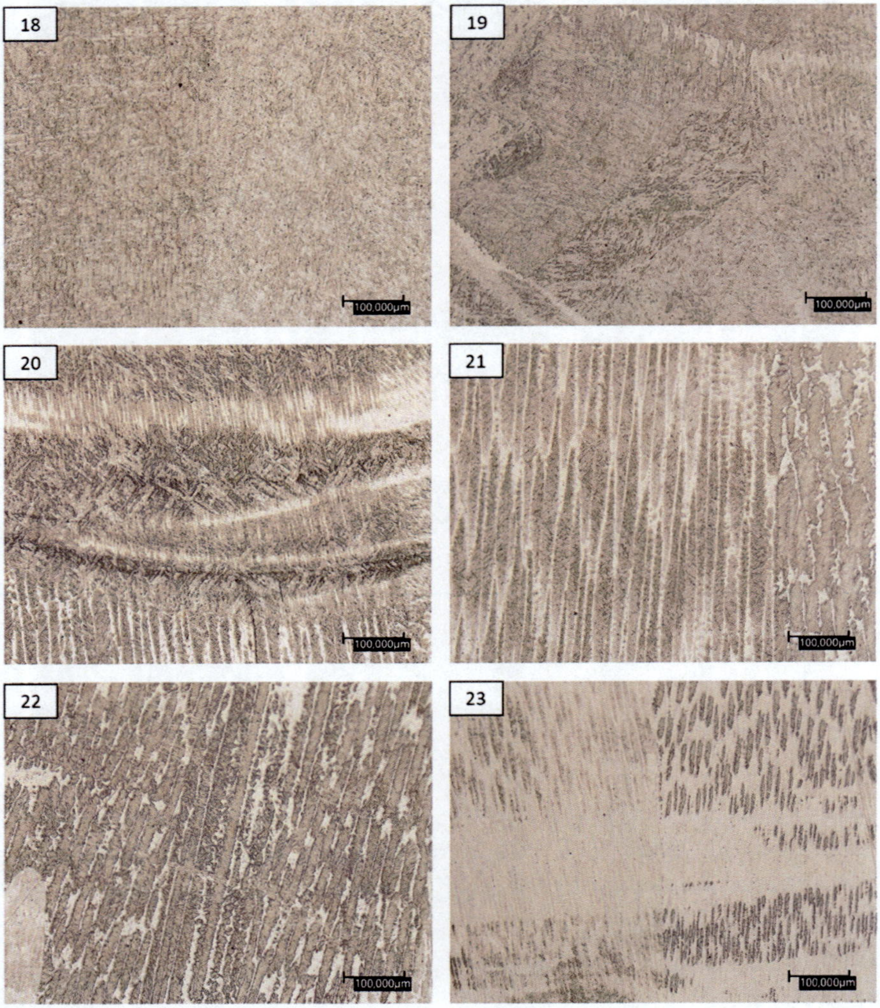

Fig. 4. Microscopic images of double wire samples 18–23, taken from the center of the sample.

design. Future research will deal with developing a dynamic path planning and process parametrization strategy to enable a global LTT effect for a large scaled additively manufactured component.

Acknowledgments. The presented investigations were carried out at RWTH Aachen University Welding and Joining Institute ISF within the framework of the Collaborative Research Centre SFB1120–236616214 "Bauteilpräzision durch Beherrschung von Schmelze und Erstarrung in Produktionsprozessen" and funded by the Deutsche Forschungs-gemeinschaft e.V. (DFG, German Research Foundation). The sponsorship and support is gratefully acknowledged.

Special thanks also go to the companies ESAB AB and EWM AG for providing welding consumables and welding machines.

References

1. Pan, Z., Ding, D., Wu, B., Cuiuri, D., Li, H., Norrish, J.: Arc welding processes for additive manufacturing: a review. In: Chen, S., Zhang, Y., Feng, Z., et al. (eds.) Transactions on Intelligent Welding Manufacturing. Springer, Singapore, pp. 3–24 (2018). ISBN 978-981-10-5354-2
2. Ding, D., Pan, Z., Cuiuri, D., Li, H.: Wire-feed additive manufacturing of metal components: technologies, developments and future interests. Int. J. Adv. Manuf. Technol. **81**(1–4), 465–481 (2015). ISSN 0268-3768. Verfügbar unter: https://doi.org/10.1007/s00170-015-7077-3
3. Norsk Titanium - Unternehmenswebseite [online] [Zugriff am: 05.09.17]. Verfügbar unter: https://www.norsktitanium.com/
4. Wegener, V.: Ramlab - Unternehmenswebsite [online] [Zugriff am: 5. September 2017]. Verfügbar unter: https://ramlab.com/
5. Piehl, K.H.: Formgebendes Schweissen von Schwerkomponenten. Thyssen, Technische Berichte **1989**(21, 1), 53–71 (1989)
6. Steven, W., Haynes, A.G.: The temperature of formation of Martensite and Bainite in low-alloy steel. J. Iron Steel Inst. **183**, 349–359 (1956)
7. Kromm, A.: Umwandlungsverhalten und Eigenspannungen beim Schweißen neuartiger LTT-Zusatzwerkstoffe. Bundesanstalt für Materialforschung und -prüfung (BAM). BAM-Dissertationsreihe, Berlin. 72 (2011). ISBN 978-3-9813853-9-7
8. Kromm, A., Kannengiesser, T.: Characterizing PHASE TRANSFORMATIONS of different LTT alloys and their effect on RESIDUAL STRESSES and COLD CRACKING [online]. Weld. World **55**(3–4), 48–56 (2011). ISSN 0043-2288. Verfügbar unter: https://doi.org/10.1007/BF03321286
9. Gach, S., Olschok, S., Arntz, D., Reisgen, U.: Erratum: residual stress reduction of laser beam welds by use of low-transformation temperature (LTT) filler materials in carbon Manganese steels— in situ diagnostic: image correlation. J. Laser Appl. **30**, 032416 (2018) [online]. J. Laser Appl. **32**(1), 19901 (2020). ISSN 1042–346X. Verfügbar unter: https://doi.org/10.2351/1.5133938
10. Reisgen, U., Olschok, S., Gach, S.: Nutzung von Low-Transformation-Temperature-Werkstoffen (LTT) zur Eigenspannungsreduzierung im Elektronenstrahlschweißprozess [online]. Materialwissenschaft und Werkstofftechnik **47**(7), 589–599 (2016). ISSN 09335137. Verfügbar unter: https://doi.org/10.1002/mawe.201600549
11. Oster, L., Akyel, F., Reisgen, U., Olschok, S., et al.: Investigating plasma keyhole welding with multiple wires for fusion welding with chemically graded weld seams (2019)
12. Mochizuki, M., Matsushima, S., Toyoda, M., Morikage, Y., Kubo, T.: Study of residual stress reduction in welded joints using phase transformation behaviour of welding material. Studies on numerical simulation of temperature, microstructure, and thermal stress histories during welding and their application to welded structures (2 nd report) [online]. Weld. Int. **19**(10), 773–782 (2005). ISSN 0950-7116. Verfügbar unter: https://doi.org/10.1533/wint.2005.3491
13. Kromm, A., Kannengiesser, T.: Effect of martensitic phase transformation on stress build-up during multilayer welding [online]. Mater. Sci. Forum: 768–769, 660–667 (2013). Verfügbar unter:https://doi.org/10.4028/www.scientific.net/MSF.768-769.660
14. Kitano, H., Nakamura, T.: Distortion reduction of parts made by wire and arc additive manufacturing technique using low transformation temperature welding materials [online]. Q. J. Jpn. Weld. Society **36**(1), 31–38 (2018). ISSN 0288-4771. Verfügbar unter: https://doi.org/10.2207/qjjws.36.31

Casting

Development of an In-Situ Observation Procedure for Hot Tear Formation in Aluminum Alloys in Gravity Die Casting

Nino Wolff[1(✉)], Rahul Sharma[2], Uwe Vroomen[1], Andreas Bührig-Polazcek[1], and Uwe Reisgen[2]

[1] Foundry Institute, RWTH Aachen University, Intzestraße 5, 52072 Aachen, Germany
n.wolff@gi.rwth-aachen.de
[2] Welding and Joining Institute, RWTH Aachen University, Pontstraße 49, 52062 Aachen, Germany

Abstract. Hot cracks are a widespread defect phenomenon in metal castings. In order to understand the underlying mechanisms and to evaluate numerical models to describe hot crack formation, it is important to know how hot cracks develop during the transition from the liquid to the solid state. An experimental setup is presented, which has been designed to generate hot cracks in aluminum gravity die casting in order to trace their formation during solidification by optical means. The solidification induced shrinkage is correlated with the crack formation. The alloy systems AlSi and AlCu are examined in practically relevant compositions. In addition to the development of the in-situ investigation methodology, the hot cracks produced are examined using elctron microscopy in order to determine the underlying crack mechanisms.

Keywords: Aluminum alloy · Gravity die casting · Hot tearing · In-situ observation

1 Introduction

Distortion, residual stresses and hot cracks are decisive quality influencing characteristics of components in foundries and, depending on their characteristics, lead to corresponding reworking and, in the case of hot cracks, to scrap. Their reduction is therefore of great interest for cost-efficient production. In the casting process, the phenomena mentioned above result from the combination of solidification shrinkage coupled with the respective local self-feeding and the geometric constraints of the mold on the component. The work presented here was carried out within the framework of the DFG-funded Collaborative Research Center SFB1120 "Precision Melt Engineering". The long-term goal is to improve the precision of the component in permanent mold casting by creating a knowledge base in the areas of distortion and hot crack formation as well as the derivation of corresponding concepts for influencing them. In order to gain a deeper understanding under which conditions and at what point in time hot cracks form during

U. Reisgen et al. (Eds.): EMPOrIA 2020, LNME, pp. 181–190, 2021.
https://doi.org/10.1007/978-3-030-70332-5_16

the solidification process, an experimental set-up was developed which allows the formation of hot cracks to be assigned to exact times and local temperatures and therefore to fraction solid values during solidification by means of optical recordings.

The subject of hot cracking which occurs while solidifying during the casting of aluminum alloys is an already very extensively investigated area e. g. by J. A. Dantzig [1], J. Langlais [2], M. Rappaz [3] and many others. Different aluminum alloys show different susceptibilities to hot tearing mainly as a function of the width of the solidification range [3, 4]. The system AlSi, which has good self-feeding properties at low strength, and AlCu, which, relatively speaking, has higher strength with poorer feeding properties, are considered here in different chemical compositions. The comparison of these two systems, both in the in-situ measurements and in the subsequent considerations of the morphology, should allow conclusions to be drawn about the mechanisms of crack formation, such as possible segregation in the areas of crack formation.

In-situ observation of crack initiation using optical methods has been performed by C. Davidson et al. Using an AlCu0.5 alloy [5]. The focus of that work is on the correlation of shrinkage induced stresses and crack initiation. However, the reliable recording of the forces as a result of settlement phenomena was questioned by the authors. Yamagata H. et al. Also worked on the optical records of crack formation during solidification [6]. The less hot crack susceptible alloy AlSi7Mg0.3 in a sand mold was investigated where the crack initiation was caused only by the casting geometry. Healed internal hot cracks and persistent cracks open to the surface were observed and compared. Apart from the video recording of the crack initiation, only temperature curves were recorded, and the occurring forces were taken from simulations.

The difficulties in detecting the shrinkage induced forces [5, 7] provide motivation to correlate the crack initiation with the contraction of the cast part. In many of the previous studies, where shrinkage was detected during crack initiation, the deviations were recorded near the crack location [8] or very far away [9]. The observation that the placement of the transducers near the crack zone influences the solidification and even more the mechanical deformation, led to the approach to determine temperature-dependent shrinkage and crack formation at two different locations in the examined component.

2 Development of Experimental Setup

The starting point for the setup for in-situ hot crack observation is the measuring equipment and experimental mold used in preceding studies for investigations of the cast-mold heat transfer coefficient as well as the cast part distortion [7, 10, 11]. The basic concept is an "F" shaped sample geometry. The mold consists of steel mold modules which enclose the contour of the "F" and are oil-temperature controlled. On the upper and lower side, the mold is closed by insulating calcium silicate plates, so that heat fluxes occur exclusively in horizontal direction into the mold modules. For the in-situ hot crack observations, corresponding mold modules were manufactured, which create a hot spot between the ribs of the "F" by integrating a further insulating plate and a quartz glass window, which also acts as an insulator. In this hotspot, where the melt is the last to solidify, hot cracks are generated due to the geometry and the solidification shrinkage. The window allows the visual recording of the crack initiation and growth with a video camera. Figure 1 shows on the left side the mold already modified for hot crack observation with the lid removed from above. Temperatures are recorded by means of type “K” thermocouples in the area of crack formation and the area of unimpeded contraction. The contraction of the free end is detected by two linear variable differential transformers (LVDT), one coupled to the mold and one to the melt. Figure 1 shows on the right side the result of an accompanying Magma simulation (performed with Magmasoft® Version 5.3). The solid phase content in the casting is shown shortly before solidification starts everywhere in the part, the hotspot is clearly visible here.

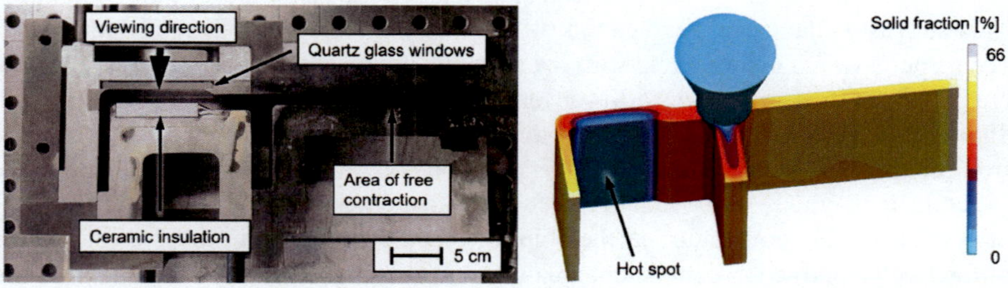

Fig. 1. The left side shows the opened mold with the hot crack area formed by the insulation and the glass window. The right side displays the calculated solid fraction during solidification and by this a hot spot where the crack is intended to form.

3 Experiments

The experiments were based on ingots, made of 99.8 grade aluminum, AlCu50 prefabrication and standard industrial A356 (AlSi7Mg0.3) and were charged according to the desired contents. Prior to casting with a superheat of 120 °C, the melt was flushed with argon for 15 min. The mold, which was coated with an insulating coating, was thermally regulated to 150 °C. For the in-situ recorded temperatures and displacements a sampling rate of 20 Hz was set and for the video 120 fps. A new quartz glass plate and new measuring rods were used for each test. In addition to the experiments, the Thermocalc software was used to calculate the liquidus and solidus temperatures (as equilibrium solidification) for the alloys under investigation.

4 Results

As expected, the tests carried out showed no crack formation for AlSi7Mg0.3 as well as for AlSi3.5Mg0.15. Crack formation only became apparent when the Si content was further halved to 1.75 wt%. Since the crack susceptibility of pure aluminum and the chosen AlCu alloys is relatively high, for these alloys hot cracks could be observed in each experiment.

4.1 In-Situ Observation

The position of the thermocouple in the observed area in the immediate vicinity of the occurring hot crack allows to determine the time between filling and the first occurrence of a hot crack and to derive the corresponding temperature. For the alloy AlSi1.75Mg the first time the crack could be visually detected was at a temperature (T_c) of 618 °C at the crack location. This lies within the calculated solidification interval of the alloy (649 – 575 °C). For the AlCu alloys with 2 and 4 wt% copper, clear hot cracks were found in both cases, each with the first appearance slightly below the solidus temperature T_{sol}. For the Al 99.8 alloy casting the crack could be visually detected for the first time at a temperature significantly below the calculated solidus temperature of pure aluminum. This suggests that strong segregation of the secondary elements in the residual melt occurs here, which in turn could shift the solidification in the last solidified area, where crack formation is observed, to lower temperatures. These results are summarized in Table 1. Also included is the time until the first visual detection of the crack, measured from the time when the area under consideration was filled with melt. The shrinkage (S) at the temperature of the first occurrence of a hot crack can be determined by means of comparing the temperature at the point of crack formation and those of the center of the freely contracting component area where the displacements are measured. In the 99.8 grade alloy the cracking starts at the lowest shrinkage of 0.056%. With AlSi1.75Mg this happens at almost double the value, which is roughly similar to AlCu2. AlCu4 shows the first signs of hot cracking at about four times the amount of shrinkage (see Table 1).

The cooling curves of the test specimen at the crack area of all four alloys showing hot cracks are depicted in Fig. 2. Marked are the time and temperature points of the first visibly indications of tearing. Using the example of the experiment with AlCu2, it can be seen that the solidification is slower in the hotspot, where crack formation takes place, compared to the area where shrinkage is measured by temperature (Fig. 3 left). The right-hand graph in Fig. 3 shows the change in length of the considered free end of the "F" shaped casting. It shows that initially the mold expands with the molten metal in it and from about 5 s on the shrinkage of the component begins, coupled with a corresponding gap formation (the difference between mold and component displacement).

Table 1. The compositions, hot crack formation, time and shrinkage (S) of the first appearance of a crack, calculated (T_{liq} and T_{sol}) and measured (T_c) temperatures of the realized experiments.

Alloy	Hot crack	Time [s]	T_{liq} [°C]	T_{sol} [°C]	T_c [°C]	S [%]
Al 99.8	Yes	55	–	660	606	−0.056
AlCu4	Yes	60	649	571	568	−0.194
AlCu2	Yes	52	655	610	604	−0.107
AlSi7Mg0.3	No	–	617	568	–	–
AlSi3.5Mg0.15	No	–	638	572	–	–
AlSi1.75	Yes	47	649	575	618	−0.085

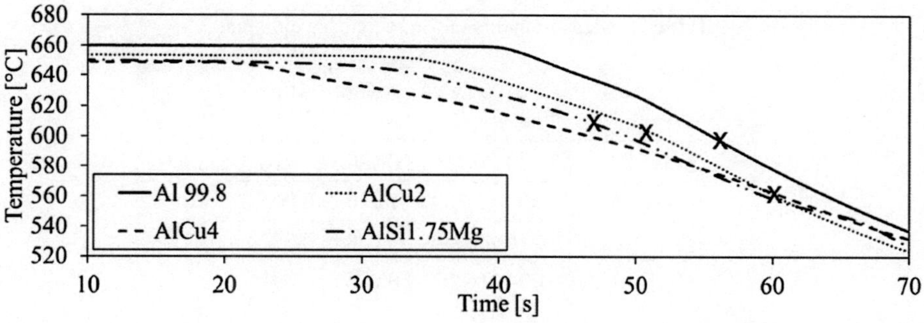

Fig. 2. The cooling curves of the test specimen at the crack area of the four samples with hot cracks. Marked are the time and temperature points of the first visible indications of tearing.

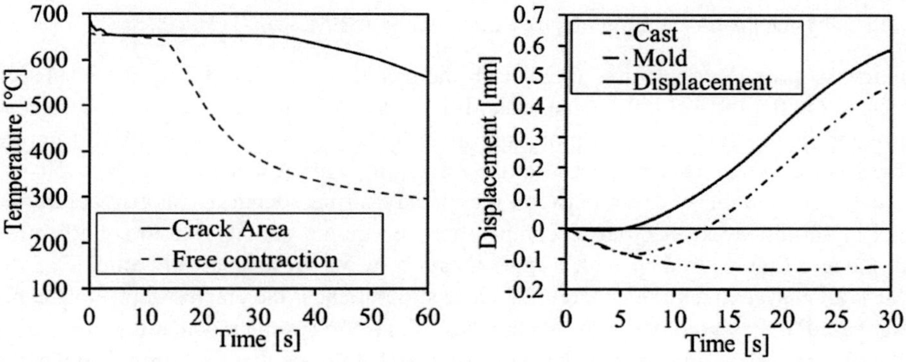

Fig. 3. The different cooling curves of the crack location and the free contracting casting area for the ALCu2 casting on the left side. The measurements of the displacements of the casting at the area with free contraction are shown on the right for this casting.

4.2 Chemical Composition

The chemical composition of the samples with cracks was determined by means of optical emission spectroscopy (OES) with spatial resolution in order to detect segregations near the crack. Measurements were taken at eight points from the outer side of the sample part towards the crack along the solidification direction. Measurements were taken with a spot diameter of 5 mm. Overlapping of the measurement spots with intermittent grinding accounted for 50% of the spot diameter. No significant concentration differences were found at different distances from the crack. It occurs that the accompanying elements of AL 99.8 are mainly Si (0.17 wt%) and Fe (0.074 wt%). The measured values are summarized for the most significant elements in Table 2.

Table 2. Measurements of the secondary elements.

	Al 99.8	AlCu4	AlCu2	AlSi1.75
Si [wt%]	0.170	0.110	0.130	1.730
Cu [wt%]	0.002	3.780	1.920	0.001
Fe [wt%]	0.074	0.080	0.078	0.076
Mg [wt%]	0.004	0.002	0.002	0.091

4.3 Morphology

After the test castings, the areas with hot cracks were cut out and examined optically as well as by means of scanning electron microscopy (SEM).

Macroscopicx. First pictures taken from the side of the casting facing the insulation plate show the last solidified area around the hot crack. In Fig. 4 it can be seen how a dendritic structure becomes more and more apparent on the surface of all samples towards the crack. This suggests that the remaining melt near the crack flows into the structure by capillary action between the already formed dendrites shortly before the end of solidification in order to compensate for volume deficits due to solidification shrinkage. Figure 5 taken with a magnification of 50 shows that for samples of the AlCu alloy system a more coherent dendrite structure near the crack front is noticeable compared to the two other alloys. What they all have in common is that it is noticeable that complete dendritic structures are clearly visible towards the crack front and that the crack runs mainly between them without dissecting them. Only the sample of Al 99.8 alloy shows torn apart dendrites in the hot crack (see Fig. 5 upper left).

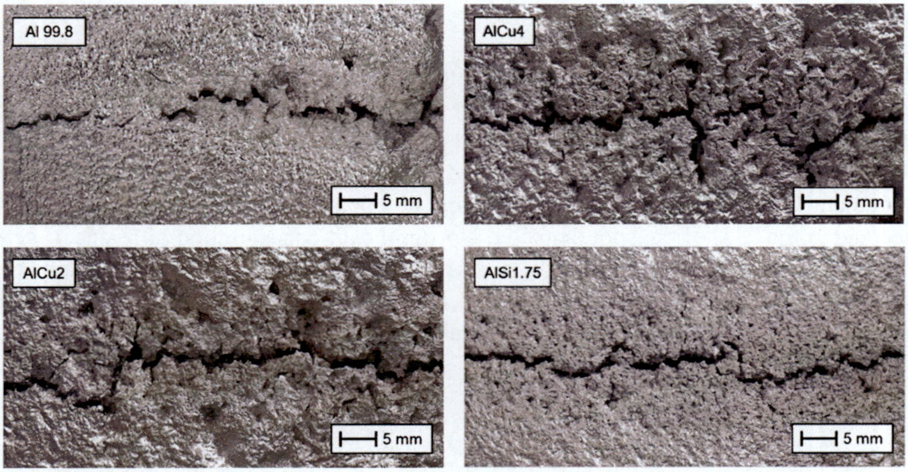

Fig. 4. Macroscopic images of the hot cracks taken with a magnification of five.

Fig. 5. Macroscopic images of the hot cracks taken with a magnification of 50. At the upper left picture of Al 99.8 the broken dendrite is encircled.

SEM Analyses. The SEM images of the prepared fracture surfaces show mostly intact dendrite structures in all samples. In Fig. 6 no ductile structures or structures clearly damaged by the crack are visible on the pictures taken with a magnification of 100. From this it can be concluded that the hot cracks were caused by separation of the remaining interdendritic residual melt in all cases. From a purely qualitative point of view, the dendrite structures appear to differ in their roundness for the various alloys. The AlSi1.75 in particular has the roundest forms. This can indicate different permeabilities of the structures during solidification and thus, via the influence on the interdendritic feeding, finally the hot cracking tendency. Clearly visible differences in the sizes of the dendrite structures were not detected. Figure 7 shows the surface of dendrites of the Al 99.8 alloy on the left. Wrinkles can be seen in the interdendritic surface, which can be explained by the solidification shrinkage of the residual melt located above the dendrites. The right picture of the AlCu4 alloy using backscatter diffraction (BSD) shows dendrite structures of primary aluminum and interdendritic a heavier phase, where the composition and the proportions will be Al_2Cu.

A significant difference between the crack appearances in the samples was found when considering the Al 99.8 sample (Fig. 8, left) in a higher magnification. In this picture, several broken dendrite arms could be seen. This indicates that the tearing occurred partly after a coherent net of dendrites has formed. Light-optical microscopy

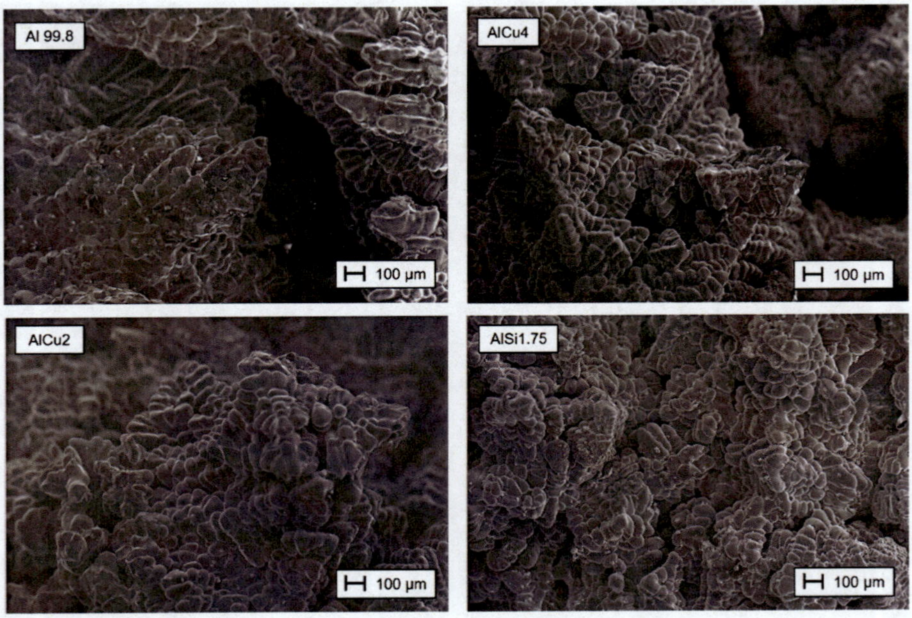

Fig. 6. SEM pictures of the dendrite structures at the crack surface for Al 99.8, AlCu4, AlCu2 and AlSi1.75.

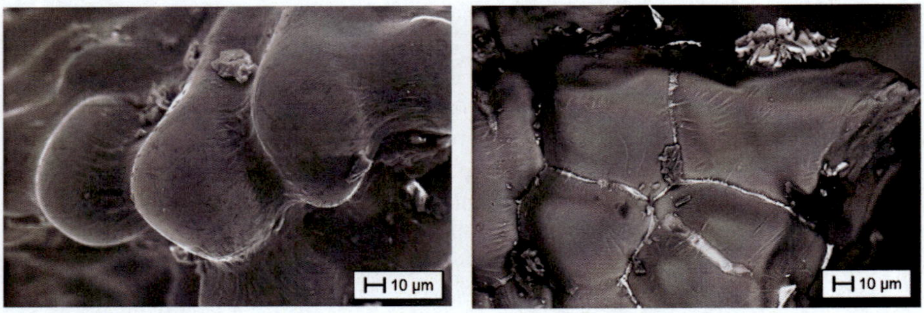

Fig. 7. On the left side a closer look at the last solidified interdendritic melt of the Al 99.8 alloy and on the right side the primary aluminum dendrite structure with interdendritic Al_2Cu in the AlCu4 alloy.

images of this sample (Fig. 8 right) show interdendritic segregations, which indicate that the small solidification interval of Al 99.8 is locally widened due to enrichment of accompanying elements (mainly Si) in the residual melt.

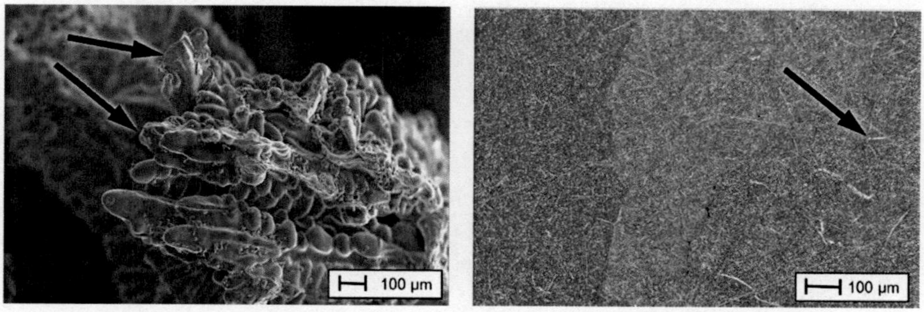

Fig. 8. Torn dendrite arms of the Al 99.8 sample (right) and segregations of interdendritic melt enriched with accompanying elements (Si) visible on a light-optical-microscopy image (left).

5 Conclusions

It can be stated that the chosen method for in-situ observation of hot cracking allows to determine the time of the beginning of tearing to be about one second. Thus, the crack formation can be clearly assigned to an associated temperature and via this to an associated shrinkage. The cast specimen allow macroscopic as well as SEM analysis of the crack locations. The observation and comparison of the four specimens with pronounced hot cracks also provides the following findings:

- The SEM images of the cracks suggest that the crack formation of all samples was comparable, mainly due to separation in the remaining interdendritic residual melt.
- The OES measurements do not show any conspicuous segregation towards the crack location, which is also consistent with the cooling curves at the crack location that matches the respective alloy compositions.
- The low temperature of the first crack occurrence in 99.8 alloy, compared to the calculated T_{sol} suits the found micro segregations of Si in the interdendritic melt and the torn apart dendrites which indicates that on this sample the crack occurred partly in the already coherent net of dendrites.

In conclusion, the observations made suggest that a more detailed quantitative investigation of the dendrite structures using automated image analysis, especially with regard to roundness and size, will provide further information. This, together with the consideration of the theoretical behavior of the solidification, promises a deeper understanding of the hot cracking susceptibility of the different alloys.

Acknowledgments. The presented investigations were carried out at RWTH Aachen University within the framework of the Collaborative Research Centre SFB1120 – 236616214 "Bauteilpräzision durch Beherrschung von Schmelze und Erstarrung in Produktionsprozessen" and funded by the Deutsche Forschungsgemeinschaft e.V. (DFG, German Research Foundation). The sponsorship and support is gratefully acknowledged.

References

1. Dantzig, J. A., Rappaz, M.: Solidification, 2nd edn., EPFL Press, Lausanne (2016)
2. Langlais, J.: Fundamental study of hot tearing mechanisms of aluminum-silicon alloys. Department of Mining, Metals and Materials Engineering McGiII University, Montreal (2006)
3. Rappaz, M., Jacot, A., Boettinger, W.J.: Last stage solidification of alloys: a theoretical study of dendrite arm and grain coalescence. Metall. Mater. Trans. A **34A**, 467–479 (2007)
4. Phillion, A.: Hot tearing and constitutive behaviour of semi-solid aluminum alloys. The University of British Columbiax (2007)
5. Davidson, C., Viano, D., Lu, L., StJohn, D.: Observation of crack initiation during hot tearing. Int. J. Cast Metals Res **1**(19), 59–65 (2006)
6. Yamagata, H., Tachibana, H., Kijima, S., Adachi, M., Koiwai, S.: Direct observation of hot tearing of Al-7%Si-0.35%Mg alloy. Adv. Mater. Process. Technol. **3**(4), 480–492 (2018)
7. Wolff, N., Vroomen U., Bührig-Polaczek, A.: Development and Evaluation of an Experimental Setup to Investigate and Influence Component Distortion in Gravity Die Casting. AFS Transactions (2020)
8. Sadayappan, K., Aguiar, A.M., Shankar, S.: Development of a hot tear test procedure for aluminum casting alloys. Mater. Sci. Technol. **2019**, 1094–1102 (2019)
9. Nasresfahani, M.R., Niroumand, B.: Design of a new hot tearing test apparatus and modification of its operation. Met. Mater. Int. **16**(1), 35–38 (2010)
10. Wolff, N., Pustal, B., Vossel, T., Laschet, G., Bührig-Polaczek, A.; Development of an A356 die casting setup for determining the heat transfer coefficient depending on cooling conditions, gap size, and contact pressure. Mater. Sci. Eng. Technol. **48**(12), 1235–1240 (2017)
11. Vossel, T., Pustal, B., Bührig-Polaczek, A.: Modellierung der Erstarrungskinetik kolumnarer und äquiaxialer Kornstrukturen zur gefügebasierten Interpolation der Werkstoffeigenschaften im Hinblick auf die Verzugsvorhersage. Gießerei Special **2**, 130–134 (2017)

Determination of the Heat Transfer Coefficient for a Liquid-Solid Contact in Gravity Die Casting Processes

Thomas Vossel(✉), Björn Pustal, and Andreas Bührig-Polazcek

Foundry Institute, RWTH Aachen University, Intzestraße 5, 52072 Aachen, Germany
t.vossel@gi.rwth-aachen.de

Abstract. The description of the heat transfer coefficient represents a core element when defining a system's heat balance. Especially in foundry processes, high temperature melts get in contact with low temperature molds. While solid-solid contacts have been the focus of numerous investigations, liquid-solid contacts are rarely described in literature. This paper will present and compare different approaches to describe the phenomena and properties necessary for modeling the contact between a liquid melt and a solid mold surface with microscopical roughness. This includes a topographical analysis of the interface and the estimation of the wettability of the surface. The resulting models will determine the heat transfer coefficient for application in thermomechanical simulations of a gravity die casting process. Identifying the model best suited for modeling the melt-mold contact in a foundry process represents the first step towards a physical model that describes the entirety of a casting process.

Keywords: HTC model · Heat transfer coefficient · Liquid-solid contact

1 Introduction

Numeric simulations are commonly used as a tool for predicting the result of gravity die casting processes. In such simulations the heat balance is of high importance to determine the local cooling conditions. Especially the heat transfer coefficient (= "HTC") for the melt-mold contact is rarely discussed in literature. A model suitable to describe the heat transfer coefficient at the contact interface of a liquid and solid material has to deliver information on two aspects of the contact:

1. Defining an expression to calculate the heat transfer coefficient at the interface with regard to its microscopically rough surface as the result of a topographical analysis
2. Describe the wetting of the surface asperities derived from a mechanical analysis in order to estimate the contact area between liquid and solid regions

This paper presents several different modeling approaches and evaluates the respective HTC predictions. These results are finally benchmarked against experimental data gathered from gravity die casting experiments and inverse simulations.

U. Reisgen et al. (Eds.): EMPOrIA 2020, LNME, pp. 191–201, 2021.
https://doi.org/10.1007/978-3-030-70332-5_17

2 Topographical Analysis

As real surfaces feature a microscopical roughness, it is important to include a topographical analysis in order to give a good representation of the respective surface. In terms of a liquid-solid contact, this is important for two reasons: On the one hand, just as for solid-solid contacts, only certain regions of the solid body will be in contact with the other one i.e. especially the surface asperities with maximum height. On the other hand, a liquid-solid contact, in contrast to a solid-solid contact, is not about just the rather few asperity contact spots, but due to the possibilities of the liquid medium to wet the surface of the solid body, the actual contact area can be significantly larger. For these reasons, the topographical analysis not only needs to come up with an appropriate description of the rough solid surface but must also describe the area in contact with the liquid material. All presented models apply the idea of wetting the solid surface up to a certain height level. The remaining gap between the wetting height and the base of the asperity is considered to be filled with gas, i.e. air.

As it is either not possible or feasible to scan the entire surface of a mold, some simplifications and assumptions have to be made in order to create a model. All presented models share the following assumptions:

1. The rough surface is represented by a continuing sequence of asperities.
2. The asperities themselves are considered to be conical in form with a constant slope along its surface line.
3. Material properties are assumed to be constant.
4. Surfaces are clean.
5. Radiation heat transfer can be neglected for liquids with low melting points and/or low emissivity.

2.1 Surface Profile Model 1

The first profile model as presented by Prasher [1] and reviewed by Somé et al. [2] further simplifies the surface structure. The asperities are considered to be of identical height and uniformly distributed. Their average height is equal to the standard deviation σ of the asperity distribution. A schematic depiction of this approach is shown in Fig. 1. In order to define the heat transfer coefficient, its basic definition is applied which divides the effective thermal conductivity λ_{eff} by a wall thickness δ. An equivalent expression to the wall thickness can be derived by evaluating the surface in contact with liquid melt. The term $\frac{A_{ratio}}{\sigma}$ applying the ratio of wetted and total asperity surfaces A_{ratio} is used as shown in Eq. (1).

$$h_{model1} = \frac{\lambda_{eff}}{\delta} = \lambda_{eff} \cdot \frac{A_{ratio}}{\sigma} \mid with\ \lambda_{eff} = \frac{2 \cdot \lambda_1 \cdot \lambda_2}{\lambda_1 + \lambda_2} \tag{1}$$

Equation (2) shows the calculation for the frustrum area i.e. the area defined by the height of the liquid wetting the conical surface roughness. This gives all the needed information for the heat transfer coefficient as shown in Eq. (3).

$$A_{ratio} = \frac{A_{wet}}{A_{total}} = \frac{[2 \cdot r_{base} - X \cdot \cot(\varphi)] \cdot \sqrt{X^2 + (X \cdot \cot(\varphi))^2}}{r_{base} \cdot \sqrt{r_{base}^2 + 1^2}} \quad | \; with \; \varphi = \arctan\left(\frac{\sigma}{r_{base}}\right) \tag{2}$$

$$h_{model1} = \frac{\lambda_{eff}}{\sigma} \cdot \frac{[2 \cdot r_{base} - X \cdot \cot(\varphi)] \cdot \sqrt{X^2 + (X \cdot \cot(\varphi))^2}}{r_{base} \cdot \sqrt{r_{base}^2 + 1^2}} \tag{3}$$

While the wetting height X will be described by the mechanical analysis of the liquid-solid contact, all other variables are defined by the material and surface properties. The dimension of r_{base} can be obtained from a topographical scan as the parameter for average groove width R_{sm} which represents the asperity diameter.

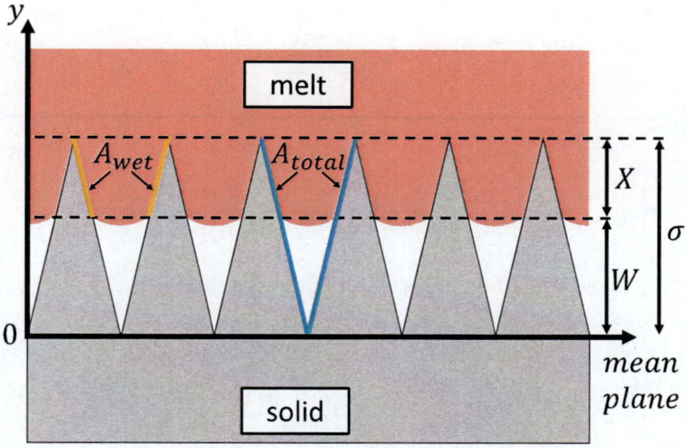

Fig. 1. Liquid-solid interface for a microscopically rough surface following profile model 1

2.2 Surface Profile Model 2

The second profile model by Hamasaiid et al. [3] adapts the modeling approach commonly used to describe solid to solid contacts and the corresponding heat transfer coefficient. As direct contact between two rough surfaces only is present at certain local spots, flux tube theory as described by Cooper et al. [4] can be applied. The final expression shown in Eq. (4) uses the size and density of circular contact spots to describe the heat transfer coefficient.

$$h_{model2} = 2 \cdot \lambda_{eff} \cdot \frac{r_{contact} \cdot \varrho_{spots}}{\left(1 - \frac{r_{contact}}{r_{base}}\right)^{1.5}} \tag{4}$$

The parameters used in Eq. (4) can be obtained from the topographical analysis. While also treating the asperities of the rough surface as a conical geometry, their height, base size and slope can be different individually. In order to give a description of such a complex surface structure the distribution of asperities is assumed to follow a Gaussian normal distribution. The average base radius of the asperities following the normal probability $\Phi(y)$ can thus be obtained as described in Eq. (6).

$$\Phi(y) = \frac{1}{\sqrt{2\cdot\pi}\cdot\sigma}\cdot\exp\left(-\frac{y^2}{2\cdot\sigma^2}\right), \quad \forall y \in (-\infty; +\infty) \tag{5}$$

$$r_{base} = \frac{1}{m}\cdot\int_{y=0}^{y=\infty} y\cdot\Phi(y)dy = \frac{\sigma}{\sqrt{2\cdot\pi}\cdot m} = \frac{R_{sm}}{2} \tag{6}$$

Here σ describes the standard deviation of the asperity heights and R_{sm} represents the average peak distance. An expression for the density of contact spots ϱ_{spots} and the contact spot radius $r_{contact}$ can be found as only peaks larger than the wetting height W, describing the thickness of the entrapped air layer, will be in contact with the liquid. Using the normal probability function describing the surface leads to the following expressions according to Hamasaiid et al. [3]:

$$\varrho_{spots} = \frac{1}{2.5\cdot\pi}\cdot\left(\frac{m^2}{\sigma^2+R_a^2}\right)\cdot erfc\left(\frac{W}{\sqrt{2}\cdot\sigma}\right) \mid with\ m = \frac{2\cdot\sigma}{\sqrt{2\cdot\pi}\cdot R_{sm}} \tag{7}$$

$$r_{contact} = \frac{1}{m}\cdot\left[\frac{\sigma}{\sqrt{2\cdot\pi}}\cdot\exp\left(-\frac{W^2}{2\cdot\sigma^2}\right) - \frac{W}{2}\cdot erfc\left(\frac{W}{\sqrt{2}\cdot\sigma}\right)\right] \tag{8}$$

Applying these equations in Eq. (4) leads to the final expression for the heat transfer coefficient for surface profile model 2:

$$h_{model2} = \frac{0.8}{\pi}\cdot\frac{\lambda_{eff}\cdot m}{(\sigma^2+R_a^2)}\cdot\frac{erfc\left(\frac{W}{\sqrt{2}\cdot\sigma}\right)\cdot\left[\frac{\sigma}{\sqrt{2\cdot\pi}}\cdot\exp\left(-\frac{W^2}{2\cdot\sigma^2}\right)-\frac{W}{2}\cdot erfc\left(\frac{W}{\sqrt{2}\cdot\sigma}\right)\right]}{\left[1-\frac{\sqrt{2\cdot\pi}}{\sigma}\cdot\left(\frac{\sigma}{\sqrt{2\cdot\pi}}\cdot\exp\left(-\frac{W^2}{2\cdot\sigma^2}\right)-\frac{W}{2}\cdot erfc\left(\frac{W}{\sqrt{2}\cdot\sigma}\right)\right)\right]^{1.5}} \tag{9}$$

2.3 Surface Profile Model 3

Another analytical model to describe the surface topology and the respective calculation for the heat transfer coefficient was developed by Hamasaiid et al. [5]. It relies on the same approach for defining the HTC coming from solid-solid contacts. In contrast to model 2, the asperities are assumed to have their base on the same zero level height as depicted in Fig. 2. The idea of leveling the asperity bases changes their slope and height distribution. Therefore, the asperity distribution function $\Phi(y)$ has to be modified to $\Phi_B(y)$ as shown in Eq. (10).

$$\Phi_B(y) = \frac{2}{\sqrt{2\cdot\pi}\cdot\sigma_B}\cdot\exp\left(-\frac{y^2}{2\cdot\sigma_B^2}\right), \quad \forall y \in [0; +\infty) \tag{10}$$

While the definition for the asperity base radius r_{base} remains identical, the contact spot radius $r_{contact}$ and density of microcontact spots ϱ_{spots} have to be adapted:

$$\varrho_{spots} = \frac{8}{\varepsilon \cdot \pi^2 \cdot R_{sm}^2} \cdot erfc\left(\frac{W}{\sqrt{2} \cdot \sigma}\right) \tag{11}$$

$$r_{contact} = \frac{1}{2} \cdot \sqrt{\frac{\pi}{2}} \cdot \frac{R_{sm}}{2 \cdot \sigma} \cdot \left[\frac{2 \cdot \sigma}{\sqrt{2 \cdot \pi}} \cdot \exp\left(-\frac{W^2}{2 \cdot \sigma^2}\right) - W \cdot erfc\left(\frac{W}{\sqrt{2} \cdot \sigma}\right)\right] \tag{12}$$

The factor ε adjusts the contact spot density to incorporate the area that exists in-between the assumed conical shapes and the potential contacts they might have.

For circular bases, a value of $\varepsilon \approx 1.5$ can be approximated statistically [5]. Applying the adapted parameters to Eq. (4) leads to the final equation for the HTC:

$$h_{model3} = \frac{8}{\varepsilon \cdot \pi^2} \cdot \frac{\lambda_{eff}}{R_{sm}} \cdot erfc\left(\frac{W}{\sqrt{2} \cdot \sigma}\right) \cdot \frac{\left[\frac{1}{2} \cdot \sqrt{\frac{\pi}{2}} \cdot \frac{R_{sm}}{\sigma} \cdot \left(\frac{2 \cdot \sigma}{\sqrt{2 \cdot \pi}} \cdot \exp\left(-\frac{W^2}{2 \cdot \sigma^2}\right) - W \cdot erfc\left(\frac{W}{\sqrt{2} \cdot \sigma}\right)\right)\right]}{\left[\frac{R_{sm}}{2} - \left(\frac{1}{2} \cdot \sqrt{\frac{\pi}{2}} \cdot \frac{R_{sm}}{\sigma} \cdot \left(\frac{2 \cdot \sigma}{\sqrt{2 \cdot \pi}} \cdot \exp\left(-\frac{W^2}{2 \cdot \sigma^2}\right) - W \cdot erfc\left(\frac{W}{\sqrt{2} \cdot \sigma}\right)\right)\right)\right]^{1.5}} \tag{13}$$

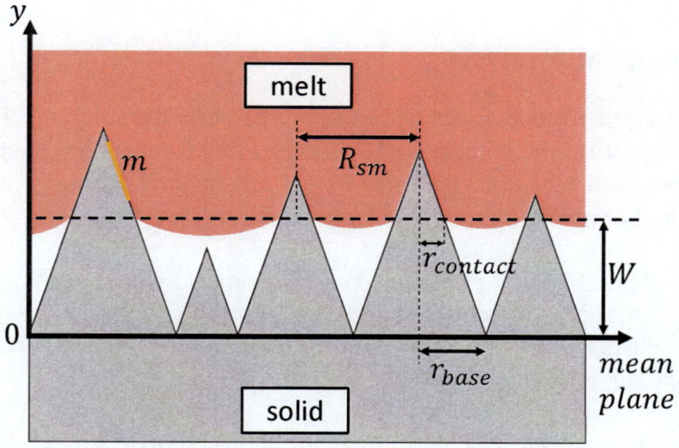

Fig. 2. Liquid-solid interface for a microscopically rough surface following profile model 3

2.4 Surface Profile Model 4

The fourth surface model proposed by Xu et al. [6] uses the modeling idea presented in model 3. When introducing a common base level height, the profile parameters R_a and R_{sm} remain the same while the slope changes. The partially significantly decreased slope this way results in an increased size of the conical asperities, leading to a changed surface profile with increased contact area as the outcome. As a result, the calculated heat transfer coefficient will be overestimated. The slope value can be preserved though by moving the asperities in parallel to the surface's mean plane while applying a leveling needed for a uniform description of the wetting height. The proposed way for achieving this is doubling the parameter R_a i.e. $R_{a_n} = 2 \cdot R_a$. This leads to a new distribution function requiring an update for contact spot density and contact spot radius:

$$\Phi_n(y) = \frac{1}{\sqrt{2 \cdot \pi} \cdot \sigma_n} \cdot \exp\left(-\frac{(y - R_{a_n})^2}{2 \cdot \sigma_n^2}\right), \quad \forall y \in [0; +\infty) \tag{14}$$

$$\varrho_{spots} = \frac{4}{\varepsilon \cdot \pi^2 \cdot R_{sm}^2} \cdot erfc\left(\frac{W - 2 \cdot R_a}{2 \cdot \sqrt{\pi - 2} \cdot R_a}\right) \tag{15}$$

$$r_{contact} = \sqrt{1 - \frac{2}{\pi}} \cdot \frac{R_{sm}}{4} \cdot \exp\left[\frac{-(W - 2 \cdot R_a)^2}{4 \cdot R_a^2 \cdot (\pi - 2)}\right] + \frac{R_{sm}}{8 \cdot R_a} \cdot (2 \cdot R_a - W) \cdot \left[1 \pm \mathrm{erf}\left(\frac{W - 2 \cdot R_a}{2 \cdot \sqrt{\pi - 2} \cdot R_a}\right)\right] \tag{16}$$

Inserting these equations in Eq. (4) gives the final description of the HTC:

$$h_{model4} = \frac{8}{\varepsilon \cdot \pi^2} \cdot \frac{\lambda_{eff}}{R_{sm}} \cdot erfc\left(\frac{W - 2 \cdot R_a}{2 \cdot \sqrt{\pi - 2} \cdot R_a}\right) \cdot \frac{\frac{1}{4} \cdot \sqrt{1 - \frac{2}{\pi}} \cdot \exp\left[\frac{-(W-2 \cdot R_a)^2}{4 \cdot R_a^2 \cdot (\pi - 2)}\right] + \frac{2 \cdot R_a - W}{8 \cdot R_a} \cdot erfc\left(\frac{W - 2 \cdot R_a}{2 \cdot \sqrt{\pi - 2} \cdot R_a}\right)}{\left[1 - \frac{1}{2} \cdot \sqrt{1 - \frac{2}{\pi}} \cdot \exp\left[\frac{-(W-2 \cdot R_a)^2}{4 \cdot R_a^2 \cdot (\pi - 2)}\right] + \frac{2 \cdot R_a - W}{4 \cdot R_a} \cdot erfc\left(\frac{W - 2 \cdot R_a}{2 \cdot \sqrt{\pi - 2} \cdot R_a}\right)\right]^{1.5}} \tag{17}$$

3 Mechanical Analysis

During the casting process, gas i.e. air is pressed into the small notches of the rough surface where it can be compressed due to external forces. The exact amount of this compression, which is crucial to the wetted height level, is the result of a pressure equilibrium. Important to the equilibrium state during wetting are:

- The initial pressure of the entrapped air inside the surface cavity p_0
- The capillarity pressure due to surface energies p_γ
- The metallostatic pressure as result of the height h of the liquid column $p(h)$
- The externally applied pressure p_{ext}

Considering the air inside a notch, Boyle-Mariotte's law applies with state 0 representing the conditions prior to casting and state 1 representing the state after pouring:

$$p_1 \cdot V_1 = p_0 \cdot V_0 \tag{18}$$

When applying the ideal gas law and when considering that the amount of substance n remains constant due to the air being entrapped, Eq. (18) can be extended including the respective temperatures and thermal effusivities b:

$$\frac{p_1 \cdot V_1}{T_1} = n \cdot R = \frac{p_0 \cdot V_0}{T_0} \mid with\ T_1 = \frac{T_{sol} \cdot b_{sol} + T_{liq} \cdot b_{liq}}{b_{sol} + b_{liq}} \quad and \quad b = \sqrt{\lambda \cdot \varrho \cdot c_p} \tag{19}$$

The following models will show different approaches to estimate the wetting height level while assuming that the solid material is formed by a non-porous medium and that the entrapped air can be described like an ideal gas.

3.1 Wetting Model A

Hamasaiid et al. [3] formed an equation based on the simplifying approach that surface tension is negligible. While for a gravity die casting process p_0 will be equal to normal barometric pressure, pressure p_1 depends on the process conditions. The respective volumes can be approximated by the following equations:

$$V_0 = \frac{\pi}{3 \cdot m_n^2} \cdot \int_{y=0}^{y=+\infty} \Phi(y) \cdot y^3 dy = \frac{\pi}{3} \cdot \sqrt{\frac{2}{\pi}} \cdot \frac{\sigma^3}{m_n^2} \tag{20}$$

$$V_1 = \frac{\pi}{3 \cdot m_n^2} \cdot \int_{y=0}^{y=W} \Phi(y) \cdot y^3 dy \approx \frac{\pi \cdot W^3}{3 \cdot m_n^2} \tag{21}$$

Combining Eqs. (21) and (22) with Eq. (19) allows for the calculation of the wetting height W as shown in Eq. (22).

$$W = \sqrt[3]{\frac{p_0 \cdot T_1}{p_1 \cdot T_0}} \cdot \sqrt[6]{\frac{2}{\pi}} \cdot \sigma \tag{22}$$

3.2 Wetting Model B

While still ension and when applying the adapted asperity distribution function Φ_B as presented in surface profile model 3, the equation for the wetting height changes to:

$$W = \sqrt[3]{2 \cdot \frac{p_0 \cdot T_1}{p_1 \cdot T_0}} \cdot \sqrt[6]{\frac{2}{\pi}} \cdot \sigma \tag{23}$$

3.3 Wetting Model C

Hamasaiid et al. [5] also presented modeling approach including surface tension influences. When modelling the surface roughness as conical structure with circular bases, the case of a circular tube with low Bond number can be assumed due to the low size of the notches. This simplifies the Laplace-Young differential equation for the pressure difference Δp across the fluid interface. The new expression depends on surface tension γ and curvature radius r_{curv} which can be approximated as a function of p_1 as shown in Eq. (24) leading to the wetting height depicted in Eq. (25).

$$p_\gamma = \frac{2 \cdot \gamma}{r_{curv}} \approx 0.87 \cdot p_1 \tag{24}$$

$$W = \sqrt[3]{2 \cdot \frac{p_0 \cdot T_1}{\left[p_1 - p_\gamma\right] \cdot T_0}} \cdot \sqrt[6]{\frac{2}{\pi}} \cdot \sigma \tag{25}$$

3.4 Wetting Model D

Prasher proposed a model for determining the wetting height based on a surface chemistry approach which includes the surface tension γ_{liq} influences. It applies the actual wetted height X instead of the thickness of the entrapped air W though. For a notch as shown in Fig. 1, the capillary pressure can be described as given in Eq. (26). The pressure equilibrium can be described by the expression in Eq. (27).

$$p_\gamma = \frac{2 \cdot \gamma_{liq} \cdot \sin(\theta + \varphi)}{r_{base} - X \cdot \cot(\varphi)} \tag{26}$$

$$p_1 + \frac{2 \cdot \gamma_{liq} \cdot \sin(\theta + \varphi)}{r_{base} - X \cdot \cot(\varphi)} = p_0 \cdot \frac{r_{base}^3}{(r_{base} - X \cdot \cot(\varphi))^3} \tag{27}$$

Equation (27) can be solved for X when assuming that for a $p_1 \approx p_0$ condition, as is the case in most gravity die casting processes, the term $X \cdot \cos(\varphi)$ will be small compared to the base radius. The wetting height X then can be calculated via Eq. (28).

$$X = \frac{2 \cdot \gamma_{liq} \cdot \sin(\theta + \varphi)}{3 \cdot p_0 \cdot \cot(\varphi)} \tag{28}$$

3.5 Wetting Model E

Yuan et al. [7] introduced an extension to the empiric Hamasaiid approach for the surface tension influence by integrating the surface chemistry approach described by Prasher which corrects the calculated wetting height W_0 for no surface tension influences as shown in Eq. (29).

$$W \approx \frac{1}{\Gamma - \frac{2 \cdot \gamma_{liq} \cdot \sin(\theta+\varphi)}{3 \cdot \Gamma \cdot p_0 \cdot W_0^3 \cdot \frac{T_1}{T_0} \cdot cot(\varphi)}} \mid \Gamma = \sqrt[3]{\sqrt[2]{\frac{p_1^2}{4 \cdot p_0^2 \cdot W_0^6 \cdot \left(\frac{T1}{T0}\right)^2} - \frac{8 \cdot \gamma_{liq}^3 \cdot \sin^3(\theta + \varphi)}{27 \cdot p_0^3 \cdot W_0^9 \cdot \left(\frac{T_1}{T_0}\right)^3 \cdot \cot^3(\varphi)}} + \frac{p_1}{2 \cdot p_0 \cdot W_0^3 \cdot \frac{T1}{T0}}} \tag{29}$$

4 Comparison of Wetting and HTC Predictions

Figure 3 left shsows the predictions of the wetting height models examined as a function of R_a as $\sigma = R_a \cdot \sqrt{\pi/2}$. All models show a linear dependency of the resulting wetting height level and, with exception of model C, predict absolute numbers of similar magnitude. Model C, which introduces an empiric surface tension influence compared to model B, predicts wetting height values that are 2–3 times larger in comparison to the other models. Even though the absolute differences between all the models are rather small, there is a strong impact on the HTC results. This can be seen in Fig. 3 right where the surface profile model predictions are examined as a function of $HTC(W)$. For model 1, a parabolic shape of its function can be seen delivering values up to 1000 W/m^{2}K in a limited range of up to a roughness of 9 μm. The model by Prasher for this reason is not suitable to describe the HTC in gravity die casting processes. All other models show an exponential behavior and are rather similar to each other with the main difference being a shift considering at which point lower wetting heights lead to very strong increases in the HTC prediction. For this reason even small changes in the wetting height prediction can lead to substantial differences in the predicted HTC which can be seen in Table 2 and 3. Here the models are applied to two different gravity die casting experiments with experimentally determined HTC by Xu [6] and Vossel [8] labeled as “GDC-A” and “GDC-B”. Table 1 shows the respective differences in process parameters. Although the wetting height predictions are all in the same magnitude and might only be different for less than 1 μm, the results in predicted HTC can be profoundly different. Models i.e. model combinations giving results close to the experimentally determined HTCs, thus suitable for an application as HTC model inside a numeric simulation, are color coded green for models with high and orange for models with moderate agreement to the experimental values.

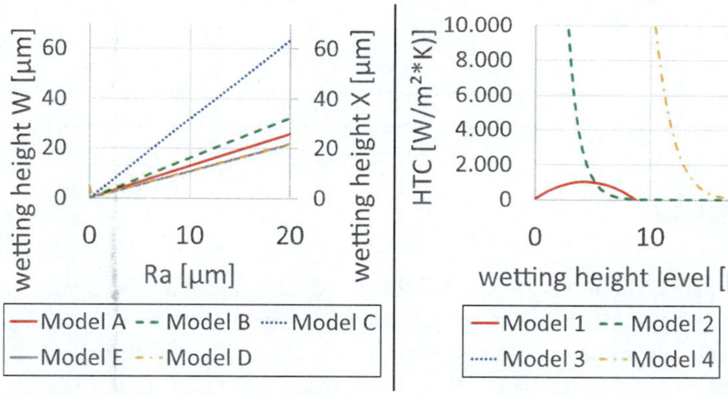

Fig. 3. Model prediction of wetting height prediction over R_a (left) and model prediction of HTC over applied wetting height level (right)

Table 1. Process parameters

“GDC-A”	$R_a\,[m]$	$\sigma\,[m]$	$R_{sm}\,[m]$	$\varphi\,[°]$
	3.52E-06	4.41E-06	1.35E-04	3.74
	$T_0\,[K]$	$T_1\,[K]$	$p_0[Pa]$	$p_1[Pa]$
	300	409	1.013E+05	1.047E+05
“GDC-B”	$R_a\,[m]$	$\sigma\,[m]$	$R_{sm}\,[m]$	$\varphi\,[°]$
	4.84E-06	6.42E-06	4.72E-04	15.22
	$T_0\,[K]$	$T_1\,[K]$	$p_0[Pa]$	$p_1[Pa]$
	300	724	1.013E+05	1.047E+05

Table 2. Resulting HTC values for "GDC-A" process parameters $\left[\frac{W}{m^2 \cdot K}\right]$

Experimentally determined HTC: 5950 W/(m²*K)

	Model 1	Model 2	Model 3	Model 4
Model A	1010	2120	23100	278000
Model B	931	694	7580	144000
Model C	-1360	1	14	6380
Model D	1010	2340	25600	296000
Model E	845	405	4420	107000

Table 3. Resulting HTC values for "GDC-B" process parameters $\left[\frac{W}{m^2 \cdot K}\right]$

Experimentally determined HTC: 2000-2500 W/(m²*K)

	Model 1	Model 2	Model 3	Model 4
Model A	238	2580	1440	392000
Model B	175	634	108	175000
Model C	-816	1	1	2050
Model D	226	26100	59900	1820000
Model E	-37	50	1	45200

5 Conclusion

The presented HTC models were applied for two different gravity die casting processes where the respective heat transfer coefficients have been determined experimentally. While some models failed to predict appropriate HTC values, the surface profile model by Hamasaiid et al. [5] and especially the approach by Xu et al. [6] in combination with a wetting model by Hamasaiid et al. [5] managed to predict HTC values close to the experimentally derived ones. Predicting the HTC for a gravity die casting process represents a challenging task as the formula are very sensitive to the surface roughness where quite strong deviations in HTC predictions could be presents within the standard deviation of a roughness measurement used as input parameter.

Acknowledgment. The presented investigations were carried out at RWTH Aachen University within the framework of the Collaborative Research Centre SFB1120–236616214 "Bauteilpräzision durch Beherrschung von Schmelze und Erstarrung in Produktionsprozessen" and funded by the Deutsche Forschungsgemeinschaft e.V. (DFG, German Research Foundation). The sponsorship and support is gratefully acknowledged.

References

1. Prasher, R.S.: Surface chemistry and characteristics based model for the thermal contact resistance of fluidic interstitial thermal interface materials. J. Heat Transfer **123**, 969–975 (2001)

2. Somé, S.C., Delaunay, D., Gaudefroy, V.: Comparison and validation of thermal contact resistance models at solideliquid interface taking into account the wettability parameters. Appl. Therm. Eng. **61**, 531–540 (2013)
3. Hamasaiid, A., Dargusch, M.S., Davidson, C., Loulou, T., Dour, G.: A Model to predict the heat transfer coefficient at the casting-die interface for the high pressure die casting process. AIP Conf. Proc. **907**, 1211 (2007)
4. Cooper, M.G., Mikic, B.B., Yovanovich, M.M.: Thermal contact conductance. Int. J. Heat Mass Transf. **12**(3), 279–300 (1969)
5. Hamasaiid, A., Dargusch, M.S., Loulou, T.: Dour, G: A predictive model for the thermal contact resistance at liquid-solid interfaces: analytical developments and validation. Int. J. Therm. Sci. **50**, 1445–1459 (2011)
6. Xu, R., Li, L., Zhang, L., Zhu, B., Bu, X.: An improved model for predicting heat transfer coefficient peak value at the casting-die interfaces. Indian J. Eng. Mater. Sci. **21**, 628–634 (2014)
7. Yuan, C., Duan, B., Li, L., Shang, B., Luo, X.: An improved model for predicting thermal contact resistance at liquid–solid interface. Int. J. Heat Mass Transf. **80**, 398–406 (2015)
8. Vossel, T., Pustal, B., Bührig-Polaczek, A.: Influence of gap formation and heat shrinkage induced contact pressure on the development of heat transfer in gravity die casting processes. In: Proceedings of Liquid Metal Processing & Casting Conference 2019 (LMPC 2019), pp. 275–284 (2019)

Check for updates

Micro-macro Coupled Solidification Simulations of a Sr-Modified Al-Si-Mg Alloy in Permanent Mould Casting

Bei Zhou[1(✉)], Herfried Behnken[1], Janin Eiken[1], Markus Apel[1], Gottfried Laschet[1], and Nino Wolff[2]

[1] Access e.V., 52078 Aachen, Germany
b.zhou@access.rwth-aachen.de
[2] Foundary Institute, RWTH-Aachen University, 52078 Aachen, Germany

Abstract. Casting simulations incorporating solidification models are commonly applied to improve the dimensional accuracy of casting components. Challenging for a predictive casting simulation is the precise description of the solidification process of the casting alloy. For Sr-modified Al-Si-Mg casting alloys (e.g. A356), the Scheil-Gulliver model fails to describe the effect of Strontium on the solidification path. In this work, a dedicated micro-macro simulation approach is applied to a permanent mould casting component with Sr-modified A356 alloy. Microstructure simulations in both 2D and 3D based on multicomponent multiphase-field method were performed to study the effect of Sr-modification, as well as the impact of cooling rate on solidification path. Casting simulations coupled with 2D and 3D microstructure simulation were compared and validated with experimental results.

Keywords: A356 alloy · Sr-modification · Phase-field simulation · Multiscale simulation

1 Introduction

In commercial Al-Si casting alloys, Strontium (Sr) is often added to improve the mechanical properties. With an amount ranging from of 100 ppm to 400 ppm, Sr transforms the eutectic morphology from coarse plate-like structure to fine coral-like fibrous networks [1, 2]. A great number of experimental and numerical studies [3–5] focused on the mechanism of Sr-modification to the Al-Si eutectic solidification and showed that Sr has impacts on both growth and nucleation of eutectic Si phases.

In practice, the Scheil-Gulliver approximations are commonly employed in thermomechanical casting simulations to estimate the solidification path, i.e., the temperature dependent fraction of solid $f_s(T)$, while experimental DSC (differential scanning calorimetry thermal analysis) data are used to describe the latent heat release. However, the Scheil-Gulliver model does not describe the effect of Sr on the nucleation and growth of eutectic Si, in particular the growth undercooling. Hence, it leads to an inaccurate description of the solidification path, latent heat release as well as temperature dependent

U. Reisgen et al. (Eds.): EMPOrIA 2020, LNME, pp. 202–211, 2021.
https://doi.org/10.1007/978-3-030-70332-5_18

effective mechanical and thermal properties. Therefore, for Sr-modified Al-Si alloy, it is necessary to get insight into the microstructure formation process and understand the effect of Sr on the solidification process.

In a previous study [6], the impact of Sr on eutectic solidification has been studied based on 2D microstructure simulations. A dedicate multiscale simulation method based on a self-consistent homoenthalpic approach [7] has been proposed and applied to simulate a permanent mould casting process component with Sr-modified A356 alloy. The spatially resolved microstructure simulations were performed using a Calphad-based multicomponent multiphase-field (MMPF) model [8]. The effect of Sr-modification was modelled by adjusting the nucleation model for eutectic Si phase, and by calibrating a growth parameter which characterizes the critical length scale of the eutectic microstructure. The 2D microstructure simulation were coupled to the thermomechanical finite element casting simulation based on a common enthalpy-temperature relationship. In this approach, the local heat extraction rate $\dot{Q}$ obtained by macroscopic casting simulation was applied as boundary conditions in microstructure simulations. Improved description of solidification path and latent heat release have been obtained by 2D microstructure simulations, and resulted in a more predictive casting simulation.

For practical applications, casting simulation combined with 2D MMPF simulations are timesaving. However, inherent simplifications and approximations in a 2D MMPF model may also lead to an inaccurate description of the enthalpy-temperature relationship. In this paper, 3D MMPF simulations coupled to the same permanent mould casting simulations will be discussed. Results for 2D and 3D MMPF simulations were compared. Moreover, in the previous study [6], 2D MMPF simulation results based on the local heat extraction rate read from one single location of the casting component were applied for all integration points in FE thermomechanical simulation. In order to identify the impact of local cooling rate on the solidification process, 3D MMPF simulations with different local extraction rates for different locations are compared.

2 The Casting Experiments

Casting experiments were performed using the experimental set-up presented in [9], see Fig. 1. It mainly consists of an inner steel core, a steel mould, a sand core on the top and the axisymmetric bowl-like casting component. The bowl component has an outer diameter of 155 mm and a wall thickness that gradually varies from 15 mm up to 30 mm. The insulation on bottom and the sand core on top of the bowl assured a radial flow of the heat flux. Cooling channels were embedded in the mould in order to control the mould temperature. The experiment was carried out with an initial melt temperature at 720 °C and a mould temperature of 30 °C. Thermocouples embedded in the mould wall as well as in melt were used to measure the temperature and the temperature gradients during the cooling process. In order to validate the heat transfer coefficient (HTC), the gap width between the mould wall and melt, as well as the contact pressure at the inner core were measured by quartz rods together with linear variable differential transformers (LVDTs).

The commercial A356 with Sr-modification and grain refinement (TiB5) has been applied as casting alloy. The grain refinement ensured a relatively homogeneous grain size distribution and an equiaxial growth of fcc-Al dendrites over all the casting component. A mean grain diameter of 400 μm and an average secondary dendrite arm spacing (SDAS) of 19.5 μm have been measured experimentally.

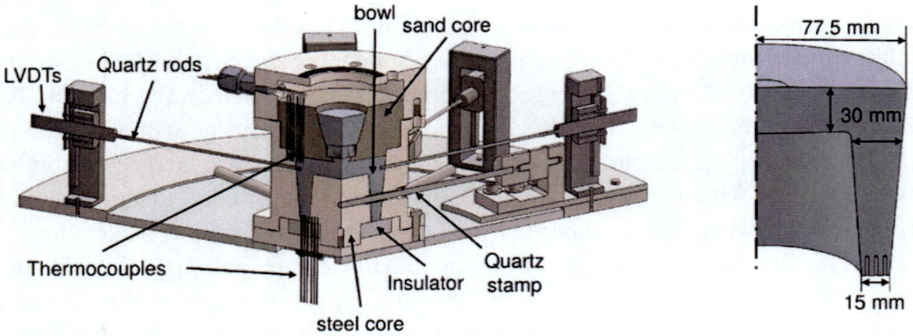

Fig. 1. The experiment set-up (left) and the geometry of the casting component (right).

3 Macroscopic Casting Simulation Model

3.1 The FE Simulation Model

The thermomechanical casting simulation were performed by using the finite element (FE) software Abaqus. The FE model, including mould set and the casting component were meshed with temperature-displacement coupled 4-node linear tetrahedron (C4D4T) elements. As the steel mould was built up with three identical 120° segments including cooling channels and thermo-elements, one third of the entire set-up was modelled for FE simulations. The bowl casting component was modelled as a thermo-elastoplastic material, whereas the mould and core parts as thermo-elastic material. According to the experiment, the melt, i.e., the bowl region had an initial temperature of 720 °C, while the mould temperature was set to be 30 °C.

Modelling of Gap Formation

As the bowl component undergoes a volume shrinkage during the solidification process, an air gap could be formed between the casting component and the mould. For the precision of the simulation, the contact condition between the bowl component and its surrounding mould distinguishes between fluid-solid contact and solid-solid contact. By modelling of the solid-solid contact condition, the impact of the formed air gap on the heat transfer coefficient (HTC) has been taken into account. In this case, the HTC is modelled as a function of the local gap width and the local contact pressure [10]. The gap conduction model was implemented in Abaqus via the GAPCON subroutine.

Modelling of Phase Transformation

In thermomechanical casting simulations, the casting alloys are treated as a liquid/solid two-phase system, which is characterized by the temperature dependent fraction solid

$f_s(T)$. During solidification, the local average enthalpy of the mixture can be expressed as a function of fraction solid

$$\langle \rho h \rangle = \int_{T_0}^{T} \left\langle \rho c_p(\tilde{T}) \right\rangle \mathrm{d}\tilde{T} + \langle \rho \rangle L_f \left[1 - f_s(T) \right], \tag{1}$$

in which ρ and h denote the density and specific enthalpy, respectively. c_p is the specific heat capacity and L_f represents the latent heat, i.e. the enthalpy of melting. Here, $\langle \cdot \rangle$ represents the volume average of a quantity in the two-phase system. According to the local energy balance in terms of specific enthalpy, the local energy balance in absence of external heat sources can be formulated as

$$\left(\langle \rho c_p \rangle + \langle \rho \rangle L_f \frac{\partial f_s}{\partial T} \right) \frac{\partial T}{\partial t} + \nabla \cdot \langle \rho h \mathbf{v} \rangle = \dot{\mathrm{Q}} \tag{2}$$

with $\mathbf{v}$ representing the velocity vector and $\dot{\mathrm{Q}}$ being the local heat flux. In the Abaqus FE model, the HETVAL subroutine in conjunction with the USDFLD subroutine were applied to define the source term related to the latent heat release in Eq. (2). Normally, the heat capacity c_p, the total latent heat L_f and the fraction solid-temperature relationship $f_\mathrm{s}(T)$ are obtained by the Scheil approximation of solidification and experimental measurements. In the micro-macro coupled approach presented here, they are obtained from the Calphad-based MMPF microstructure simulation.

As an example, the simulated temperature evolution during the solidification process in the bowl is shown in Fig. 2. The grey colour indicates temperatures higher than liquidus temperature, whereas the black colour in the right picture represents temperature lower than the solidus temperature.

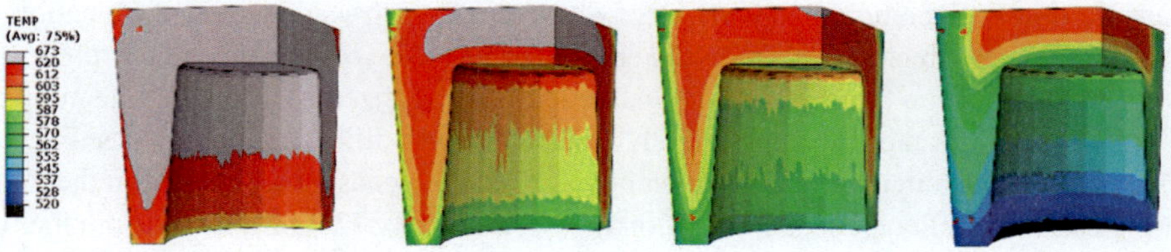

Fig. 2. Temperature (°C) evolution during solidification with the grey and black coloured region indicating pure liquid and pure solid, respectively.

4 Microstructure Simulation

Microstructure simulations were performed using the software MICRESS®, which is based on the multicomponent multiphase-field model described in [8]. The Calphad-database are coupled via the TQ-interface. The alloy system is defined by 92.7 wt% Al, 7 wt% Si and 0.3 wt % Mg. Further chemical elements normally also present in commercial A356 alloys, such as Mn and Fe, are not considered. During the simulation, the nucleation of primary fcc-Al, eutectic Al-Si and Mg_2Si phase were considered. According to the experimentally measured average grain size, a simulation domain with a dimension of 320 μm × 320 μm × 320 μm and the grid resolution of 2 μm has been used. Compared to primary fcc-Al dendrites, the interdendritic eutectic has a much finer microstructure, which cannot be resolved with a 2 μm grid resolution. Therefore, the two-phase eutectics was handled as an effective single phase. The simulation started from 100% liquid phase at 615 °C. Heat extraction rate was applied as the boundary condition. As shown in Fig. 3, an equiaxed single fcc-Al dendrite growths at the beginning of the solidification. At around $f_s = 0.5$, the first interdendritic eutectic phases were formed between primary fcc-Al dendrites and residual melt, leading to a rather complex microstructure at the end of the solidification process.

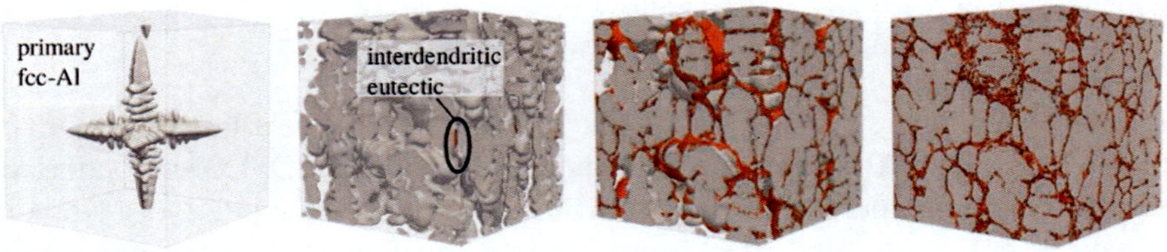

f_s=0.02, T=600°C f_s=0.52, T=571°C f_s=0.71, T=570°C f_s=0.99, T=521°C

Fig. 3. Results of a 3D microstructure simulation

4.1 Effect of Sr-Modification on the Eutectic Solidification

The modelling of Sr-modification in the course of the eutectic solidification were achieved by applying appropriate nucleation and growth parameters for the effective eutectic phase. Specifically, the effect of Sr on the growth was simulated by calibrating a parameter which characterize the critical length scale of the eutectic morphology, whereas its effect on the nucleation of eutectic Si was realized by adjusting the number of initial nuclei for the eutectic phase. In the case of unmodified A356 alloy (see Fig. 4), a potentially unlimited number of nuclei for the eutectic phase were seeded in the RVE at the beginning of eutectic solidification ($f_s = 0.52$, $T = 572.3$ °C), i.e., the number of grains is governed by the interplay of cooling rate and latent heat release during growth. The microstructure size parameter was set to be 0.1 μm. For the Sr-modified case (see Fig. 5), only 8 nuclei for the eutectic phase were allowed to grow up. The microstructure size parameter was set to be 0.01 μm. As illustrated in Fig. 6(a), in the unmodified case the eutectic growth temperature is very close to the equilibrium temperature obtained by Scheil approximation, while by Sr-modified alloy it is nearly 5 °C lower. Additionally, recalescence has been observed in the primary solidification (fcc-Al). Consequently, the fraction solid curve differs to the Scheil approximation, see Fig. 6(b).

4.2 Impact of Cooling Rates on Solidification Path

For coupling to the macroscopic casting simulation, 3D MMPF simulations were performed by taking the local heat extraction rate $\dot{Q}$ obtained in casting simulations as

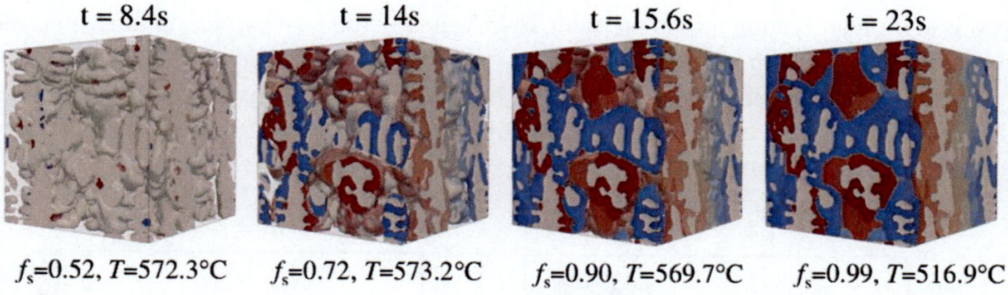

Fig. 4. Growth of effective eutectic grains during solidification of unmodified A356 alloys.

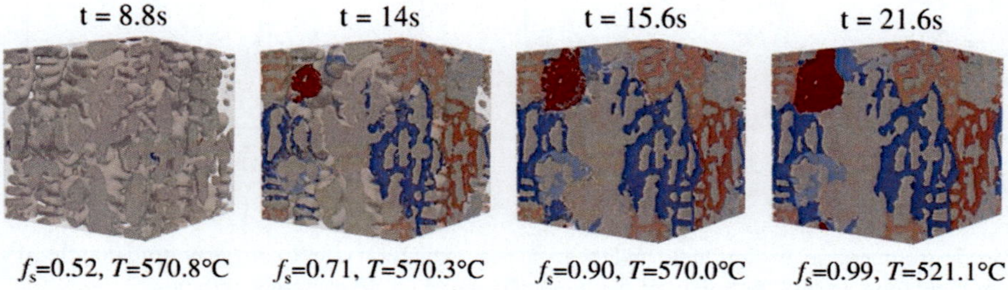

Fig. 5. Growth of effective eutectic grains during solidification of Sr-modified A356 alloys.

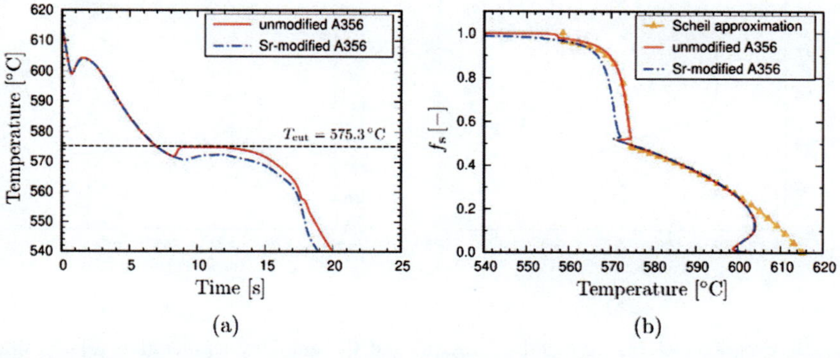

Fig. 6. Comparison between simulation results of Sr-modified A356 and unmodified A356 alloy: (a) temperature evolution; (b) fraction solid curve.

boundary condition. The resulting fraction solid, latent heat release and heat capacity in dependency of temperature can then be applied for improving the macroscopic process simulation. However, in the casting simulation, the local heat extraction rates are not identical in the entire casting component. It is very expensive to perform MMPF simulation on each integration point of the FE model. In order to study the impact of cooling rate variation on the solidification, MMPF simulations were performed with heat extraction rate obtained on two locations, i.e., on the bottom middle (BM) and the top middle (TM) of the bowl component. As shown in Fig. 7(a), the cooling rate on point

BM is significantly higher than on point TM. Although the heat extraction rates on these two points differ from each other, see Fig. 7(b), the fraction solid curve and enthalpy-temperature relationship obtained from both MMPF simulations are quite similar, as shown in Fig. 8. Hence, in the present casting simulation, a uniform fraction solid curve and enthalpy-temperature relationship were applied for the entire component.

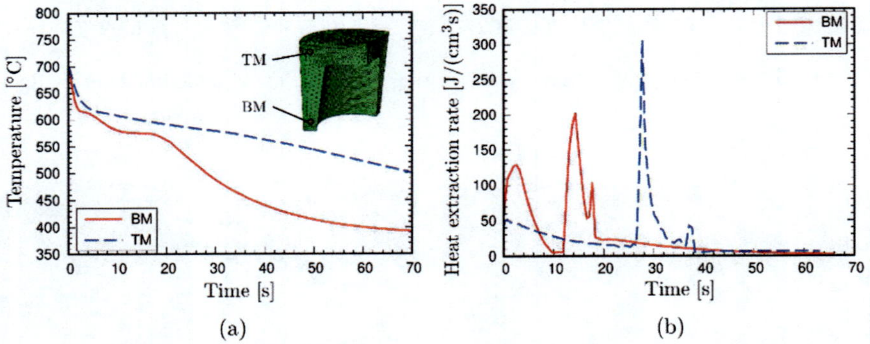

Fig. 7. Temperature evolution (a) and the local heat extraction rate (b) on the bottom middle (BM) point and top middle (TM) point of the bowl calculated by FE casting simulation.

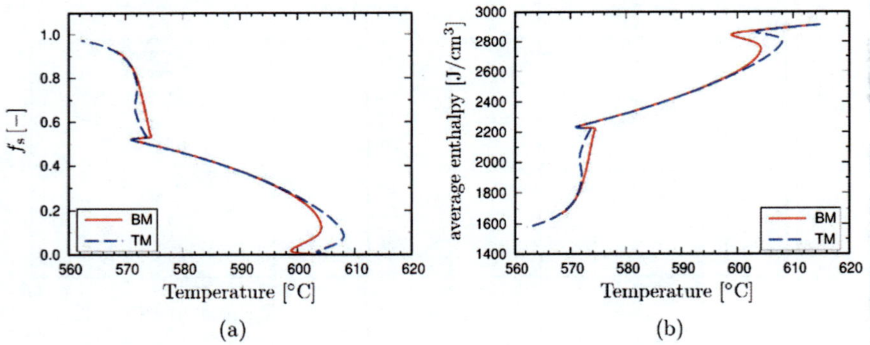

Fig. 8. Comparison of faction solid curves (a) and the enthalpy-temperature relationship (b) obtained by MMPF simulations on two locations.

5 Results of Micro-macro Coupled Simulation

The micro-macro coupled simulations were performed in two iterative loops. It began with a FE casting simulation with a fraction solid curve approximated by the Scheil solidification model. One should note that the result from the Scheil model does not consider different cooling rates, i.e. the Scheil curve is independent from the local temperature history. The local heat extraction obtained on the point BM was then applied as boundary condition for the temperature calculation in the 2D or 3D MMPF simulation. In Fig. 9, the 3D MMPF simulation results are compared with the 2D MMPF simulation using the same boundary condition and nucleation model. In the second FE simulation, the new fraction solid curve and latent heat release obtained from the microstructure simulation was applied. For validation, the simulated temperature evolutions in four different bowl locations have been compared with the experimental measurements, see Fig. 10.

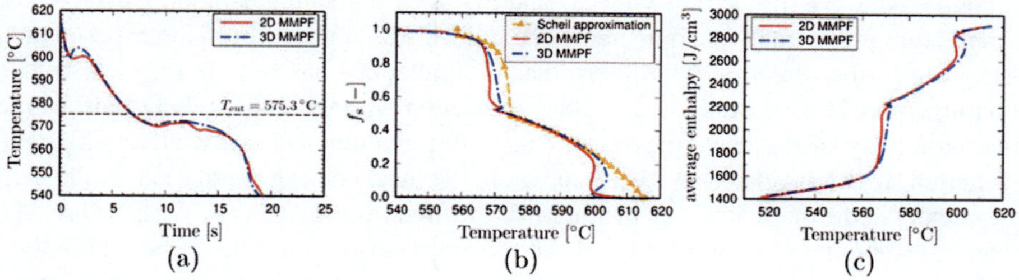

Fig. 9. 2D and 3D MMPF simulation results: (a) temperature evolution; (b) fraction solid curve; (c) enthalpy-temperature relationship.

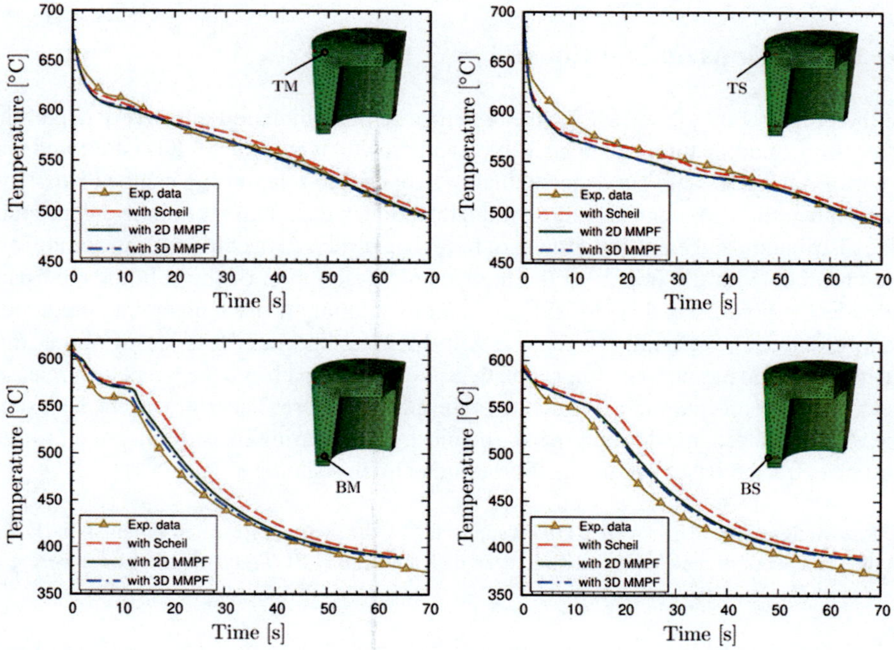

Fig. 10. Temperature evolutions obtained by FE casting simulations on four experimentally observed locations, which are labelled as TM (top middle), TS (top surface), BM (bottom middle) and BS (bottom surface).

As illustrated in Fig. 10, the FE simulation with the Scheil approximation supplied good temperature predictions in locations TM and TS. Whereas on the bottom of the bowl (locations BM and BS) where the cooling rate is higher, the predicted temperature was significantly higher than in the experiment, especially at the later stage of solidification. By applying the fraction solid-temperature relationship and the latent heat release from the MMPF simulation, a better temperature prediction was obtained on locations BM and BS. Although the 2D MMPF simulation results show a slightly different recalescence and eutectic growth undercooling compared to the 3D simulation (see Fig. 9), the results of coupled casting simulations with 2D and 3D MMPF are quite similar. However, the temperature predictions for some regions, e.g., BS, still deviated from the experimental data. One further iteration with micro-macro simulations has been performed, but no obvious improvement in the temperature prediction was achieved. Likely reasons are the complexity of the casting geometry including thermomechanical effects like gap formation, which is related to a wide spread in the local cooling conditions, or the high complexity in the microstructure evolution of the Sr modified Al-Si alloy. Therefore, one potential solution for minimization of the remaining error could be further calibration of the number of potential nuclei and the microstructure size parameter of the eutectic phase. Moreover, instead of a uniform fraction solid curve and enthalpy-temperature relationship, locally different relationships could also lead to a more precise temperature prediction in casting simulation.

6 Conclusions and Outlooks

In the present work, the MMPF based microstructure simulation effectively depict the effect of Sr-modification on the nucleation and growth of eutectic Si. It is confirmed that a uniform fraction solid curve and enthalpy-temperature relationship can be to apply to every integration point in FE casting simulation model, if a relatively homogeneous grain size distribution and equiaxial growth of fcc-Al dendrites all over the casting component are ensured. With the improved thermodynamic description of the solidification path provided by both 2D and 3D MMPF simulations, a more precise temperature prediction can be obtained, especially in regions with high cooling rate. It is worth noting that Sr-modification has not only impact on the solidification path and the latent heat release, but also on the morphological structure of the eutectic phase. Therefore, as the next step, the effective mechanical properties of the mushy zone during the solidification process will be extracted based on the fine fibrous eutectic morphology.

Acknowledgments. This work is kindly supported by the German Research Foundation (DFG) in the framework of the Collaborative Research Centre SFB1120 "Precision Melt Engineering".

References

1. Sigworth, G.K.: The modification of Al-Si casting alloys: important practical and theoretical aspects. Int. J. Metalcast. **2**(2), 41 (2008)
2. Timpel, M., Wanderka, N., Schlesiger, R., Yamamoto, T., Lazarev, N., Isheim, D., Schmitz, G., Matsumura, S., Banhart, J.: The role of strontium in modifying aluminium-silicon alloys. Acta Mater. **60**(9), 3920–3928 (2012)

3. Lu, S.Z., Hellawell, A.: Growth mechanisms of silicon in Al-Si alloys. J. Cryst. Growth **73**(2), 316–328 (1985)
4. Dahle, A.K., Nogita, K., McDonald, S.D., Dinnis, C., Lu, L.: Eutectic modification and microstructure development in Al-Si Alloys. Mater. Sci. Eng. A **413–414**, 243–248 (2005)
5. Eiken, J., Apel, M., Liang, S.M., Schmid-Fetzer, R.: Impact of P and Sr on solidification sequence and morphology of hypoeutectic Al-Si alloys: combined thermodynamic computation and phase-field simulation. Acta Mater. **98**, 152–163 (2015)
6. Zhou, B., Laschet, G., Eiken, J., Behnken, H., Apel, M.: Multiscale solidification simulation of Sr-modified Al-Si-Mg alloy in die casting. In: IOP Conference Series: Materials Science and Engineering, vol. 861 (2020)
7. Böttger, B., Eiken, J., Apel, M.: Phase-field simulation of microstructure formation in technical castings – a self-consistent homoenthalpic approach to the micro-macro problem. J. Comput. Phys. **228**(18), 6784–6795 (2009)
8. Eiken, J., Böttger, B., Steinbach, I.: Multiphase-field approach for multicomponent alloys with extrapolation scheme for numerical application. Phys. Rev. E Stat. Nonlinear Soft Matter Phys. **73**(6), 1–9 (2006)
9. Wolff, N., Pustal, B., Vossel, T., Laschet, G., Bührig-Polaczek, A.: Development of an A356 die casting setup for determining the heat transfer coefficient depending on cooling conditions, gap size, and contact pressure. Materialwiss Werkstofftech **48**(12), 1235–1240 (2017)
10. Laschet, G., Jakumeit, J., Benke, S.: Thermo-mechanical analysis of cast/mould interaction in casting process. Zeitschrift für Metallkunde **95**(12), 1087–1096 (2004)

Molding

Analysis of Radial Heat Transfer in an Injection Mold with Highly Dynamic Segmented Mold Tempering

Christian Hopmann[1], Cemi Kahve[1](✉), and Cheng-Long Xiao[1,2]

[1] IKV Institute for Plastics Processing in Industry and Craft, RWTH Aachen University, Seffenter Weg 201, 52074 Aachen, Germany
cemi.kahve@ikv.rwth-aachen.de

[2] Department of Mechanical Engineering, University of South China, Hengyang 421001, China

Abstract. The demand for precise injection molded parts is continuously increasing. One of the main effects that negatively influence the geometrical part properties is warpage due to inhomogeneous shrinkage. By homogenizing the specific volume and thus the shrinkage potential, warpage is expected to be reduced. The following work addresses the homogenization of the specific volume by controlling the local part temperature with 18 individual tempering zones in the mold, which are capable of rapid heating and cooling.

To control the tempering elements precisely, a model predictive control (MPC) approach has been developed, which predicts mold and melt temperature for each tempering zone based on a discretized one-dimensional heat transfer equation. Due to radial heat transfer processes in the mold cavity and the part, the temperature of one tempering zone is expected to have an influence on neighboring tempering zones. By additional consideration of the radial heat transfer in the mold and the part, the heating and cooling output of each tempering zone can be adapted according to the heat input from the neighboring tempering zones.

To quantify the reciprocal thermal influence of the tempering zones, heating trials have been performed and recorded with a thermographic camera.

Keywords: Injection molding · Precision molding · Mold tempering · Process control · PvT-optimization · Modeling

1 Introduction

The process of injection molding offers many advantages regarding cost and efficiency for mass production. However, the manufactured parts always undergo shrinkage and often warpage, negatively affecting the geometrical part properties. One of the main material properties influencing shrinkage and thus the warpage is the pressure-specific volume-temperature behavior (pvT-behavior). According to the pvT-behavior, illustrated in Fig. 1 for a semi-crystalline polymer, melt temperature and pressure during the molding process have a major impact on the specific volume and thus the shrinkage [1, 2].

U. Reisgen et al. (Eds.): EMPOrIA 2020, LNME, pp. 215–224, 2021.
https://doi.org/10.1007/978-3-030-70332-5_19

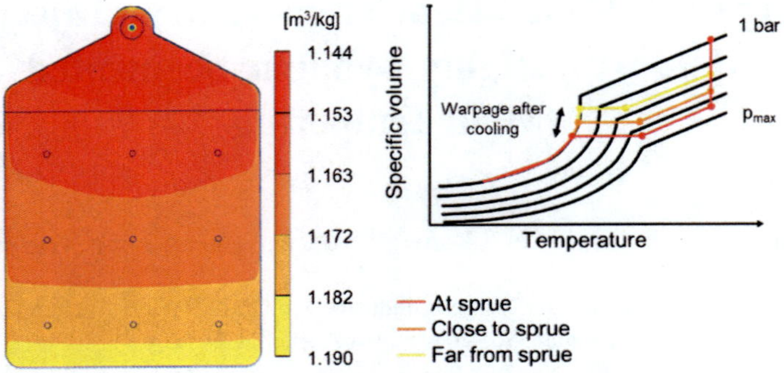

Fig. 1. Specific volume distribution of plate geometry at the packing phase (left); inhomogeneous shrinkage and warpage due to locally varying pressures and temperatures (right) [3].

As illustrated in Fig. 1 for a platen shaped geometry, the specific volume varies depending on the relative position to the sprue, due to locally varying melt pressures and temperatures. This leads to inner stress during the solidification and ultimately results in warpage. To minimize part warpage, the specific volume needs to be homogenized prior to ejection. According to the pvT-behavior, the specific volume can mainly be manipulated by controlling the local melt pressure or temperature [1, 2]. The local pressure can be influenced by compression zones, which alters the local part geometry and is therefore not suited for most applications. This project therefore aims to manipulate the local specific volume without altering the local part geometry, by controlling the local melt temperature.

2 Injection Mold with High Segmented Temperature Control

To achieve highly dynamic temperature changes during the injection molding process, a mold has been developed, which consists of 18 individually controllable tempering zones. An overview of the mold and a cross-sectional view of one tempering zone is given in Fig. 2.

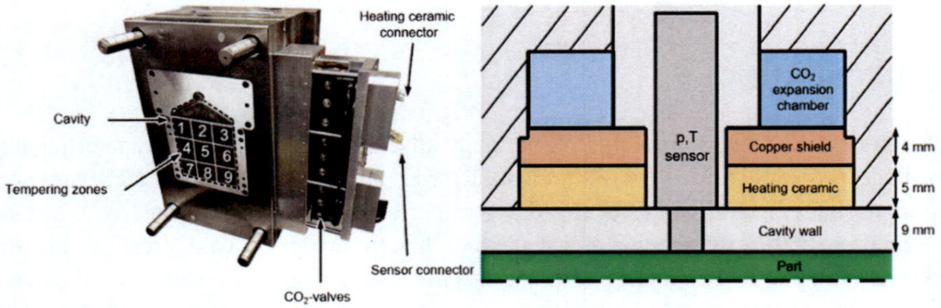

Fig. 2. Movable side of the mold with segmented temperature control (left); cross-sectional view of one tempering zone (right) [3].

In the first setup, a platen-shaped geometry (170 mm × 170 mm × 3 mm) is used and segmented into 9 tempering zones. As shown in Fig. 2, right, each tempering zone consists of a ceramic heating element with a maximum power of 1300 W and a CO_2 expansion chamber for cooling. These evaporation chambers are filled with CO_2 by magnetic valves which can be turned either fully on or off. When the liquid CO_2 reaches the evaporation chamber, it evaporates due to the rapid decrease in pressure and thus cools the tempering zone with a temperature minimum of approximately −78 °C [4]. The ribbed copper shield between CO_2 expansion chamber and ceramic heating element dampens mechanicals loads on the ceramic element and enhances the heat transfer between CO_2 and copper.

In the center of each tempering zone, an infrared temperature sensor is located to measure the local part temperature [5]. Infrared temperature sensors excel regarding the reaction time, the sample rate as well as the precision. The sensors in tempering zones 2, 5 and 8 are combined infrared temperature and piezoelectric pressure sensors to measure the in-mold pressure at three different positions along the flow path. Due to the coat-hanger shaped melt distributor, a nearly straight flow front is expected, such that the measured pressures can be extrapolated for neighboring zones.

As indicated in Fig. 2, right, the distance between cavity surface and CO_2 evaporation chamber is 18 mm. Therefore, heat transfer processes underlie thermal delays, meaning that approximately three seconds elapse between the actuation of a tempering zone and the actual change of the cavity surface temperature. Because of this delay, the process control strategy that actuates the heating and cooling elements need to contain a predictive element. Furthermore, each tempering zone is expected to have a thermal influence on adjacent zones. A controller which considers the heat transfer mechanisms from the tempering elements to the cavity surface therefore allows for an optimal control of each zone.

3 Model Predictive Control Approach

The model predictive control approach (MPC) is able to predict the temperature in the mold and the part by calculating the heat transfer from the part center up to the tempering element. This allows a precise actuation of the heating and cooling elements while considering the thermal delay in the mold.

3.1 Prediction and Control Mechanism

The temperature prediction is based on five discrete heating/cooling scenarios that are simultaneously calculated for all 18 tempering zones every 100 ms. The prediction horizon is set to three seconds since the thermal delay inside the mold is approximately three seconds. For all scenarios, the temperature T_i^t at the corresponding tempering element is calculated depending on the according power output or the CO_2 valve position. To calculate the temperature from timestep t to the next timestep $t + \Delta t$, a physical model is needed that describes the thermal processes in the mold and the part. In a first step, heat transfer processes are simplified to one dimension. For this the one-dimensional heat equation derived from *Fourier's law* is used:

$$\frac{\partial T}{\partial t} = a_{eff} \cdot \frac{\partial^2 \mathrm{T}}{\partial x^2} \quad (1)$$

The thermal diffusivity a_{eff} describes the effective material specific rate of heat transfer dependent on the thermal conductivity λ, the density ρ and the heat capacity c. The heat conduction within each material is currently calculated by using individual non-temperature-dependent material properties. The heat transfer is simplified to heat conduction with a reduced conductivity. To solve Eq. 1 for each timestep and tempering zone, the equation is discretized using the Crank-Nicolson discretization scheme.

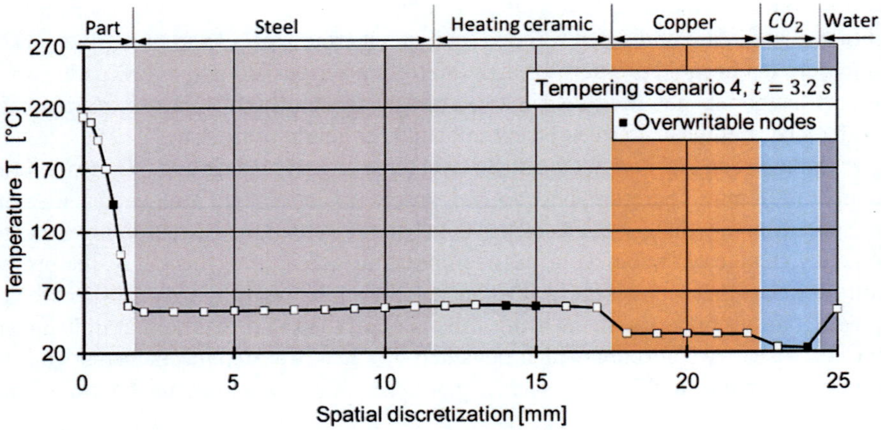

Fig. 3. Temperature prediction based on one-dimensional temperature nodes [3].

Equation 2, allows the calculation of the temperature in the part and the mold using discretization nodes. Figure 3 shows the calculated temperature distribution for one tempering zone at 3.2 s. The black discretization nodes allow to dynamically overwrite the current temperature according to the measured melt temperature from the infrared sensors, the calculated heating ceramic temperature, or a fixed temperature at the CO_2 expansion chamber of −78 °C. Using this one-dimensional finite difference approach allows a precise calculation of the individual part temperature. However, this setup cannot consider the reciprocal influence of adjacent tempering zones.

3.2 Simulation and Modelling of Radial Heat Transfer

In the current configuration, the calculated temperature for each tempering zone is assumed to be uniformly distributed. Furthermore, the reciprocal influence of neighboring tempering zones cannot be considered, which is expected to be significant. To qualitatively examine both hypothesis in a first step, a finite element simulation has been conducted using all heating ceramics in the movable mold side at a power of 650 W.

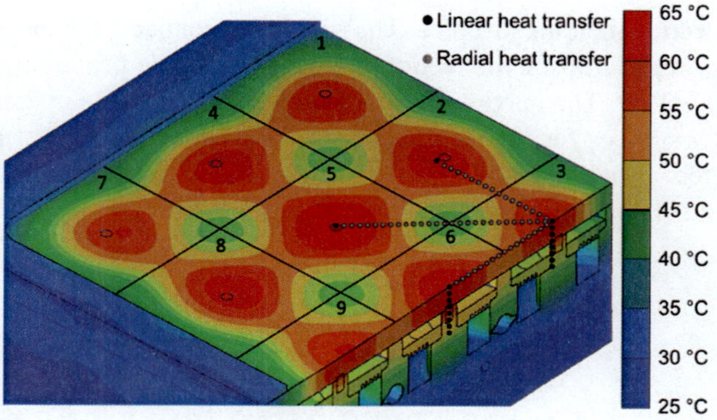

Fig. 4. Distribution of discretization nodes for radial heat transfer on the cavity surface for tempering zone 3.

Figure 4 shows the cavity surface temperature and illustrates that the temperature distribution of one tempering zone is not uniform as assumed with the one-dimensional approach. Furthermore, the cavity plate allows radial heat transfer between adjacent tempering zones. By combining the current linear heat transfer model with radial heat transfer, both phenomena can be considered. Rewriting *Fourier's law* for the case that heat is transferred through a cylindric wall with the thickness r yields the equation for radial heat transfer [7]:

$$\frac{\partial T}{\partial t} = a_{eff}\left(\frac{1}{r}\frac{\partial T}{\partial r} + \frac{\partial^2 T}{\partial r^2}\right) \tag{2}$$

with a_{eff} being the effective thermal diffusivity, and r radial the distance. As with the linear heat transfer equation, Eq. 2 needs to be discretized using the Crank-Nicolson discretization scheme, such that it can be connected with the linear heat transfer nodes as indicated in Fig. 4. The connection of linear and radial heat transfer nodes creates a three-dimensional heat transfer model. Using the three-dimensional discretization scheme will allow to consider the reciprocal thermal impact of the neighboring tempering zones as well as the uninform temperature distribution in each tempering zone.

4 Heating Trials to Quantify the Impact of Radial Heat Transfer

The development of an optimal discretization scheme for the cavity surface and the part requires knowledge about the exact temperature distribution for a given tempering scenario. Furthermore, the impact of radial heat transfer on the temperature zone needs to be examined quantitatively. Within the scope of these trials, the MPC was not used to activate the heating elements. Each element was activated manually to have full control over each zone.

In order to address both issues, heating trials have been performed using the ceramic heating elements in the movable mold side. The cavity surface temperature was recorded by a thermographic camera FLIR A645sc, FLIR Systems AB, Sweden. Due to space restrictions resulting from the thermographic measuring setup, the mold could not be setup on an injection molding machine. The trials were conducted without general mold tempering. During all trials in the sections below, the ceramic heating power was set to approximately 650 W. The maximum power the MPC can actuate is limited to 650 W, since it already suffices for high heating rates. The heating duration was set to 10 s. These parameters are additionally listed in Table 1 for a better overview and are equally used throughout all scenarios.

Table 1. Important parameters for the heating trials.

Parameter	Value	Unit
Mold temperature	23	°C
Heating ceramic power	650	W
Heating duration	10	s
Cycle time	50	s

4.1 Calibration of the Ceramic Heating Elements

In a first approach, it was examined whether the temperature of each tempering zone is similar for a given voltage. The power of the ceramic heating elements can be expressed as a function of the voltage U and the electrical resistivity $R(t)$ according to Eq. 3:

$$P(t) = \frac{\mathrm{U}^2}{\mathrm{R(T)}} \tag{3}$$

For a rising resistivity, more voltage is needed to hold the power level. The resistivity of the ceramic elements generally tends to rise with higher element temperature decreasing the maximum power output for a given voltage.

More importantly, the ceramic heating elements that were used in the cavity have varying resistivities. According to the manufacturer datasheet, the resistivity R_{20} at 20 °C of each heating element can vary up to ±25% [8]. Assuming the same applied voltage, this would result in significantly varied power output and temperature distribution for different tempering zones.

As shown in Fig. 5, left, the temperature of all 9 tempering zones varies significantly for the same applied voltage due to variations in the individual resistivity. Thereby, the average zone temperature is 49.3 °C with a standard deviation of approximately 5.6 °C across all zones.

The voltage therefore has been adjusted with the aim to homogenize the power output and with that the temperature of all tempering zones. The resulting temperatures as well as the average zone temperature and the standard deviation are illustrated in Fig. 5 (center and right). As it can be seen in the thermographic recording, the temperature distribution is visually homogenized. The average zone temperature is 48.9 °C at a standard deviation across all zones of 1.6 °C. Adapting the voltage of all heating elements significantly reduced the standard deviation of the average zone temperature.

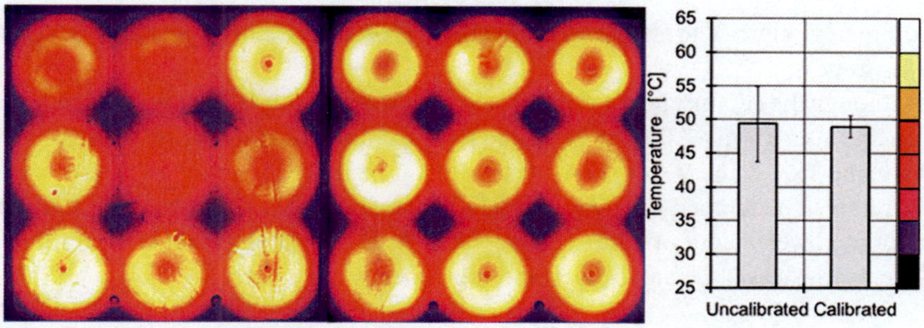

Fig. 5. Recorded cavity temperature distribution without adjusted voltages (left) and with adjusted voltages (center) as well as the average zone temperature with standard deviation (right).

4.2 Analyzing the Impact of Radial Heat Transfer

After successfully adjusting the heating output for the ceramic heating elements, the impact of radial heat transfer can be analyzed. For this, investigations on multiple heating scenarios have been conducted using a varying amount of simultaneously activated heating elements. For the following sub sections, two distinct tempering scenarios are selected and presented.

Heating Scenario 1: Impact of One Activated Heating Element on Cavity Temperature In a first approach tempering zone 5 is solely activated for 10 s to examine the impact of one activated heating element on the temperature of the adjacent tempering zones. Figure 6 shows the recorded temperature for the movable mold side as well as the corresponding temperature legend. The cavity surface temperature is shown for two different recording times to highlight the heat flux. Figure 6 shows the temperature distribution after 10 s, which also equals the heating time.

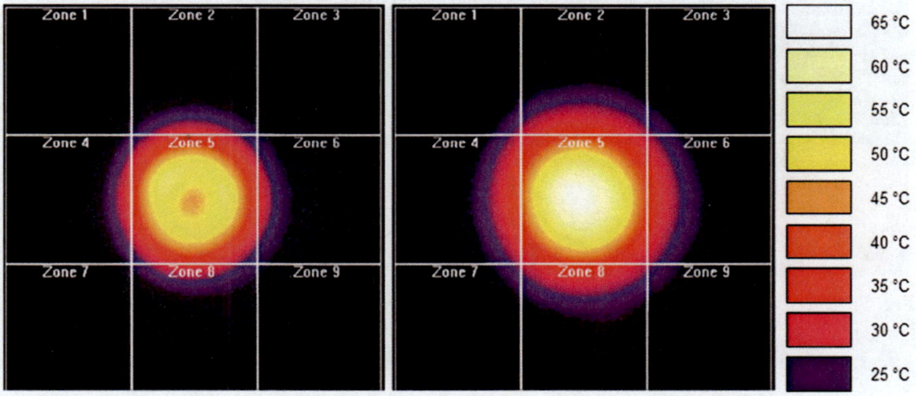

Fig. 6. Recorded cavity surface temperature for heating scenario 1 after 10 s (left); and 20 s (right).

After 10 s, there is no significant impact on tempering zones 1, 3, 7 and 9. The average temperature of the remaining tempering zones raised approximately 1 °C and is therefore negligible. The results show that there is no short-term impact for one activated heating zone.

Although the ceramic heating element is turned off after 10 s, the average temperature of zone 5 raises from 42.8 to 48.8 °C due to linear and radial diffusion processes in the mold. Zone 1, 3, 7 and 9 still show no significant increase regarding their average zone temperature. The temperature of the remaining zones raised approximately 2 °C for the maximum power output of approximately 650 W. For this configuration, the short- and long-term impact on the adjacent tempering zones is negligible.

Heating Scenario 2: Impact of Two Activated Heating Elements on the Cavity Temperature. In a next step, tempering zones 7 and 9 were activated simultaneously. Based on the first results, tempering zones 1, 2 and 3 are expected to be uninfluenced. Regarding tempering zones 4, 5 and 6, the average temperature increase is assumed to be around 1–2 °C. Zone 8 is expected to experience the most influence, since it is located between zones 7 and 9. In this heating scenario, the cavity surface temperature will be recorded for 120 s, in order to investigate the long-term thermal impact for the complete injection cycle.

Figure 7 illustrates the cavity surface temperature for heating scenario 2 after 10 and 50 s, respectively. In Fig. 7 left, the temperature of zone 1 and 3 is slightly elevated, which can be attributed to prior infrared recordings. Similar to heating scenario 1, the short-term impact on the temperatures of the adjacent zones is not large. At 10 s, the average temperature of zones 4 and 6 raised 0.5 °C. The temperature of tempering zone 8 raised 1 °C.

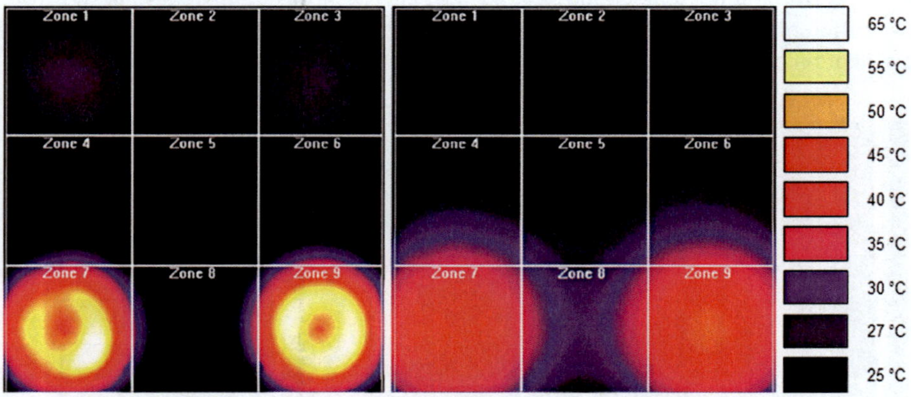

Fig. 7. Recorded cavity surface temperature for heating scenario 2 after 10 s (left); and 50 s (right).

At 50 s, which equals to the cycle time, the long-term impact of radial heat transfer becomes evident. The average temperature of zones 4 and 6 raised approximately 2 °C. The temperature of zone 8 raised approximately 5 °C.

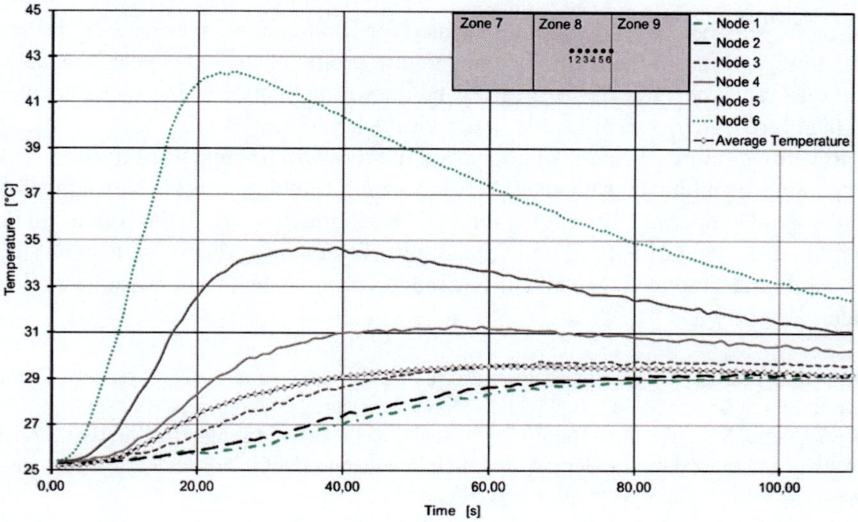

Fig. 8. Temperature distribution of tempering zone 8 for heating scenario 2.

In order to highlight the temperature distribution in zone 8, the temperature curves of six temperature nodes are shown in Fig. 8. Node 1 is located at the center of zone 8 while node 6 is on the border to zone 9. The nodes in-between 1 and 6 are evenly distributed as indicated.

Based on the local temperatures of tempering zone 8, node 6 reaches its peak temperature at 18 s with 42 °C. The temperature of nodes 1 and 2 on the other hand keep rising even after the cycle time of 50 s and plateau at 29 °C for another 60 s, such that the cavity surface temperature is altered in the second injection cycle. Similar behavior can be seen with the average zone temperature. This long-term alteration of a non active tempering zone is currently not considered by the MPC and will significantly influence the temperature control precision.

5 Conclusion and Outlook

For a precise and on-time actuation of the tempering elements, a model predictive controller was developed based on a one-dimensional heat equation. To describe the reciprocal thermal impact of neighboring tempering zones, radial heat transfer needs to be considered.

To validate and calibrate the radial heat transfer model, multiple heating scenarios were analyzed, of which two were presented. It could be seen that the short-term impact on the adjacent tempering zones can be neglected due to the low thermal diffusivity of the cavity plate. However, it could be shown that the long-term impact was not negligible. The temperature of non-active and directly neighboring tempering zones rose significantly after the actual heating time. For a tempering zone located between two active tempering zones, which were activated for 10 s at a power of 650 W, the long-term zone temperature rose 5 °C. This rise in temperature plateaued for additional 60 s, exceeding the actual cycle time, and thus influencing the temperature profile of the next cycle. Considering additional radial heat transfer in the MPC model will help in detecting such phenomena and therefore lead to a significantly better local temperature control.

Regarding future research, additional simulations will be conducted to consider further parameters influencing the thermal balance of the mold such as the melt temperature and the mold tempering. The simulation will be calibrated to match the conducted heating trials such that the thermal simulation during the molding process has a high validity. The additional simulated results will be used to optimize the future radial heat transfer model.

Acknowledgments. All presented investigations were conducted in the context of the Collaborative Research Centre SFB1120 "Precision Melt Engineering", subproject B3 "Self-optimizing Process Control Strategies for a Highly Segmented Injection Mould Tempering" at RWTH Aachen University and funded by the German Research Foundation (DFG). For the sponsorship and the support, we wish to express our sincere gratitude.

References

1. Wang, J., Hopmann, Ch., Röbig, M., Hohlweck, T., Kahve, C., Alms, J.: Continuous two-domain equations of state for description of the pressure-specific-volume-temperature behavior of polymers. Polymers **12**, 409 (2020)
2. Johannaber, F., Michaeli, W.: Handbuch Spritzgießen. Carl Hanser Verlag, München (2004)
3. Hopmann, C., Kahve, C.E., Schmitz, M.: Development of a novel control strategy for a highly segmented injection mold tempering for inline part warpage control. Polym. Eng. Sci. **60**, 2428–2438 (2020)
4. Berghoff, M.: Perspektiven bei der Temperierung von Problemzonen im Werkzeug, https://www.isk-iserlohn.de/fileadmin/medien/Dokumente/co2_temperierung_fachbeitrag.pdf. Accessed 11 Feb 2020
5. Giese E.: Infrared temperature sensor for plastics molds, Datasheet. https://www.fos-messtechnik.de/MTS%20408_IR_BTS_STS_XSR_2017.pdf. Accessed 28 Jan 2020
6. Thomas, J. W.: Numerical Partial Differential Equations: Finite Difference Methods (1995). ISBN: 978-1-4899-7278-1.
7. Kneer, R: Wärme-und Stoffübertragung I/II, RWTH-Aachen (2014)
8. NN: Heating elements, Details of Standard. https://www.bachrc.de/print.html?category=ceramic-8&group=1&product=0&language=en. Accessed 23 Apr 2020

Evaluation and Transport of the Crystallization Heat in an Iterative Self-consistent Multi-scale Simulation of Semi-crystalline Thermoplastics

Christian Hopmann[1], Jonathan Alms[1](✉), and Gottfried Laschet[2]

[1] Institute for Plastics Processing, RWTH University, Seffenter Weg 201, 52074 Aachen, Germany
Jonathan.Alms@ikv.rwth-aachen.de
[2] ACCESS e.V., Intzestraße 5, 52072 Aachen, Germany

Abstract. In the injection molding process of semi-crystalline thermoplastics, the melt is subjected to a complex deformation and temperature history. This leads to an inhomogeneous microstructure over the component and thus to local inhomogeneities in the component's effective properties. To predict the component properties precisely, a multi-scale simulation is used, which couples the filling simulation at the component level with a microstructure simulation at the scales of microns (SphaeroSim). The influence of the crystallization heat is considered in the filling simulation with an averaged empirically determined degree of crystallization. To achieve higher precision in the microstructure simulation the influence of the crystallization heat is considered at the microscale of SphaeroSim. SphaeroSim is extended by the calculation of a local crystallization degree, which is used to calculate the amount of local crystallization heat and heat transport calculations.

Keywords: Semi-crystalline polymers · Polymer crystallization · Injection molding · Multiscale simulation

1 Introduction

The manufacturing of high-quality injection molded parts requires a deep understanding of material properties, process parameters and product design [1]. The material behavior of injection molded components highly depends on the formation of microstructures during manufacturing, which in turn is linked to the macroscopic heat and material flow during manufacturing [2]. In semi-crystalline thermoplastics two types of microstructures must be distinguished at different scales: the formation of molecular chain folded lamellae at the nanoscale and the formation of superstructures, named spherulites, at the microscale. Each spherulite contains lamellae structures originating from a single nucleus. The lamellae are surrounded by amorphous material, which in the case of isotactic polypropylene (iPP) makes up 47% of the solidified material [3, 4]. The form and distribution of superstructures as well as crystalline-amorphous composition of each spherulite has a high impact on the macroscopic thermal and mechanical properties [5,

U. Reisgen et al. (Eds.): EMPOrIA 2020, LNME, pp. 225–235, 2021.
https://doi.org/10.1007/978-3-030-70332-5_20

6]. A precise calculation of superstructures provides deep insights into the thermal and mechanical properties of an injection molded component.

Due to the huge computational effort to calculate the molecular behavior at the scale of a component, a 4 step multiscale simulation chain for semi-crystalline thermoplastics is developed within the collaborative research center SFB 1120, "Precision Melting Engineering" (see Fig. 1) [4]. The simulations are performed at three different levels of scale: The macroscale, a coupled filling, heat transfer and solidification simulation is performed using COMSOL Multiphysics, COMSOL Multiphysics GmbH, Göttingen [3]; at the microscale, an cellular-automata solver, SphaeroSim, is developed to calculate the form and distribution of superstructures and at the nanoscale, at which a Representative Volume Element (RVE) of the bi-lamella formed from crystalline and amorphous phases, is discretized and homogenized within HOMAT [7]. As both phases co-exist always, their pure phase properties are calculated by molecular-dynamics simulations [7]. HOMAT simulates the crystalline-amorphous composition of the bi-lamellae formation within the superstructures of iPP. This is combined with the properties resulting from the bi-lamella homogenization, to calculate the effective thermal and mechanical properties of the superstructures. Using the distribution of superstructures on the RVE, its effective mechanical and thermal properties are evaluated by adopting a special 3D spherulite model [2]. The simulation on the three different scales and the homogenizations allow the computation of effective properties in a reasonable time.

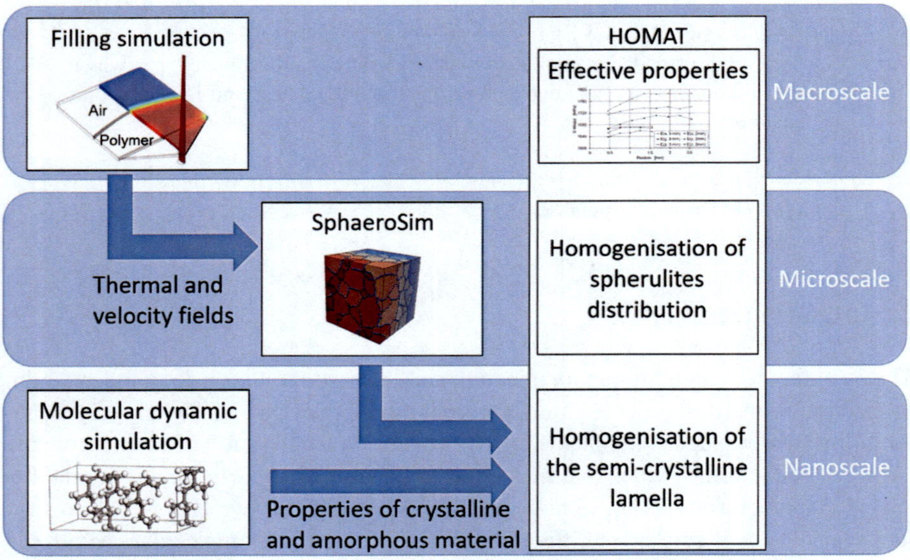

Fig. 1. The multiscale simulation scheme developed within the collaborative research center SFB 1120

Currently, SphaeroSim uses thermal and velocity fields calculated by filling simulation (FS) as starting conditions without calculating internal thermal diffusion during solidification. This leads to neglecting the crystallization heat at the point of solidification at the microscale. The consideration of the crystallization heat at the microscale was found to impact the microstructure and thus the effective properties [8, 9].

Böttger et al. [10] developed a self-consistent multiscale simulation scheme for metallic alloys, which correlates the thermal fields calculated at the macro- and microscale in an iterative way, to ensure that the thermal fields and the cooling behavior are consistent at both scales. To incorporate a similar approach to the multiscale simulation chain for semi-crystalline thermoplastics, parameters to link microstructure simulation (MS) and FS are determined in Sect. 2. Since the consideration of crystallization heat at the microscale impacts the superstructure formation, the release of crystallization heat (Sect. 4) and its propagation (Sect. 3) are implemented in SphaeroSim.

2 The Self-consistent Approach Within the Multiscale Simulation Chain

2.1 Self-consistency of the Multiscale Simulation Scheme

Böttger et al. [10] discovered that crystallization heat at the microscale cannot be averaged from the macroscale and the cooling behavior must be the same at both scales otherwise a completely different microstructure is formed. Since the crystallization heat and its thermal diffusion influences the microstructure of polymers, this approach is suitable to improve the multiscale simulation chain for semi-crystalline thermoplastics [8, 9].

In the original formulation of self-consistent iterative multiscale simulation, the FS uses process-related boundary conditions (e.g. geometric, thermal) applied on the RVE for the microstructure nucleation and growth simulation. In turn, MS is used to provide an accurate calculation of solid fraction, crystallization heat release and the temporal enthalpy change at a much smaller scale than FS. From MS, the solid fraction and thermal calculations are fed back to the FS and adjustments are done to reduce the differences between both simulations. Updated thermal and velocity fields based on the adjustments are used in another microstructure evolution simulation run. This iterative feedback-loop between FS and MS is performed until consistency within thermal fields, solid fraction and release of crystallization heat is reached. A much more realistic temperature-time curves was reached in comparison to widely used macroscale models [10].

This approach is transferred here to the multiscale simulation chain of semi-crystalline thermoplastics. Here, both simulations benefit from the iterative self-consistent framework, where MS cannot consider geometric or thermal boundary conditions (e.g. mold wall, melt temperature) on the scale of the component, the FS is equipped to provide the necessary calculations. At selected locations, the FS benefits from solidification calculations at a 1,000,000 times finer grid within the MS.

To implement the iterative self-consistent approach both simulations must provide a description of crystallization degree, solid fraction and crystallization heat release including thermal diffusion. Those parameters can be calculated within SphaeroSim to a much higher accuracy than in FS. An improvement of the FS is expected if the FS parameters are adjusted accordingly.

2.2 Macroscale Filling Simulation

The FS performs the filling simulation and heat transfer calculations within the cavity in the injection molding process. This takes into account many boundary conditions set by production-related parameters. Phase change, cooling and unsteady flow behavior of the melt are considered during the injection molding process.

The melt flow description is provided by incompressible viscous Navier-Stokes flow based on the conservation equations of mass, momentum and energy. This is implemented in a dedicate finite element model, which is implemented in the multiphysics solver, COMSOL [3, 11].

The solidification is calculated using a simple Avrami-equation to predict the solid fraction ξ, which depends on time t and the non-isothermal kinetic rate constant K_a:

$$\xi = 1 - \exp\{[K_a t]^{n_a}\} \tag{1}$$

where n_a is the Avrami-exponent, which can take values in the interval between 1 and 4 and characterizes the spherulitic growth mechanism [2, 12–14]. The solid fraction states the relative amount crystallized material in relation to the maximum reachable crystallization degree. The Avrami-equation therefore yields the amount of crystallized material with time depending on the non-isothermal kinetic rate constant and time. The change in solid fraction is then used within the heat transfer equations to calculate the crystallization heat release $\dot{Q}$ during solidification in the form of:

$$\dot{Q} = \xi_{max}^{abs} \Delta h_{m,0} \frac{\partial \xi}{\partial t} \tag{2}$$

where ξ_{max}^{abs} is the maximum reachable crystallization degree (here 53%); and $\Delta h_{m,0}$ the change in enthalpy for ideal 100% crystallization [3, 15]. Overall, this calculates the emitted local heat by the solidification processes for local, temporal and thermal evolution. However, the maximum reachable crystallization degree is assumed to be fixed and equal throughout the component. Here, SphaeroSim provides a significant improvement by calculating a local crystallization degree depending on crystallization kinetics and on a much finer grid.

This leaves the identification in SphaeroSim, how the crystallization heat is calculated and a description of the solid fraction within a simulation volume.

2.3 Microscale Superstructure Simulation (SphaeroSim)

SphaeroSim is a cellular-automata solver, which solves the Avrami-equation modified by the Hoffman-Lauritzen nucleation and growth theory. To simulate the microstructure within a support location, the area is usually discretized into voxels of $1 \times 1 \times 1\ \mu m^3$, since this is the actual maximum fineness within SphaeroSim [4]. The starting conditions (thermal and velocity fields) from FS are interpolated to the finer grid for each time step calculated in SphaeroSim. Currently this setup allows disregarding of all thermal behavior on the microscale since the considerations of crystallization heat and thermal diffusion are provided by the microscopic local temperature. However, due to the finer grid size and the findings of Böttger et al., the local impact of crystallization heat on the formation of spherulites is high. Therefore, the calculation of crystallization heat released during solidification at the microscale and thermal diffusion are now implemented.

3 Implementation of the Thermal Transport in the Microstructure Simulation

The three-dimensional heat transfer equation is derived from Fick's second law of diffusion and describes the diffusion of heat over time and space [16]:

$$\frac{\partial T}{\partial t} = D\left(\frac{\partial^2 T}{\partial x^2} + \frac{\partial^2 T}{\partial y^2} + \frac{\partial^2 T}{\partial z^2}\right) \tag{3}$$

where x, y, z are the three-dimensional relative coordinates. The thermal diffusion coefficient D needs to be treated for anisotropic material, since every voxel calculated by SphaeroSim has a unique crystallization degree k_i based on its temperature history. Additionally, the thermal conductivity λ, the specific heat capacity c_p and the density ρ depend on the local crystallization degree [7].

$$D_i(k_i) = \frac{\lambda(k_i)}{c_p(k_i)\rho(k_i)} \tag{4}$$

A unique D_i must be calculated for each voxel i to ensure precise thermal calculations.

To simplify the determination of the local diffusion coefficient, it is assumed that the material properties in an undercooled melt do not change until solidification. Once solid, a voxel can neither remelt, nor change its crystallization degree. These assumptions allow the initialization of D_i with material parameters of only molten material. At the time of state change the corresponding crystallization degree is calculated from T_c and D_i is updated. It is to note, that in the initial formulation of the iterative self-consistent multiscale scheme, Böttger et al. used a one-dimensional description of heat diffusion in their simulation [10]. This is not possible in the case of three-dimensional unique diffusion coefficients and a 3D-heat transfer is implemented.

An analytical solution for the heat-equation is obtained by Fourier analysis in the form of:

$$T(x, y, z, t) = \frac{1}{\sqrt{4\pi Dt}^3}\exp\left(-\frac{x^2 + y^2 + z^2}{4Dt}\right) \tag{5}$$

Equation (5) is the so-called fundamental solution of the heat-equation. The function can be used to find a general solution for the heat-equation over certain domains with simple boundary conditions [16]. Since the starting conditions received from the FS are discrete and heat sources (crystallization heat) are not covered in this solution, the usage of the fundamental solution must be adopted.

The fundamental solution is used for the purpose of propagating temperature within SphaeroSim by treating each voxel as a separate heat source calculating unhindered thermal diffusion within a small timeframe t. The superposition of the thermal diffusion of all voxels lead to the temperature diffusion throughout the whole simulation volume. The timeframe t must be chosen small enough to ensure a negligible difference of the local diffusion coefficients D_i in the affected neighborhood.

The continuous formulation of the fundamental solution allows the calculation of the temperature influence of each voxel to all other voxels in the simulation area. The calculation of temperature influences to all voxels is CPU time consuming, however it allows to choose the size of the neighborhood by variation of the time t for the diffusion to propagate. The discretization, once a time step size is chosen, is performed by integration of the fundamental solution over each voxel in the neighborhood, where the neighborhood is here defined as each voxel within 99.5% of the distributed heat. This ensures the validity and accuracy of the thermal diffusion throughout the simulation area, while keeping the calculation time reasonable.

4 Crystallization Heat in SphaeroSim

4.1 Implemented Crystallization Kinetics Model

As well as the FS, the Avrami-equation is used within SphaeroSim to determine solidification of single voxels, but in a modified form. The non-isothermal kinetic rate constant K_a can be further split into a geometric factor, a term connected to the nucleation rate and a growth function determined by the type of spherulitic growth (see Eq. (6)) [17].

At first, the type of spherulitic growth is determined to be spherical, which demands an Avrami-exponent of at least 3. The addition of a nucleation rate $\dot{N}$, adds another degree of freedom, which sets the Avrami-exponent to 4. The final form of the kinetic rate constant used in SphaeroSim:

$$K = \frac{\pi}{3}\dot{N}(T, \Delta g)G^3(T, \Delta g) \tag{6}$$

where $G(T)$ is the growth rate function of the spherulites, which is derived from the Hoffman-Lauritzen nucleation and growth theory [17, 18] (explicit forms for $\dot{N}$ and G can be found in [4] p. 31). The nucleation rate as well as the growth rate are dependent on the temperature T and the change in Gibbs free energy Δg, which in turn is dependent on the temperature and material constants.

The implementation of SphaeroSim splits Δg into a melt flow dependent term and a melt flow independent term $\Delta g_{v,q}$, which states the change of Gibbs free energy in quiescent conditions. In order to determine the crystallization heat, the flow dependent term can be ignored, since it only contributes to the nucleation rate (shear induced nucleation) but not to the growth rate, in the current model. The change in Gibbs free energy in quiescent conditions is implemented from the Hoffman-Lauritzen theory as:

$$\Delta g_{v,q} = \Delta h_{m,0} f \frac{T_{m,0} - T}{T_{m,0}} \tag{7}$$

where $\Delta h_{m,0}$ is the crystallization heat for an ideal crystallization degree of 100%; $T_{m,0}$ is the equilibrium melting temperature; and f is a correction factor proposed in Hoffman et al. to widen the application of $\Delta g_{v,q}$ to a large range of undercooling [19, 20].

During cooling the iPP melt passes $T_{m,0}$ (194 °C) and becomes undercooled. In the undercooled melt the formation of nuclei and the growth of spherulites is possible. Since variations in cooling rate of injection molding between 60 K/min at the core of the component and up to 3,000 K/min at the mold walls is common, large undercooling must be considered [21]. Measurements using a Flash-DSC 2+, Mettler-Toledo, Columbus,

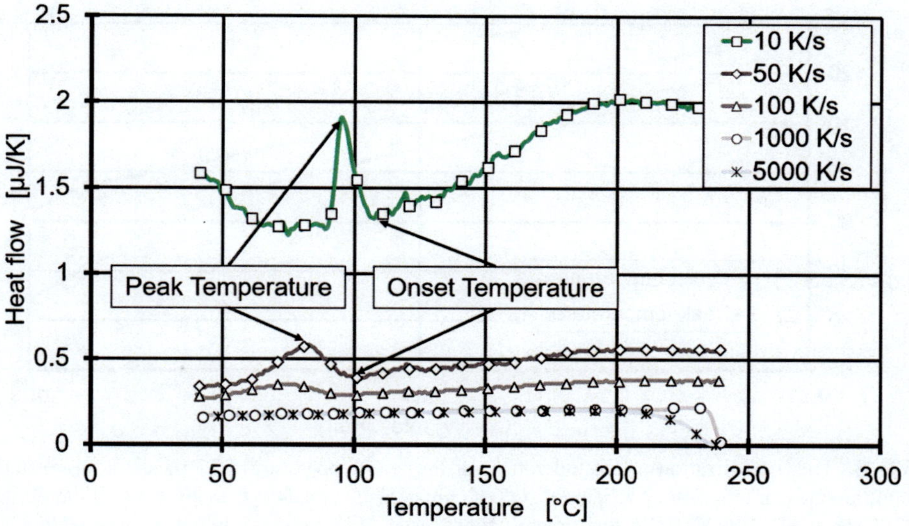

Fig. 2. The heat flow normalized with the cooling rate of each cooling cycle of iPP using the Flash-DSC 2+, Mettler-Toledo, Columbus, USA. The sample is heated to 240 °C using a constant heating rate of 100 K/s, followed by rapid cooling down to 40 °C for each measurement.

USA are performed to measure the crystallization heat during crystallization at very fast cooling rates from 600 K/min to 300,000 K/min (see Fig. 2).

The crystallization process is described by the onset temperature and the peak temperature. The onset temperature shows the point at which the first nucleus forms and the solidification starts. The peak temperature shows the highest solidification rate per time. Both temperatures are dependent on the cooling rate (see Fig. 3), however measurements taken with cooling rates of 60,000 K/min and 300,000 K/min do not show any crystallization peak. The rapid cooling suppresses molecular movement to the point where no molecular chain folding is possible, and the melt solidifies purely in an amorphous phase. This points out, that the increasing cooling rates lead to a reduction in crystallization degree and therefore to a reduced crystallization heat.

In order to compensate for the reduction in crystallization heat at high undercooling, the correction factor f is introduced in the Hoffman-Lauritzen theory.

$$f = \frac{2T}{T_{m,0} + T} \tag{8}$$

f describes the reduction in crystallization heat released into the melt, therefore it is interpreted as the factor defining the crystallization degree during the simulation.

In the implementation of SphaeroSim, voxels are either in a molten state or in a solid state and no solid fraction within a voxel is considered. The crystallization degree of a voxel is therefore calculated from the factor $f(T_c)$ at the point of phase change and the solidification temperature T_c. The point of phase change of a voxel is calculated from the Avrami-equation reaching a solid fraction of 50%. This approximation simplifies the assignment of a voxel to a spherulite to the point of half crystallization time and

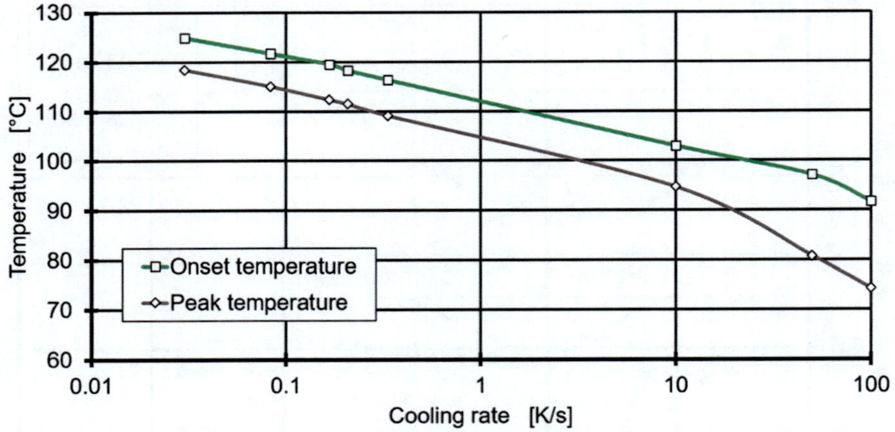

Fig. 3. The crystallization onset temperature and the peak temperature of iPP measured at constant cooling rates ranging from 2 K/min to 6,000 K/min cooled from 240 °C to 40 °C using the Flash-DSC 2+ and DSC Q2000, TA Instruments, New Castle, USA for cooling rates up to 20 K/min.

reduces the amount of data generated during simulation significantly [4]. Furthermore, all crystallization heat of a voxel is released at the time of phase change. This simplifies the treatment of the temperature change to a simple addition to the solidified voxel.

4.2 Comparison of the Calculation of Crystallization Heat at micro- and Macroscale

The crystallization heat release during solidification plays a major role within the self-consistent approach. To compare the amount of crystallization heat released in both simulations, MS and FS, thermal calculations of a simulation volume of step-plate component are performed. The selected volume is 3 mm thick, spanning from one mold wall to the other one with an area of 0.5×0.5 mm^2.

To calculate the crystallization heat released throughout the solidification within FS is performed by temporal integration of Eq. (2):

$$\int_0^\infty \dot{Q}dt = \xi_{max}^{abs} \Delta h_{m,0} \tag{9}$$

The integration of the change in solid fraction $(\partial\xi/\partial t)$ limits to 1 for a completely solidified component; with the crystallization heat for 100% crystallinity $\Delta h_{m,0} =$ 0.14 J/mm^3 the released heat is calculated to $Q = 55.6$ mJ.

The calculation of crystallization heat in SphaeroSim cannot be performed by integration since no gradual increase of crystallinity is calculated, but a sudden phase change from melt to solid at 50% relative crystallinity. It is therefore assumed, that all crystallization heat is released at this point in time and temperature. From Eq. 7 the crystallization heat can be derived in the form of $f\,\Delta h_{m,0}$. The comparison of both formulations suggests that f should resemble the reached crystallization degree of each voxel. However, the solidification temperature for $f = 0.53$ is -104.6 °C. This temperature is well below the glass transition temperature of iPP, therefore the assumption of a crystallization process starting at that temperature is wrong. The average crystallization degree calculated by ShpaeroSim via $f\,\Delta h_{m,0}$ is 88% and crystallization heat released within the simulation volume is 92.3 mJ.

At this point some problems of the currently implemented correction factor f must be pointed out: The Hoffman-Lauritzen theory from which the factor originates, was developed assuming isothermal solidification conditions with small undercooling. The formulation in general was proven to be valid for several non-isothermal cases, but the Avrami-equation in which the growth-function G is embedded only predicts a relative crystallinity. In the formulation of solidification in both simulations the relative crystallinity is assumed to be a solid fraction. This disregards the crystallization degree reached at a given solidification temperature. The factor f is used to compensate for the reduction in crystallization degree at higher undercooling but shows that additional parameters must be introduced to model solidification of semi-crystalline polymer in injection molding processes.

It is necessary to introduce a function that calculates the crystallization degree depending on the cooling rate up to the point at which formation of lamellae is completely suppressed. This ensures, that the crystallization heat can be correctly calculated even for extreme cooling rates, which are found in injection molding processes. This still leaves the Avrami-model valid to determine solid fraction, but it would result in the amount of crystallized phase present in a voxel, disregarding amorphous material as part of the solid.

5 Conclusion and Outlook

The formation of microstructures during the solidification of semi-crystalline thermoplastics has a major impact on the effective thermal and mechanical properties of the final component. The multiscale simulation chain was developed to predict the microstructure in a simulation chain across macro-, micro- and nanoscale. Since the macro- and microscale simulation both simulate solidification but on different scales, the iterative self-consistent approach is proposed. This uses the accuracy of the microscale simulation to improve the accuracy of the macroscale simulation results. In order to link both simulations, the thermal fields, crystallization degree and solid fraction are determined.

Since the microscale simulation relied purely on the calculated thermal fields provided by the macroscale simulation, thermal calculations are implemented into the microscale simulation. The diffusion of temperature during solidification is calculated via the fundamental solution of the heat diffusion equation. The fundamental solution allows the choice of the time step size for the thermal diffusion, due to its continuous formulation.

The thermal diffusion is implemented to consider crystallization heat emitted during the solidification at the microscale, since the crystallization heat has a major impact on the microstructure formation. The comparison of the crystallization heat from the mirco- and macroscale showed significant differences in scale and thermodynamic treatment of the calculation of the crystallization heat. The microscale simulation uses the correction factor f from the Hoffman-Lauritzen theory to determine the amount of crystallization heat released to the system, which is not sufficient to determine the crystallization in according to experimental data and should be replaced.

The correction factor f has certain shortcomings, such as the overestimation of crystallization heat release. For semi-crystalline thermoplastics, a functional relationship between the crystallization degree reached and cooling rate of a sample will be developed in future works. This function will enable the Avrami-model to describe the amount of crystallized material at a given time and temperature, which is preferable, since it neglects the amorphous material as part of the solid.

Acknowledgments. The depicted research was funded by the Deutsche Forschungsgemeinschaft (DFG) as part of the collaborative research center SFB 1120 "Precision Melt Engineering". We would like to extend our thanks to the DFG. The authors wish to thank Prof. R. Spina (Politecnico di Bari, Italy) for providing the simulation results of the injection molding process with COMSOL.

References

1. Spina, R., Spekowius, M., Hopmann, C.: Analysis of polymer crystallization with a multi-scale modeling approach. In: Key Engineering Materials, vol. 611, pp. 928–936. Trans Tech Publications Ltd (2014)
2. Laschet, G., Apel, M., Wipperfürth, J., Hopmann, C., Spekowius, M., Spina, R.: Effective thermal properties of an isotactic polypropylene (α-iPP) injection moulded part by a multiscale approach. Materialwiss. Werkstofftech. **48**(12), 1213–1219 (2017)
3. Spina, R., Spekowius, M., Hopmann, C.: Multi-scale thermal simulation of polymer crystallization. Int. J. Mater. Form. **8**(4), 497–504 (2015)
4. Spekowius, M.: A new microscale model for the description of crystallization of semi-crystalline thermoplastics. Verlagsgruppe Mainz GmbH, Aachen (2017)
5. Zhao, S., Xu, N., Xin, Z., Jiang, C.: A novel highly efficient nucleating agent for isotactic polypropylene. J. Appl. Polym. Sci. **123**(1), 108–117 (2012)
6. Drogelen, M., Erp, T., Peters, G.: Quantification of non-isothermal, multi-phase crystallization of isotactic polypropylene: the influence of cooling rate and pressure. Polymer **53**(21), 4758–4769 (2012)
7. Laschet, G., Spekowius, M., Spina, R., Hopmann, C.: Multiscale simulation to predict microstructure dependent effective elastic properties of an injection molded polypropylene component. Mech. Mater. **105**, 123–137 (2017)
8. Foks, J.: The influence of latent heat release on polymer morphology. In: Crystallization of Polymers, pp. 337–343. Springer, Dordrecht (1993)
9. Raimo, M., Cascone, E., Martuscelli, E.: Review Melt crystallisation of polymer materials: the role of the thermal conductivity and its influence on the microstructure. J. Mater. Sci. **36**(15), 3591–3598 (2001)
10. Böttger, B., Eiken, J., Apel, M.: Phase-field simulation of microstructure formation in technical castings–a self-consistent homoenthalpic approach to the micro–macro problem. J. Comput. Phys. **228**(18), 6784–6795 (2009)
11. Spina, R., Spekowius, M., Dahlmann, R., Hopmann, C.: Analysis of polymer crystallization and residual stresses in injection molded parts. Int. J. Precis. Eng. Manuf. **15**(1), 89–96 (2014)
12. Mubarak, Y., Harkin-Jones, E., Martin, P., Ahmad, M.: Modeling of non-isothermal crystallization kinetics of isotactic polypropylene. Polymer **42**(7), 3171–3182 (2001)
13. Hao, W., Yang, W., Cai, H., Huang, Y.: Non-isothermal crystallization kinetics of polypropylene/silicon nitride nanocomposites. Polym. Testing **29**(4), 527–533 (2010)

14. Isayev, A.I., Chan, T.W., Shimojo, K., Gmerek, M.: Injection molding of semicrystalline polymers. I. Material characterization. J. Appl. Polym. Sci. **55**(5), 807–819 (1995)
15. Galera, V., Marinelli, A.: Determination of non-isothermal crystallization rate constant for pseudo-experimental calorimetric data. Mater. Res. **12**, 151–157 (2009)
16. Evans, L.C.: Partial Differential Equations, vol. 19. American Mathematical Society, Rhode Island (2010)
17. Hammami, A., Spruiell, J., Mehrotra, A.: Quiescent nonisothermal crystallization kinetics of isotactic polypropylenes. Polym. Eng. Sci. **35**(10), 797–804 (1995)
18. Lauritzen, J.I., Hoffman, J.D.: Theory of formation of polymer crystals with folded chains in dilute solution. J. Res. Natl. Bureau Stand. Sect. A Phys. Chem. **64**(1), 73 (1960)
19. Hoffman, J., Davis, G., Lauritzen, J.I.: The rate of crystallization of linear polymers with chain folding. In: Treatise on Solid State Chemistry, pp. 497–614. Springer, Boston (1976)
20. Hoffman, J., Miller, R.: Kinetic of crystallization from the melt and chain folding in polyethylene fractions revisited: theory and experiment. Polymer **38**(13), 3151–3212 (1997)
21. Wang, J., Hopmann, C., Röbig, M., Hohlweck, T., Kahve, C., Alms, J.: Continuous two-domain equations of state for the description of the pressure-specific volume-temperature behavior of polymers. Polymers **12**(2), 409 (2020)

Thermal Optimisation of Injection Moulds by Solving an Inverse Heat Conduction Problem

Tobias Hohlweck(✉) and Christian Hopmann

Institute for Plastics Processing (IKV), RWTH Aachen University, Seffenter Weg 201, 52074 Aachen, Germany
tobias.hohlweck@ikv.rwth-aachen.de

Abstract. The thermal design of injection moulds is a complex process and is often conducted manually by the mould maker. However, manufacturing of highly precise parts has become more and more important over the last years. The focus is laid especially on part warpage and reproducible results in the context of a stable process.

In this paper, a simulative approach for an automatic cooling channel layout via an inverse heat conduction problem is presented. Based on previous results, the methodology is extended using an improved objective function. This approach minimizes the thermal inhomogeneity in the part during the holding and cooling phase by evaluating the heat flux distribution in the mould. In this project, a virtual methodology is developed for an efficient thermal mould design. Exemplary simulations with two different geometries and materials validate this approach. For a thick-walled lens geometry produced with an amorphous material, a significant improvement of the resulting temperature distribution can be shown. For the second, technical geometry with a semi-crystalline thermoplastic, the results are very close to the solution with the previous objective function.

Keywords: Injection moulding · Thermal mould design · Thermal optimisation

1 Introduction

Injection moulding is widely used for the large-scale production of complex thermoplastic parts with short cycle time. In order to manufacture high quality parts, the injection moulding process needs optimal process conditions as well as a high reproducibility. Quality criteria such as weight, density, shrinkage and warpage determine the mechanical and dimensional integrity of the product and directly correlate to the corresponding process conditions. In this process, the design phase is the most time consuming and most costly step of the construction of an injection mould. Especially the thermal design phase requires a lot of experience by the mould designer. Every mould is as unique as the parts, which are produced. Therefore, it is highly complex to develop a thermal design with reproducible quality assuring a minimum part warpage at optimal cycle time. Conventional, analytical thermal mould design relies on the calculation of the cooling error at the cavity surface wall [1]. This calculation method is easy and suitable for conventionally drilled cooling channels but it does only give an estimation of the temperature

U. Reisgen et al. (Eds.): EMPOrIA 2020, LNME, pp. 236–245, 2021.
https://doi.org/10.1007/978-3-030-70332-5_21

at the cavity surface rather than the actual temperature distribution in the part. Hot spots in small areas can only be detected very difficultly. It becomes even more complex for cooling channels manufactured with the freedom of the additive manufacturing (AM). Nowadays, simulation tools such as Moldflow, Moldex3D, Sigmasoft etc. are the state of the art in industrial mould-making and heavily needed to identify possible challenges already in the design phase. Consequently, process simulation requires experience of mould designers on the one hand for setting up correctly the simulation and on the other hand for interpreting correctly the calculated results. It is not assured that an optimal solution can be found [2, 3].

This is why currently research focusses on the investigation how to design cooling channels for injection moulds in an automatic and reproducible way. After a discussion of the current state of the art on automatic cooling channel generation, this paper presents a revised method to evaluate objectively the quality of a thermal mould design by solving an inverse heat conduction problem and an improved objective function, which has been developed in previous research.

2 State of the Art on Automatic Cooling Channel Generation

Putting cooling channels close to the cavity, increases the cooling capacity but increases also mechanical weaknesses of the mould steel. *Jahan et al.* developed an elaborate approach and proposed a thermo-mechanical design optimisation of cooling channels. This way it is ensured that the injection mould is mechanically and thermally stable and is able to take up the arising heat fluxes. The cooling channels cannot bear any mechanical forces, so that channels close to the cavity create high stresses in the remaining material. This lead to the conclusion that not every cooling channel design that would create an optimal thermal result for the part is suitable for a production process with high pressure in the filling phase. Therefore, further factors as the created stress in the mould material were considered. *Jahan et al.* parameterize the geometry: the diameter and the pitch distance between the centreline of the cooling channel and the simple cup-shaped part were varied. By a design of experiments (DoE) several cooling channel designs were tested and the best design was chosen by the criteria of minimum yield stress and minimum cooling time. In conclusion, the outcome of this study heavily depends on the weighting of the input parameters of the user, which does not assure that a global optimum can be found. Also, the resulting part warpage is not considered but only the achievable cooling time [4].

Several authors have taken the cooling demand of the part into the focus of their research. Using an objective function to evaluate the cooling quality, they developed a quality criterion at the end of the cooling phase such that an optimisation software can solve the thermal problem:

Agazzi et al. decoupled the optimization process from the generation of the cooling channel system in a new approach. By solving an inverse heat conduction problem, the heat dissipation was determined on an outer contour surrounding the part. Within that part, there was a homogeneous starting temperature distribution related to a selected reference temperature. A uniform demoulding temperature is the goal to achieve on the surface of the moulded part at the defined point of ejection. The homogeneity of the temperature distribution was evaluated by a objective function that adds up the local differences in temperature. This sum was minimized iteratively in a thermal optimization. The temperature control system was then derived based on isothermal surfaces of the calculated temperature field. In contrast to an analytical design, this method takes into account the heat conduction within the part and the mould as well as the heat transfer between these two components. However, no starting positions for the location of the cooling channels have to be defined and thus the boundary conditions are free of user influences. In contrast, the method developed by *Agazzi et al.* neglects the physical density properties of thermoplastics, which show a pronounced dependence on pressure and temperature [5, 6]. Therefore, the proposed objective function of *Agazzi et al.* considering only thermal effects has been extended with a second term taking into account the local difference in density by *Nikoleizig et al.*

$$Q(T_C) = \sum_{i=1}^{m} \int \left(\frac{T_{Ej} - T_{Pl}\left(x_i, t, T_{CC_0}\right)}{\omega_1} \right)^2 dA_{part} + \sum_{j=1}^{k} \int \left(\frac{\bar{\rho}_{Pl} - \rho_{Pl}\left(x_j, t, T_{CC_0}\right)}{\omega_2} \right)^2 dA_{off} \quad (1)$$

This objective function $Q(T_C)$ addresses a quick cooling through the first term, where a desired ejection temperature T_{Ej} for the surface A_{part} of the part is given and compared to the actual local temperatures $T_{Pl}(x_i, t, T_{CC_0})$ of the part. The second term addresses the density homogeneity, with the differences of local density $\rho_{loc}(x_i, t, T_{CC_0})$ compared to a average density $\overline{\rho_{Ejec}}$, which should be reached on an offset surface A_{off} within the part. Both terms are summed over their respective areas and can be weighted with the variable $\omega_{1/2}$. The temperatures T_C on the outer mold contour are varied to minimize the objective function. This approach is purely thermal, which means that no filling effects are calculated. *Nikoleizig et al.* have also added a realistic starting temperature distribution to the methodology, which shows better results. Starting from the temperature distribution after the end of filling the optimal temperature distribution in the mould during the holding and cooling phase is iteratively calculated, such that at the end of the cooling phase the above-mentioned differential function is minimal and the part has a homogeneous temperature and density distribution [7, 8].

This approach has been used and optimised for the injection moulding process in recent years at IKV. In the following, this approach is further developed to a tempering system for optical lenses.

3 Thermal Optimization – Revising the Objective Function

The methodology of *Nikoleizig et al.* shown in the precedent chapter is further investigated here. The core aspect of the thermal optimisation of the mould is the definition of the objective function. This function returns a specific value to the optimisation software such that the thermal state around the mould can be changed for an optimal heat flux from part to mould. This definition of the objective function is now revised:

To take into account the compressible character of the plastic melt, *Nikoleizig et al.* have added the consideration of the density into the objective function (see Eq. 1). The two terms can be weighted individually by the terms ω_1 and ω_2. These weighting factors represent an absolute number and needs to be adapted to every moulded part. A series of thermal optimisations have to be performed in order to find the best weighting for the specific part and process. Exemplary calculations have shown, that the value of the two terms can differ by several orders of magnitude. Changes in density are in the range of 0.1 g/cm^3 whereas Temperatures may differ up to 100 K from each other over the part's surface. Therefore, it is proposed to normalise the two terms of the function by T_{ejec} and $\overline{\rho_0}$ respectively. This causes the two terms to be in the same order of magnitude and the whole term becomes unitless. In a second step, this weighting can now be changed and adapted by $\omega_{1/2}$, but in a way that the optimal weighting needs to be found only one time.

The optimisation function proposed by *Nikoleizig et al.* does also weight implicitly more complex regions by not taking into account mesh effects. Regions that have a lot of e.g. edges and complex geometry features are usually meshed more densely than areas with low geometric complexity. As in the proposed objective function (1) the integral becomes numerically a sum. This is why in this function the sum operator is used. The respective node density from the mesh leads to an implicit weighting of the optimisation function. Based on this effect, the following revised objective function is proposed:

$$Q(T_C) = \sum_{i=1}^{m}\left(\frac{T_{Ej} - T_{Pl}\left(x_i, t, T_{CC_0}\right)}{T_{ej} * \omega_1}\right)^2 \cdot \frac{A_{el,i}}{A_{part}} + \sum_{j=1}^{k}\left(\frac{\overline{\rho}_{Pl} - \rho_{Pl}\left(x_j, t, T_{CC_0}\right)}{\overline{\rho}_{pl} * \omega_2}\right)^2 \cdot \frac{A_{el,j}}{A_{Off}} \quad (2)$$

This equation is normalised in the described way such that the two terms are now in the same order of magnitude and do not implicitly affect the cooling quality through numerical effects. A_{part} represents the surface of the part, whereas A_{Off} is the offset surface inside the part where the density function is evaluated $A_{el,i/j}$ represents the share of the surface of each element *i/j* on this surface. The dependency of the thermal optimisation of the mesh size should be reduced by this new formulation. In general, the methodology is not dependent on the mesh size as long as enough mesh elements are chosen because this affects the precsion of the heat transfer calculations. Over the part's thickness, at least three elements are chosen. Over the part's surface, differences in mesh density are now evened out.

To show the advantages of this new formulation of the objective function firstly thermal optimisations are performed with a lens geometry with very few features and a thick-walled mid-section. Secondly, a more complex part containing freestanding ribs, deep boxes and changes in wall thickness are investigated.

4 Experimental - Thermal Optimisations

As described in Sect. 2, in this approach only the holding and cooling phase without flow effects are considered as these two phases make up the most of the cooling time of the injection moulding cycle. The melt is already in full contact with the cavity and the flow effects are neglected here. The temperature distribution after the injection phase is calculated with commercial injection moulding software. A constant mould temperature close to the final solutions is chosen and no cooling channels are used. A heat up process of the mould over a series of injections is calculated to use a starting temperature distribution, which is close to the final solution. This distribution is mapped into the software environment Comsol, Comsol AB, Stockholm, Sweden. This starting temperature distribution is used for the part as well as for the surrounding mould contour. In order to save calculation time, a constant offset around the part of 10 mm is used. The temperature distribution directly around the part influences the part quality critically. The resulting temperature distribution is then approximated by a derived cooling channel design.

The first geometry is a typical convergent lens. Due to the low refractive index of plastics the part is considerably thick, with a maximum wall thickness of 10.21 mm. The part is visualised in Fig. 1. The used material is a Polymethyl methacrylate (PMMA) of the type Plexiglas 7N by the manufacturer Roehm GmbH, Weiterstadt, Germany. The moulding parameter can be found in Table 1. The temperatures are taken from the calculation of the commercial software calculation.

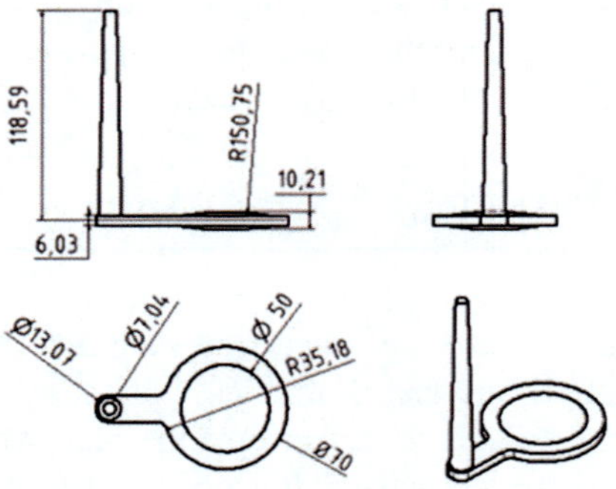

Fig. 1. Exemplary lens geometry and moulding parameters

Table 1. Parameter of the thermal optimisation for the lens geometry

Moulding parameter	Value
Cooling time	240 [s]
Melt temperature distribution	198–241 [°C] (depending on distance to the gate)
Mould temperature	83–93 [°C]

In a second step, a reference geometry that has been used in earlier publications [9], is investigated to show the improvement by the new objective function. This geometry is visualised in Fig. 2. The geometry has specific elements, which are common for plastic parts, such as boxes, freestanding ribs and changes in wall thickness. These features are potentially prone to severe warping. With this exemplary geometry for technical parts, a comparison between the existing and the revised objective function is performed. For this part a Polybutylene terephthalate (PBT) B4520 from the company BASF SE, Ludwigshafen, Germany, is used. The calculation parameters are summarized in Table 2.

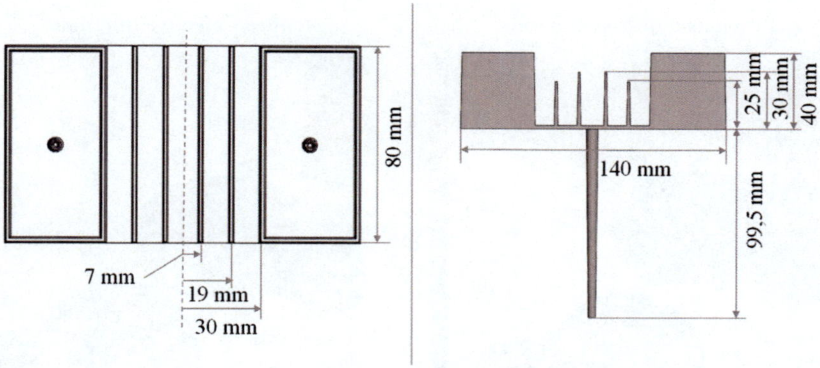

Fig. 2. Complex reference geometry

Table 2. Process parameters for the thermal optimisation of the complex geometry

Moulding parameter	Value
Cooling time	10 s
Melt temperature	168–281 °C (depending on distance to the gate)
Mould temperature	88–65 °C
Ejection temperature	110 °C

5 Results and Discussion

After finishing the thermal optimisations, an optimal temperature distribution can be extracted from the Comsol environment. In Fig. 3, a comparison between the new and the previous objective function is made. In this figure, the temperature on the mould contour can be seen. According to the calculation using the previous objective function, a warm ring on the upper side is necessary to cool down more slowly the outer area of the lens while the thicker parts need to be cooled more quickly. With the revised objective function, this ring of higher temperature is also visible and the inside belonging to the thick-walled lens area is cooled more strongly showing a more realistic result. Furthermore, in the mid-plane, a warm area can be seen, indicating a slow cooling of the side plane of the lens.

Comparing the temperature distribution on and inside the lens shows a similar result. In Fig. 4, the resulting temperature distribution can be seen. Using the previous objective function, an overheated section in the middle of the part occurs with temperatures above 130 °C, which is highly above glass transition temperature and the outer areas of the lens are already too cold. The whole temperature level is closer to the required ejection temperature $T_{ejec} = 110°C$. The effects of the mesh size and the unequal order of magnitude of the two terms are now considered. This should result in a better optical quality due to less part warpage, which will be validated in future work.

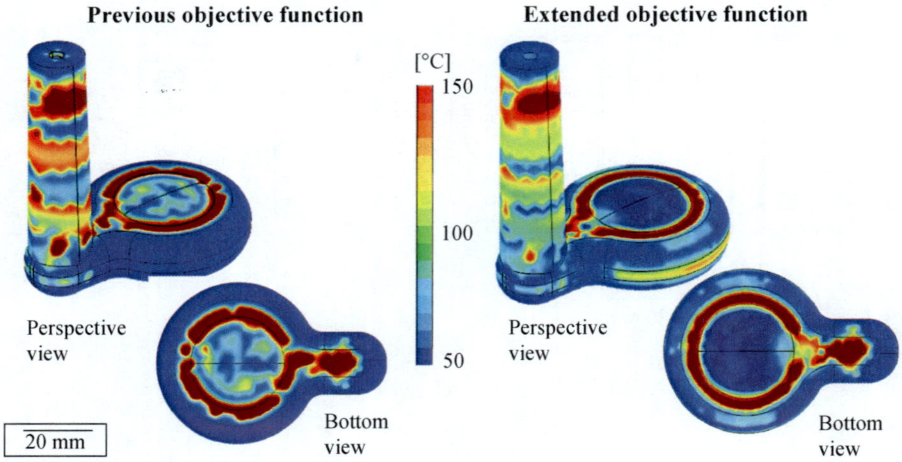

Fig. 3. Comparison of the temperature distribution on the mould contour

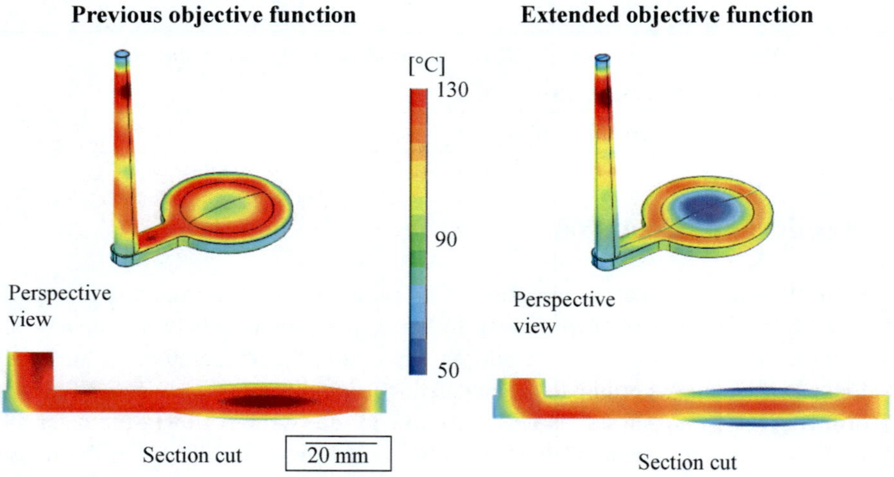

Fig. 4. Temperature distribution of the lens at the point of ejection

In the next step, the second geometry is investigated. In this case, the analysis of the resulting temperature distribution is not as obvious because the temperature distribution cannot be analysed easily due to many geometric features blocking the vision on the part. The overview on the temperature distribution (cf. Fig. 5) on the mould contour shows for the previous objective function a set of complex patterns with a slightly higher temperature level at the end of the flow path. This higher temperature distribution at the end of the flow path is more visible for the extended objective function. This takes into account, that the wall thickness is lower due to the drafting angle and cools down more quickly at the same mould temperature.

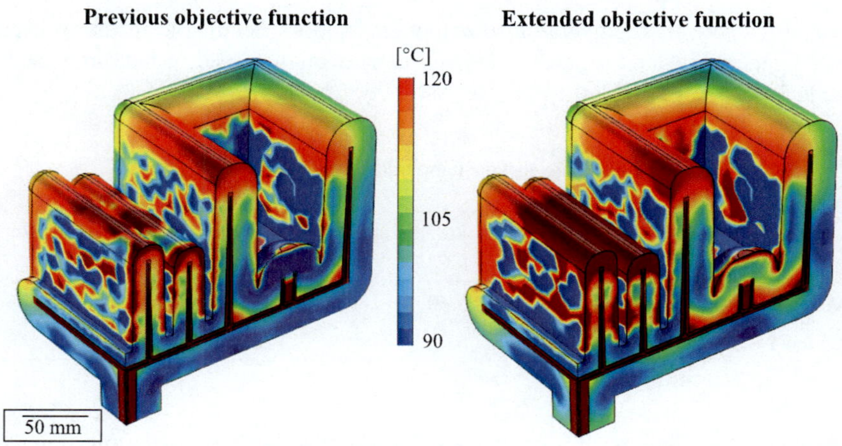

Fig. 5. Comparison of the temperature distribution in the mould contour

These differences in the mould temperature distribution are barely visible on the part. In Fig. 6, the comparison of the temperature distributions on the part are shown. In contrast to the lens geometry, it can be seen, that the two temperature distributions are quite close to the required ejection temperature of 110 °C. The extended objective function shows slightly higher gradients on the ribs, whereas the temperature gradients in the box section are more distinctive for the previous objective function. Both optimisations have problems to cool sufficiently corners of parts. These elevated gradients especially in these areas prone for warpage may lead to future unwanted deviations of the geometry after the derivation of the cooling channel design.

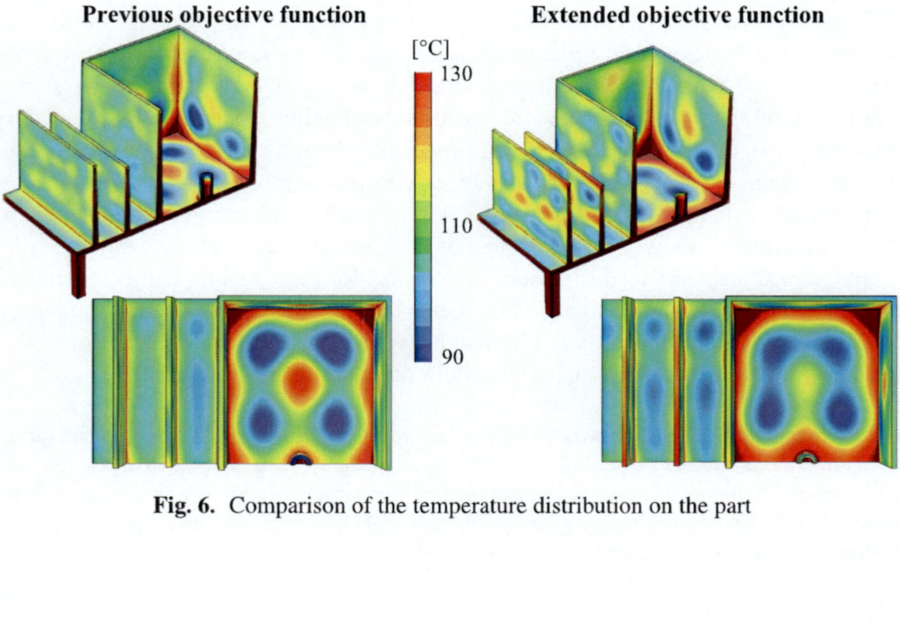

Fig. 6. Comparison of the temperature distribution on the part

To investigate this temperature distribution in more detail, the resulting average surface temperatures on the part and average density on the offset surface are visualised in Table 3.

Table 3. Average surface temperature and density of the part

Quality function	Av. temperature [°C]	Std. deviation [°C]	Av density [kg/m^3]	Std. deviation [kg/m^3]
Previous	110.19 °C	5.30	1219.56	10.04
Extended	110.87 °C	6.23	1219.37	10.02

By comparing these results, it becomes obvious, that the thermal optimisation works quite well for both objective functions and the objective of an optimal homogeneity can be reached with both objective functions. The marginal differences can be neglected. So far it is not obvious, why for this part the differences are this small for this box geometry and very obvious for the lens geometry. Generally, the extended objective function shows better or same results and should therefore be used in future works. The reason for the different behaviour needs to be investigated in further research.

6 Conclusion and Outlook

In this paper, a methodology for an automatic thermal optimisation of a mould for the injection moulding process has been discussed. Based on previous research, a revision of the objective function of the thermal optimisation has been proposed. This objective function now includes an automatic weighting of the different objectives by normalising the respective term. Furthermore, the numerical effects of different mesh sizes are evened out by taking only into account the respective part of every element of the whole surface. This revised objective function has been tested at two different geometries representing two main sectors of injection-moulded products. The lens geometry has shown a significant improvement as especially the mid-section of the lens is now more homogeneously cooled down. The part with ribs is more difficult to evaluate as the graphical evaluation can be misleading. The analysis of the average temperature and density distribution shows a slight decline of the homogeneity of the two objectives.

In the next steps, an analysis about the reasons for these results will be conducted in order to receive further insights on the behaviour of this optimisation methodology. Furthermore, an extensive study will be conducted to investigate the influence of further weighting factors on the resulting temperature distribution and part warpage.

Acknowledgement. The presented investigations were carried out at RWTH Aachen University within the framework of the Collaborative Research Centre SFB1120–236616214 "Bauteilpräzision durch Beherrschung von Schmelze und Erstarrung in Produktionsprozessen" and funded by the Deutsche Forschungsgemeinschaft e.V. (DFG, German Research Foundation). The sponsorship and support is gratefully acknowledged.

References

1. Hopmann, Ch., Menges, G., Michaeli, W., Mohren, P.: Spritzgießwerkzeuge, 7th edn. Hanser, Munich
2. Baur, E., Osswald, T.A., Rudolph, N.: Saechtling Kunststoff Taschenbuch, 31st edn. Hanser, Munich
3. Karlinger, P., Hiken, F.: Mold design. In: Mennig, G., Stoeckhert, K. (eds.) Mold-Making Handbook, vol. 3, pp. 301–331. Hanser, Munich
4. Jahan, S.A., Wu, T., Zhang, Y., Zhang, J., Tovar, A., Elmounayri, H.: Thermo-mechanical design optimization of conformal cooling channels using design of experiments approach. Procedia Manuf. **10**(5), 898–911 (2017)
5. Agazzi, A., Sobotka, V., Legoff, R., Garcia, D., Jarny, Y.: A methodology for the design of effective cooling system in injection moulding. Int. J. Mater. Form. **3**(1), 13–16 (2010)
6. Agazzi, A., Sobotka, V., Legoff, R., Jarny, Y.: Uniform cooling and part warpage reduction in injection molding thanks to the design of an effective cooling system. Key Eng. Mater. **556**(31), 1611–1618 (2013)
7. Nikoleizig, P.: Inverse thermische Werkzeugauslegung. Dissertation. RWTH, Aachen (2018)
8. Hohlweck, T., Fischer, T., Hopmann, C.: Simulative and experimental validation of an inversed cooling channel design for injection molds. In: Proceedings of the SPE ANTEC Conference, Detroit, pp. 1–6 (2019)
9. Hopmann, C., Wehr, H., Schmitz, M., Schneppe, T., Theunissen, M.: Digitalisierung der Werkzeugentwicklung im Spritzgießen. In: Internationales Kolloquium Kunststofftechnik. Shaker Publishers Verlag, Düren (2018)

Check for updates

Reduction of Internal Stresses in Optics Through a Demand-Oriented Cooling Channel Layout in Injection Moulding

Christian Hopmann and Jonas Gerads(✉)

IKV Institute for Plastics Processing in Industry and Craft, RWTH Aachen University, Seffenter Weg 201, 52074 Aachen, Germany
jonas.gerads@ikv.rwth-aachen.de

Abstract. The production of plastic optics creates great challenges for the injection moulding process. The lens geometry is designed to fulfil the optical requirements and often contradict the plastic design guidelines. Especially thick-walled lenses with changes in wall thickness are challenging not only in the classic injection moulding process but also in injection-compression moulding processes, as the compression pressure can lead to internal stresses and thus to reduced optical properties. In order to increase the optical performance of thick-walled plastic lenses, a methodology developed at the Institute of Plastics Processing is being adapted for the inverse calculation of the cooling requirements of plastic components. Based on this, a cooling channel design is derived and validated in injection moulding simulations. With this method, lower peak values of residual stress and birefringence can be achieved. Furthermore, it can be shown that the developed cooling channel layout significantly reduces the cooling time required to reach the glass transition temperature.

Keywords: Injection moulding · Cooling channel design · Plastic optics

1 Introduction

The manufacturing of optical plastic parts using injection moulding or injection compression moulding is a challenge for mould, machine and process, as the geometry of the optical components often contradicts any conventional, process-oriented design guidelines. Conventional injection moulding processes often reach their limits when it comes to the high demands on moulding accuracy and internal properties, which directly influence the optical properties.

Initially during the injection phase of the injection moulding process, the cavity is completely filled and then during the holding pressure phase the material shrinkage is compensated by further transport of melt through the sprue channel into the cavity. The melt movement causes inhomogeneous internal component properties such as orientation and internal stresses. The pressure distribution drops along the flow path towards the end of the part and is only effective in the areas close to the sprue. The pressure gradient induces further stresses in the moulded part, which have a negative effect on the optical

U. Reisgen et al. (Eds.): EMPOrIA 2020, LNME, pp. 246–255, 2021.
https://doi.org/10.1007/978-3-030-70332-5_22

properties. As soon as the sprue has completely solidified (sealing point), the holding pressure phase is over and shrinkage processes can no longer be compensated [1].

This is why injection compression moulding has become established in many optical applications [2, 3]. Depending on the process variant, the compression phase is used to mould the part and to compensate for shrinkage. At the beginning of the compression process, the cavity is reduced at a defined velocity until a desired compression force or pressure is reached. Afterwards, the system switches over to a force- or pressure-controlled compression process. During the compression phase, the pressure is introduced over the entire surface of the component, so that, in contrast to injection moulding, the pressure distribution over the component is significantly more homogeneous.

However, for thick-walled lenses with large wall thickness differences, which contradict a plastic-compatible design, the homogeneous pressure distribution during injection-compression moulding represents a challenge. Injection-compression moulded parts with large wall thickness variations sometimes have poorer properties than injection-moulded parts [4]. Due to the long compression phase which is necessary for the thick-walled area to compensate for the local shrinkage. This introduces stresses in the already solidified thin-walled areas (Fig. 1). These stresses and can lead to birefringence, whereby a lens is not correctly focused. The residual stresses therefore lead to a reduction in the optical performance of the moulded lens [5, 6].

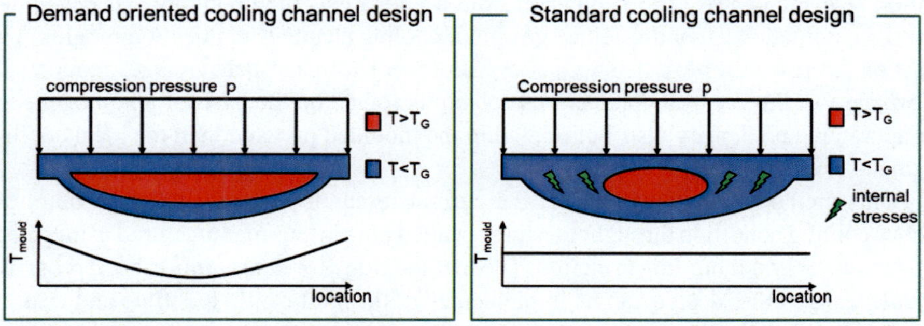

Fig. 1. Optical properties of a biconvex lens with different measuring range

In order to avoid deformation of already solidified areas of the moulded part, a compression pressure profile would be preferable, which applies a higher pressure to thick-walled areas than to thin-walled areas. However, a profiled compression pressure cannot be realised in terms of tooling. A uniform mould cavity pressure over the lens surface, as is present in injection compression moulding over the entire lens surface, would be appropriate if the same cooling conditions were present for thick-walled and thin-walled areas. This can be achieved by means of cooling channels adapted to requirements. In this way, thick-walled and thin-walled areas of the lens can be cooled to different intensities.

2 Inverse Thermal Cooling Channel Design

The current thermal mould design method is a forward iterative process based on empirical experience. Up to now, thermal mould design has usually been designed in such a way that a mould designer adapts the design of the cooling channels in order to create a homogeneous temperature distribution at the cavity. By means of digital process simulation to calculate the temperature distribution, the forward iterative method could be used more effectively and at a lower cost. In the next step, the mould designer changes the position of the cooling channels, based on personal experience, so that hot spots are minimized. This procedure is highly dependent on the experience of the individual mould designer [7].

For this reason, current process simulation software such as Sigmasoft (SIGMA Engineering GmbH, Aachen, Germany) or Moldflow (Autodesk Inc., San Rafael, USA) offer an optimization of the mould design through virtual test plans (DoE) [8, 9]. This optimization is based on the degrees of freedom given by the designer.

By choosing different geometries or process parameters, the designer can define the optimization space. Several designs can be calculated automatically one after the other. Although an optimised draft is found in the end, it is not guaranteed that it is generally an optimum.

At the Institute of Plastics Processing (IKV), within the framework of the collaboration research centre 1120 "Precision from Melt", a new methodology was developed for the automatic, demand-oriented design of cooling channels in injection moulds. The aim of the research project is to achieve uniform shrinkage behaviour to reduce part warpage. An inverse heat conduction problem is solved on the basis of a homogeneous temperature and density distribution within the moulded part and then the local cooling demand is determined [9].

This methodology is adapted in the current research project to create a homogeneous cooling condition throughout a thick-walled plastic lens manufactured using injection compression moulding principle. The methodology for determining the appropriate mould temperature design can be implemented with significantly less time and cost.

In this paper the automatic thermal mould design for optical parts is considered and validated with injection moulding simulations. The focus will be on the birefringence caused by residual stresses.

2.1 Calculation of an Inverse Demand-Oriented Cooling Channel Design

The methodology for inverse cooling channel design was originally developed by Agazzi et al. and further developed by Nikoleizig et al. and Hohlweck et al. [9–11]. In a first step, the method calculates the optimum temperature distribution in the mould so that after the cooling phase a thermally homogeneous component is obtained. To evaluate the temperature distribution, a function was developed to analyse the weighted temperature and density distribution in the part. In addition, the temperature distribution on a mould envelope around the component is calculated and optimised. It is assumed that a homogeneous temperature and density distribution leads to minimal warpage.

$$Q(T_{KK0}) = \sum_{i=1}^{m} \int_{0}^{t_{cool}} \frac{(T_{ejec} - T_F(x_i, t; T_{KK0})}{\omega_T} dt + \sum_{j=1}^{n} \int_{0}^{t_{cool}} \frac{(\overline{\rho_F} - \rho_F(x_j, t; T_{KK0})}{\omega_\rho} dt \quad (1)$$

The temperatures $T_{KK01}; \ldots; T_{KK0n}$ are calculated on the mould envelope Γ_W surrounding the moulded part, in a way that the error square sum of the quality function Q is minimized for all $T_{KK0i} \in [a, b]^n$ for $0 < t < t_{cool}$ [9].

In the first term, the difference between the desired demoulding temperature T_{Ejec} and the local moulded part temperature T_F over the surface of the part Γ_{Fo} and the cycle time is evaluated. In the second term, the difference between averaged density $\overline{\rho_F}$ and local density ρ_F over the surface inside the part Γ_{Fi} and the cycle time is evaluated. Through $\omega_{T;\rho}$ a different weighting of the terms is possible. Then, the temperature distribution on Γ_W is iterated via a conjugated gradients method for high number of degrees of freedom. Finally, the calculated optimized temperature distribution is analysed and iso-surfaces are generated by the following equation:

$$0 = \int_0^{t_{cool}} (h_{M \to CC}(T_{loc} - T_{coolant}) - |\lambda_M \, gradT_{loc}|)dt \tag{2}$$

This equation expresses the equilibrium between the locally existing absolute heat flux ($|\lambda_M \, gradT_{loc}|$) and a theoretical heat flux inside the mould steel induced by a cooling channel at a specific position with a coolant temperature [9]. By this method "iso-flux" surface are derived that indicate the position of the cooling channels. In the second step, a cooling channel is then designed based on the computation results. This is then designed on the basis of the "iso-flux" surfaces, taking into account the component and a possible parting plane.

3 Investigation of a Demand-Oriented Cooling Channel Layout for Thick-Walled Optics in Injection Moulding

The method has so far only been used for thin-walled plastic components with a maximum thickness of up to 3 mm. Plastic optics often have significantly greater wall thicknesses, which also significantly extends the cooling time. Due to the high temperature gradients in the components, a targeted and optimized mould temperature control is therefore important.

3.1 Geometry of the Plastic Lens

An application-oriented demonstrator was selected in cooperation with an industrial consortium. The optics consists of five areas, whereby the three middle ones additionally have a structure for light scattering (Fig. 2). The component is a thick-walled plastic lens with wall thickness ranging from 10 to 15 mm within the optically relevant area. The methodology was extended, such that it is suitable for thick-walled components with wall thickness differences.

For the following investigation, the injection moulding process is considered in a manner that a constant heat flux from the plastic melt into the mould is assumed.

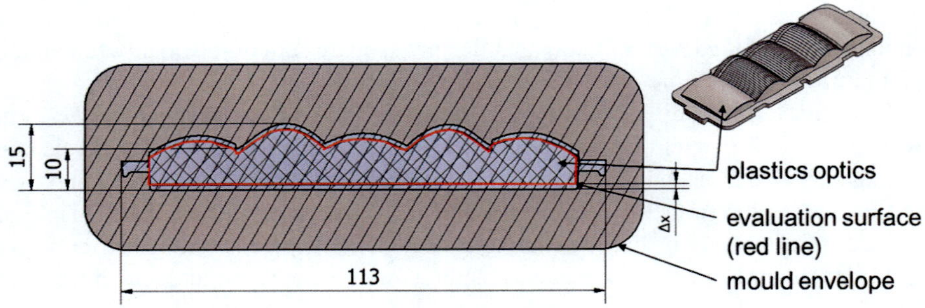

Fig. 2. Application-oriented demonstrator with max. and min. thickness and the evaluation surface within the part

3.2 Adaption of the Methodology

In the presented approach only one area is defined for both, the temperature and density evaluation. The simulation evaluates the quality function for a surface inside the plastic lens that follows the outer contour by the constant distance of Δx. This evaluation surface was created using CAD software so that there is a constant distance to the outer contour (Fig. 2).

With only one evaluation surface for the thermal and density term, the quality function needs to be adjusted for m = n. Continuously the quality is determent by the homogeneity of the solidification.

$$Q(T_{KK0}) = \sum_{i=1}^{n} \int_0^{t_{cool}} \frac{(T_{ejec} - T_F(x_i, t; T_{KK0})}{\omega_T} dt + \sum_{i=1}^{n} \int_0^{t_{cool}} \frac{(\bar{\rho_F} - \rho_F(x_i, t; T_{KK0})}{\omega_\rho} dt \quad (3)$$

The two terms are independent from each other and the first term describes the difference in temperature. In the previous methodology the temperature distribution is evaluated over the part surface. The area is changed to the quality area, which also applies to the second term, which describes the local homogeneity of the density. The thermal term thereby evaluates the internal temperatures in the component more strongly, while the focus is no longer on the outermost surface layer. This change is made to take greater account of the temperature conditions inside the thick-walled lens.

3.3 Designing of Demand-Oriented Cooling Channels

To calculate the cooling behaviour and to determine the optimised iso-flux areas, a start temperature and pressure distribution within the component must be available. Therefore a filling simulation with the simulation software Sigmasoft (SIGMA Engineering GmbH, Aachen, Germany) using realistic injection moulding parameter for thick-walled lenses is carried out. In this first injection moulding simulation no mould cooling system is implemented so that only the melt heat flow is present at the end of the filling phase.

The temperature and pressure data on the individual mesh node elements are transferred to the multiphysics software COMSOL (COMSOL AB, Stockholm, Sweden). The cooling phase can then be calculated within this software using the existing data and the used pvT data, which are implemented in the simulation using the Tait Model. An amorphous thermoplastic, the polymehtyl methacrylate Plexiglas 7N (Röhm GmbH, Darmstadt, Germany) is used. The result of the iterative optimization of the thermal system, consisting of the lens and the mould envelope, for the demonstrator is shown in Fig. 3. This shows in which areas (blue) of the injection mould a cooling system is required due to the quality function computation. The iso-flux surfaces can then be generated from this information (Fig. 3). For this purpose the temperature surfaces are selected which correspond to a specific coolant temperature.

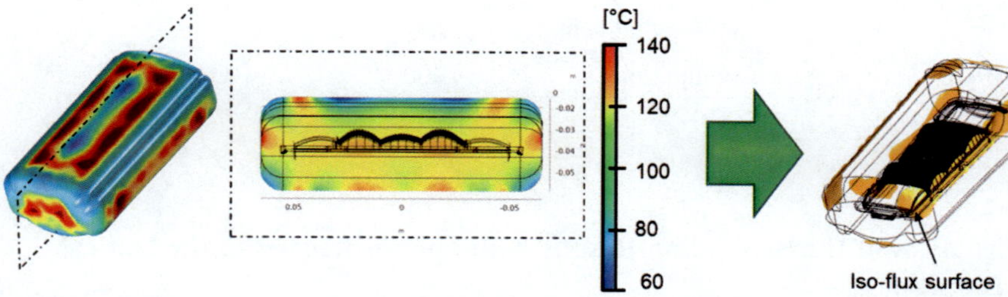

Fig. 3. Thermal System in the optical lens and mould envelope and the derived iso-flux Surfaces

The iso-flux surfaces are transferred into a cooling channel design using CAD software. The derived cooling channel design is then adapted for the injection moulding simulation, so that a parting plane is present and inlets and outlets have realistic position in the mould (Fig. 4, left). A simple standard cooling channel system determined by the simulation tool is used to validate the demand oriented cooling system (Fig. 4 right).

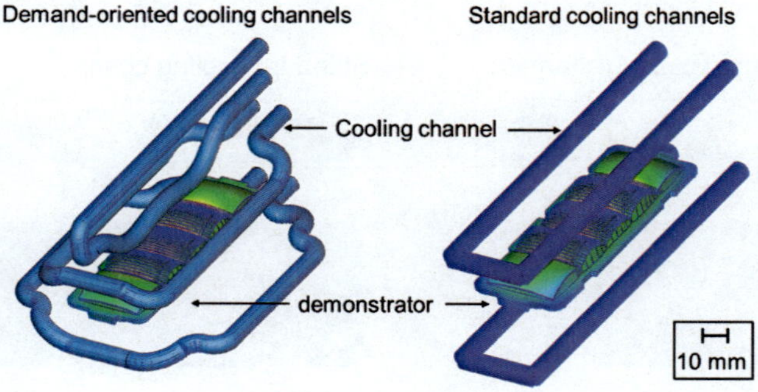

Fig. 4. Cooling channel designs

3.4 Implementation of the Injection Moulding Simulations

The demand-oriented cooling channel layout is analysed for residual stress and birefringence by means of an injection molding simulation using Moldex3D (CoreTech System Co., LTD., Chupei City, Taiwan). The analysis of both cooling channel designs is performed with the injection moulding parameters shown in Table 1. The cooling time has been calculated using the standard cooling formula to ensure that the glass transition temperature is reached.

Table 1. Simulation process cooling parameters.

Parameter	Value	Unit
Melt temperature	240	°C
Flowrate	120	cm^3/s
Cooling fluid temperature	80	°C
Cooling time	400	s

3.5 Analysis of the Simulation Results with Special Regard to the Optical Properties

The results of the injection moulding simulations with the standard and demand-oriented cooling channel designs are shown and analysed subsequently. The cavity is filled uniformly in both designs without differences in filling and temperature at the end of the filling time (EOF). This phase is mainly dependent on the part geometry and the injection parameters, neither of which have been changed, so these results were to be expected. The influence of demand-oriented cooling is seen in the simulations at the end of the cooling time (EOC), where the maximum temperature inside the lens is ~5% lower with the demand-oriented cooling channel design.

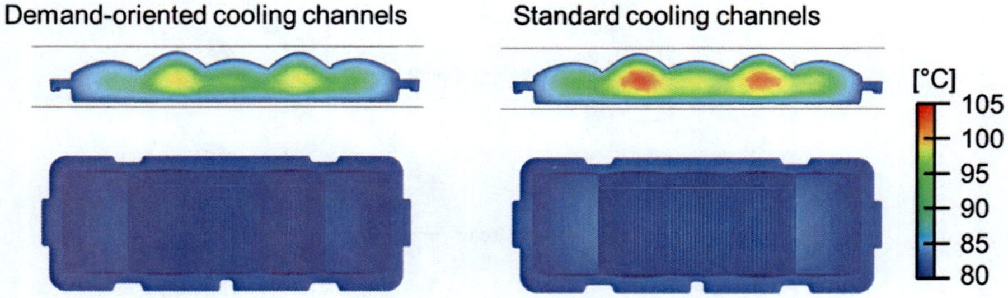

Fig. 5. Temperature distribution within the lens at EOC

When comparing the temperature distribution in the component at EOC with the simulation results from Comsol, it becomes clear that the distribution of temperature is similarly distributed over the component, but there is a shift in the temperature range between Moldex3D and Comsol. The average temperature in Comsol is 110,65 °C and in the Molex3D simulation 86 °C (Fig. 6). This may be due to the fact that Moldex3D also simulates cooling of the mould by air (25 °C) during the injection moulding process with demoulding to determine more realistic product results.

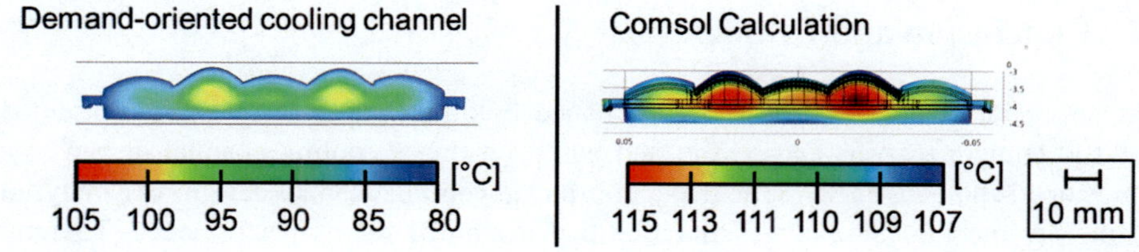

Fig. 6. Temperature distribution comparison between simulation results of the demand-oriented cooling channel and COMSOL calculation

The optical properties of the plastic lens can be analysed by the stresses and birefringence in the part. An improvement can be seen in the thermally induced stresses (Fig. 7). In this case, the maximum stress is reduced by 5.3% whereas the average stress is only reduced by 1%. This means that a more homogeneous stress distribution is present in the component. This is also reflected in the birefringence, which is reduced by 2.3% on average due to the optimized cooling channel design. If only the maximum birefringence is considered, a reduction of 5.5% is achieved.

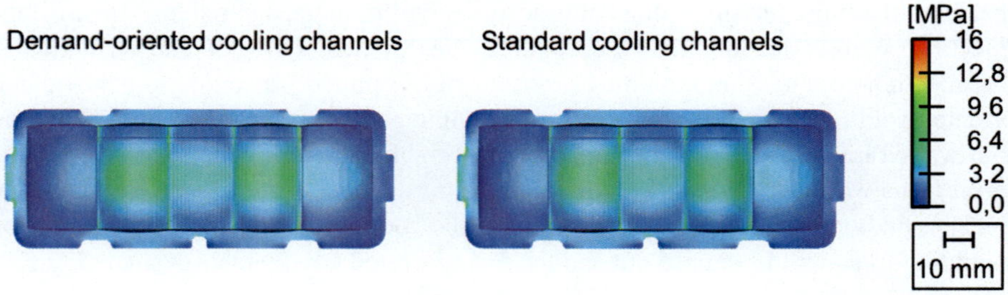

Fig. 7. Thermally induces stresses

In the injection moulding simulation the cooling time was estimated conservatively. The results of the simulations show that after 360s at the latest, the lens has reached the glass transition temperature even inside the lens. With the cooling channel design oriented to cooling requirements, a significantly earlier solidification of the innermost core is shown. The glass transition temperature in the lens center is reached ~53 s earlier (Fig. 8). This indicates that, in addition to lower peak values in stress and birefringence, a more profitable production of thick-walled plastic lenses can be achieved with the demand-oriented cooling channel design.

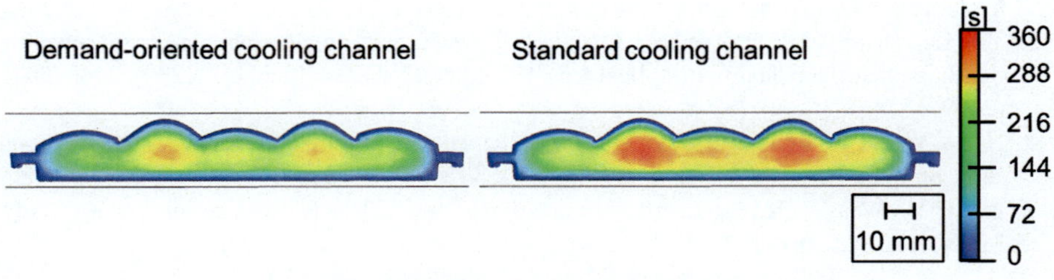

Fig. 8. Time to reach glass transition Temperature

4 Conclusion and Outlook

In the conducted investigation a cooling channel optimization for an application-oriented plastic component was carried out and based on this a cooling channel design was developed and constructed. The demand-oriented cooling channel design was analysed using an injection moulding simulation and examined for its performance. Thereby the optical properties were analysed in particular. The result was benchmarked with a standard cooling system. It could be shown that an improvement and homogenization of the optical property, the birefringence, could be achieved. An improvement can be found mainly in the reduction of the peak values of stress and birefringence within the part. The demand-oriented cooling was able to significantly reduce the temperatures within the thick-walled plastic optics. As a result, the glass transition temperature within the component can be reached one sixth and a cycle time reduction with improved optical properties can be achieved.

For future research, additional calculations must be carried out to determine the influence of the weighting and quality areas. To take the internal temperature of the component into more effective consideration, a volumetric approach is being investigated that takes the entire internal volume into account. Furthermore, an implementation of the injection-compression moulding process in the present methodology is carried out. Thus, the injection-compression moulding process can be applied, since due to the demand-oriented cooling channel design, the same cooling conditions are present in thick and thin-walled areas. In addition, the impact of the weighting factors, which was investigated in previous work for thin-walled components, must be re-analysed and optimized for thick-walled optical components. To validate the cooling channels, practical injection moulding trials will be carried out and the manufactured components evaluated.

Acknowledgments. All presented investigations were conducted in the context of the Collaborative Research Centre SFB1120 "Precision Melt Engineering", subproject T3 "Injection compression moulding of thick-walled plastic optics with demand-oriented mould temperature control taking into account local cooling demands" at RWTH Aachen University and funded by the German Research Foundation (DFG). For the sponsorship and the support, we wish to express our sincere gratitude.

References

1. Bürkle, E., Klotz, B., Schnerr, O.: Der gläserne Innendruck. Kunststoffe **101**, 71–74 (2011)
2. Isayev, A.I.: Tsui-HsunLin: Frozen-In Birefringence an Anisotropic Shrinkage in Optical Moldings: 1 Theory and Simulation Scheme. Polymer **1**(51), 316–327 (2010)
3. Tsai, K., Hsieh, C.-Y., Lo, W.-C.: A study of the effects of process parameters for injection molding on surface quality of optical lenses. J. Mater. Process. Technol. 1(209), 3469–3477 (2009)
4. Hsien, W., Su, Y.-L.: Optimization of wedge-shaped parts for injection compression molding. Int. Commun. Heat Mass Transf. **2**(30), 215-224 (2003)
5. Lin, C.M., Wang, C.K.: Processing optimization of optical lens in the injection molding. Adv. Mater. Res. **813**, 161–164 (2013)

6. Chung, C.-Y.: Integrated optimum layout of conformal cooling channels and optimal injection molding process parameters for optical lenses. Appl. Sci. **9** (20), 4341(2019)
7. Hopmann, C., Menges, G., Michaeli, W., Mohren, P.: Spritzgießwerkzeuge. Anwendung. Carl Hanser Verlag, München, Auslegung, Bau (2018)
8. SIGMASOFT. https://www.sigmasoft.de/en/press/pressreleases/improved-part-properties-via-virtual-doe/. Accessed 04 May 2020
9. AUTODESK. https://knowledge.autodesk.com/support/moldflow-insight/learn-explore/caas/CloudHelp/cloudhelp/2018/ENU/MoldflowInsight/files/GUID-5EC36B17-F7B4-4083-8872-E86AA04BA849-htm.html. Accessed 04 May 2020
10. Nikoleizig, P.: Inverse thermische Spritzgießwerkzeugauslegung auf Basis des lokalen KühlbedarfsRWTH Aachen, Dissertation: Betreuer: Prof. Dr.-Ing, Christian Hopmann (2017)
11. Agazzi, A., Sobotka, V., LeGoff, R., Jarny, Y.: Optimal cooling design in injection moulding process – a new approach based on morphological surfaces. Appl. Therm. Eng. **52**(1), 170–178 (2013)
12. Hopmann, C., Fischer, T., Hohlweck, T.: Simulative and Experimental validation of an inversed cooing channel design for injection molds. In: Conference 2019, SPE ANTEC, vol. 1, pp. 339-344. Curran Associates, Inc., Red Hook (2020)

Check for updates

Inverse Design Method for Injection Molding Cavity Shapes

Florian Zwicke(✉) and Stefanie Elgeti

Institute of Lightweight Design and Structural Biomechanics, Vienna University of Technology, Gumpendorfer Straße 7, 1060 Vienna, Austria
{zwicke,elgeti}@ilsb.tuwien.ac.at

Abstract. We present a method that uses simulation models to automatically design suitable cavity shapes for injection molding. Due to constrained solidification and inhomogeneous cooling behavior, the molding shapes resulting from the process deviate from the cavity shapes that are used. While it is possible to optimize certain process conditions to reduce these deviations, our approach is to find a cavity shape that, despite such deviations, will produce the desired molding shape.

Such an approach presents several difficulties. The simulation of the injection molding process has to model a great number of phenomena, such as viscoelasticity and solidification. This complex simulation naturally leads to more complex inversion or optimization procedures. In our specific approach, we use a combination of forward and inverse simulations to produce a cavity design method that is efficient while still taking into account the important physical phenomena.

Keywords: Injection molding · Shape optimization · Inverse methods · Finite element method

1 Introduction

In the process of injection molding, the shape of a produced molding can differ from the shape of the mold cavity. This is the result of several physical phenomena that occur during the process.

One cycle of the injection molding process starts with the filling of the mold cavity with a hot polymer melt. The cavity walls are colder than the melt and cause the melt to cool down. This cooling, in turn, starts the solidification of the melt. During the solidification, stresses that exist in the material may be frozen in, as they are no longer able to relax (cf., e.g., [1, 2]). The motion of the solidified material may also be geometrically constrained. This leads to further residual stress in the material. When the molding has cooled down sufficiently, it is ejected from the cavity. As a result of the residual stresses and the inhomogeneous temperature distribution that remains in the material, the part deforms as it cools down further to the environment temperature and reaches its equilibrium state. These process steps and the change in shape are illustrated in Fig. 1. For more detailed information on the injection molding process we refer to, e.g., [3].

U. Reisgen et al. (Eds.): EMPOrIA 2020, LNME, pp. 256–263, 2021.
https://doi.org/10.1007/978-3-030-70332-5_23

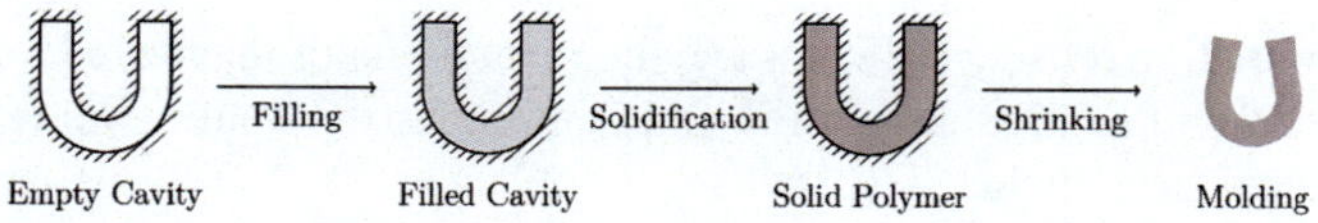

Fig. 1. Simplified illustration of injection molding process.

There are several aspects that influence the amount of shrinkage and warpage that occur. For instance, changes in the cooling behavior may be helpful in reducing defects in the molding shape. However, it is the shape of the mold cavity that has the greatest influence on the molding shape. Therefore, adjustments of this cavity shape can help keep shape defects to a minimum, even with shrinkage and warpage still occurring.

We address, in this paper, the task of finding such an improved cavity shape that will ensure a suitable molding shape. The inverse problem that results from predicting the molding shape and determining a corresponding cavity shape is depicted in Fig. 2.

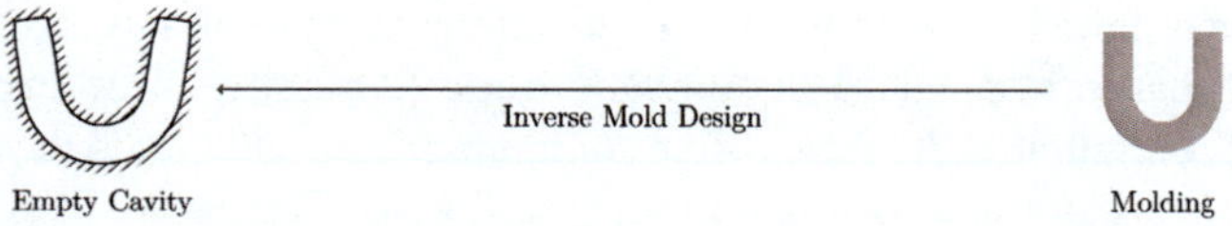

Fig. 2. Inverse problem of injection molding cavity design.

As a basis for the inverse method, we apply a finite-element simulation method for the injection molding process. By creating an improved cavity shape with a computational method, costly iterations in injection mold design can be saved.

2 Injection Molding Simulation

Since the injection molding process involves multiple material states and a number of physical phenomena, a reliable simulation method can become quite complex. To keep the complexity of the simulation at a level that is suitable for optimization methods in terms of efficiency and computational cost, we split the simulation into two separate parts. Each of these simulation parts models a specific part of the injection molding process with the most important physical phenomena.

The first part of the simulation, which we call the filling and solidification simulation, is a transient fluid simulation from the Eulerian viewpoint. This simulation part models the viscous and elastic behavior of polymers at a wide temperature range from the hot liquid melt to a colder solidified state. Both viscoelastic stress relaxation and simplified solidification behavior are represented in the model. Overall, given a cavity shape, the first simulation part determines a temperature and stress field at the time of ejection, after the material has solidified (see Fig. 3).

The second part of the simulation is called the shrinkage and warpage simulation.

For this part of the simulation, we make the simplification of assuming that solidification has been completed. This means that only a stationary thermoelasticity simulation is required to simulate how the material cools down to the environment temperature. This part of the simulation is using the stress and temperature fields that were transferred from the filling and solidification simulation and determines the final molding shape along with the residual stress field.

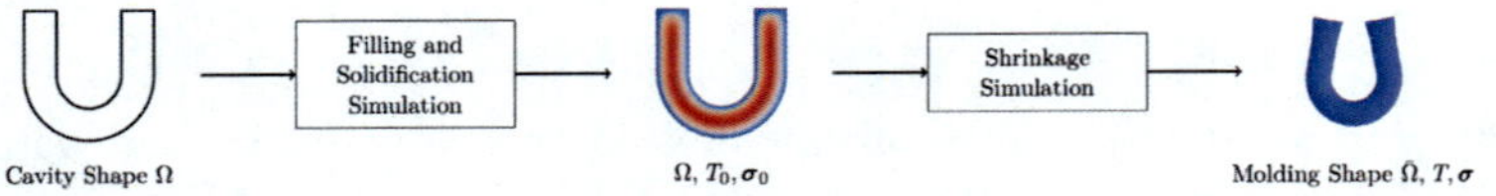

Fig. 3. The simulation is split into one transient filling and solidification part and one stationary shrinkage and warpage part.

We will now focus on the shrinkage and warpage simulation since this will be the basis of our inverse formulation. The principal requirements on this simulation are that both nonlinear elastic material behavior as well as thermal expansion can be modeled.

An additional requirement, however, is the prescription of an initial stress field. The existence of initial stresses means that the reference configuration, i.e., the material configuration that corresponds to the simulation mesh, is not stress-free. Instead, we need to add a third configuration, the stress-free configuration, in addition to the reference and equilibrium (i.e., deformed) configurations (see Fig. 4). This stress-free configuration differs from the other two in the fact that there may not exist a coordinate system that corresponds to this configuration. The reason for this is that the initial stresses result from the solidification process and not the deformation of a solid body. To simplify the transfer of the initial stress, we represent this stress by means of a deformation tensor.

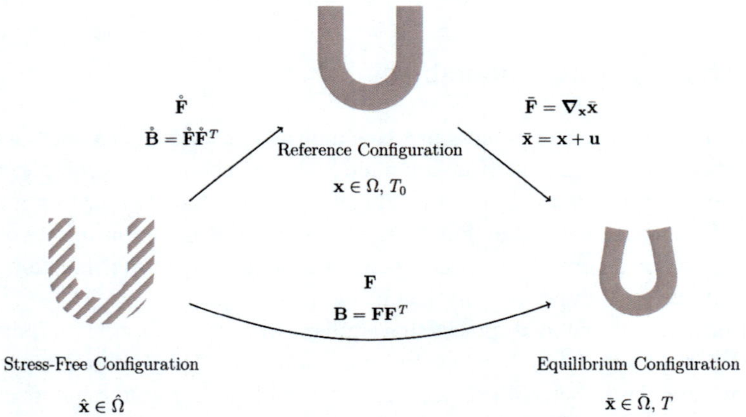

Fig. 4. In addition to the two configurations that are common in Lagrangian structural simulations, we use a stress-free configuration. The relations are demonstrated in terms of deformation gradient tensors **F**, left Cauchy-Green deformation tensors **B**, coordinates **x**, and the displacement field **u**.

We arrive at the following system of partial differential equations:

$$
\begin{aligned}
-\nabla \cdot \mathbf{P} &= 0, \quad \text{(momentum conservation)} \\
-\nabla \cdot \mathbf{q} &= 0, \quad \text{(energy conservation)}
\end{aligned}
$$

where **P** is the first Piola-Kirchhoff stress tensor and **q** is the heat flux vector. We have extended two nonlinear elasticity laws to allow for initial deformation and thermal expansion (cf. [4]),

$$\mathbf{P} = {}^{\circ}J^{-1}\left(\frac{\lambda}{2}\mathrm{tr}(\mathbf{B}-\mathbf{I})\mathbf{I} + \mu(\mathbf{B}-\mathbf{I}) - \alpha(T-T_0)\mathbf{I}\right)\overline{\mathbf{F}}\overset{\circ}{\mathbf{B}}, \qquad \text{(based on St. Venant – Kirchhoff)}$$

$$\mathbf{P} = {}^{\circ}J^{-1}\left(2D_1 J(J-1)\mathbf{I} + 2C_1 J^{-\frac{2}{3}}(\mathrm{dev}\,\mathbf{B}) - \alpha(T-T_0)\mathbf{I}\right)\overline{\mathbf{F}}^{-T}, \qquad \text{(based on Neo - Hooke)}$$

where thermal expansion is modeled in a simplified manner that allows small temperature changes. Finally, the heat flux follows Fourier's law:

$$\mathbf{q} = -\kappa \nabla T.$$

3 Inverse Elasticity Formulation

In the previous section, we have described all the ingredients for a forward injection molding simulation. Such a simulation already allows us to make predictions about the shape of the molding that results when a certain cavity shape is used. However, our intention, as stated earlier, is to create an inverse method for the design of a suitable cavity shape.

Figure 5 depicts the comparison of the forward and inverse approaches. Since the second part of the simulation, the elastic shrinkage and warpage simulation, is a stationary simulation, it can be inverted. If the material behavior was assumed to be plastic rather than purely elastic, such an inversion would be more difficult. The main difficulty in the elastic case stems from the need to prescribe inhomogeneous temperature and stress fields on the initial shape, which is to be determined by the simulation. We cannot know what temperature and stress fields would result at ejection time for a different cavity shape, without completely rerunning the costly filling and solidification simulation.

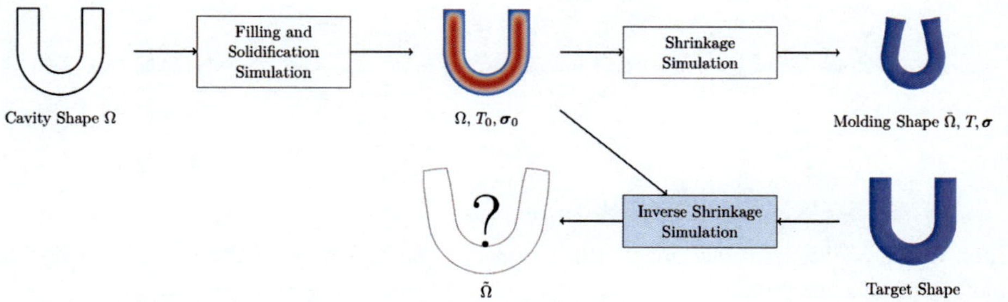

Fig. 5. For the inverse variant of the shrinkage and warpage simulation, we need to prescribe temperature and stress fields on an unknown initial shape.

In order to enable the inverse solution of the shrinkage and warpage simulation we need to make assumptions about the changes in the auxiliary fields (i.e., stress and temperature fields) that are caused by the change in the shape boundary. We achieve this by borrowing an idea from the Elastic Mesh Update Method (EMUM) [5]. We introduce a displacement field $\mathbf{r}$ in addition to $\mathbf{u}$. While the displacement field $\mathbf{u}$ describes the coordinate movement from the reference to the equilibrium configuration, the initial displacement field $\mathbf{r}$ describes the movement from the reference to the initial configuration. In order to smoothly adjust the interior node positions, we solve the equation of linear elasticity in the interior of the domain (see Fig. 6).

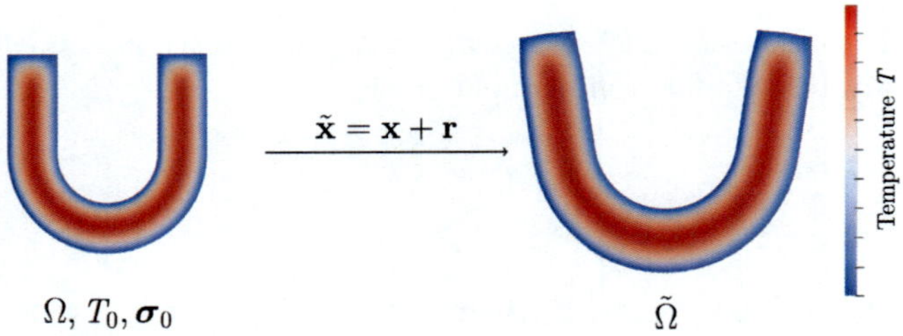

Fig. 6. An additional displacement field is introduced to map the auxiliary fields smoothly to the initial shape. Here, this is demonstrated on an arbitrary temperature field.

As a result of this additional displacement field, we need to introduce yet another material configuration. This initial configuration is located between the stress-free and reference configurations (see Fig. 7).

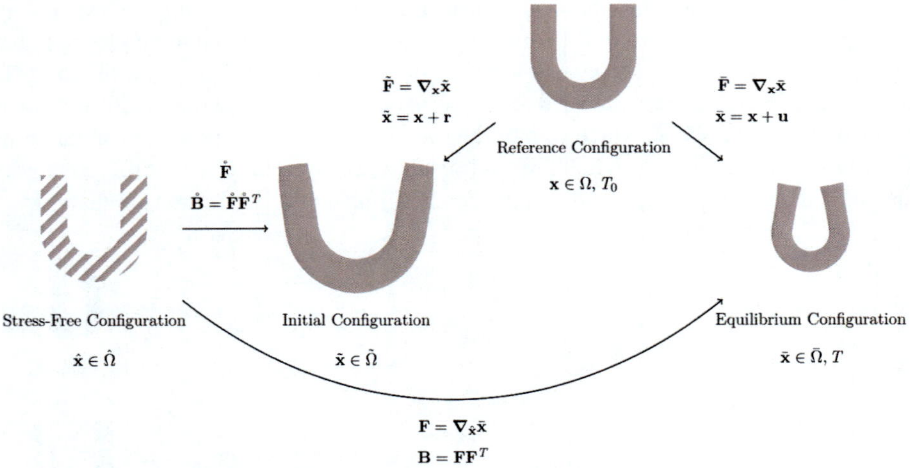

Fig. 7. The additional displacement field introduces a fourth configuration between the stress-free and reference configurations.

We now arrive at a system of three differential equations,

$$
\begin{aligned}
-\nabla \cdot \mathbf{P} &= 0, && \text{(momentum conservation)}\\
-\nabla \cdot \boldsymbol{\sigma}^{\varphi} &= 0, && \text{(smoothing conservation)}\\
-\nabla \cdot \mathbf{q} &= 0, && \text{(energy conservation)}
\end{aligned}
$$

where an additional equation – the smoothing equation – has been added for the initial displacement field $\mathbf{r}$. The expressions for the first Piola-Kirchhoff stress $\mathbf{P}$ need to be adjusted to account for the deformation between the initial and reference configurations (cf. [4]):

$$\mathbf{P} = \tilde{J}^{\circ} J^{-1}\left(\frac{\lambda}{2}\mathrm{tr}(\mathbf{B}-\mathbf{I})\mathbf{I} + \mu(\mathbf{B}-\mathbf{I}) - \alpha(T-T_0)\mathbf{I}\right)\overline{\mathbf{F}}\tilde{\mathbf{F}}^{-1}\overset{\circ}{\mathbf{B}}\tilde{\mathbf{F}}^{-T}, \quad \text{(based on St. Venant – Kirchhoff)}$$

$$\mathbf{P} = \tilde{J}^{\circ} J^{-1}\left(2D_1 J(J-1)\mathbf{I} + 2C_1 J^{-\frac{2}{3}}(\mathrm{dev}\,\mathbf{B}) - \alpha(T-T_0)\mathbf{I}\right)\overline{\mathbf{F}}^{-T}, \quad \text{(based on Neo - Hooke)}$$

Finally, the stress for linear elasticity – or the smoothing equation – is given by

$$\boldsymbol{\sigma}^{\varphi} = \lambda^{\varphi}(\nabla\cdot\mathbf{r})\mathbf{I} + 2\mu^{\varphi}\,\mathrm{sym}\,\nabla\mathbf{r}.$$

These equations allow the usage of an adjusted initial shape, but we still have not made the switch from a forward to an inverse simulation. In fact, the equations that are required are identical in both cases. It is only the boundary equations that are adjusted (see Fig. 8).

For the forward simulation, Dirichlet boundary conditions are used on the boundary for the initial displacement field $\mathbf{r}$ to prescribe the boundary coordinates of the initial shape. To switch to the inverse formulation, these boundary conditions are removed and Dirichlet boundary conditions are added to the displacement field $\mathbf{u}$, where the boundary coordinates of the target molding shape are prescribed. In both cases, Neumann boundary conditions are used on the momentum equation to achieve an equilibrium state.

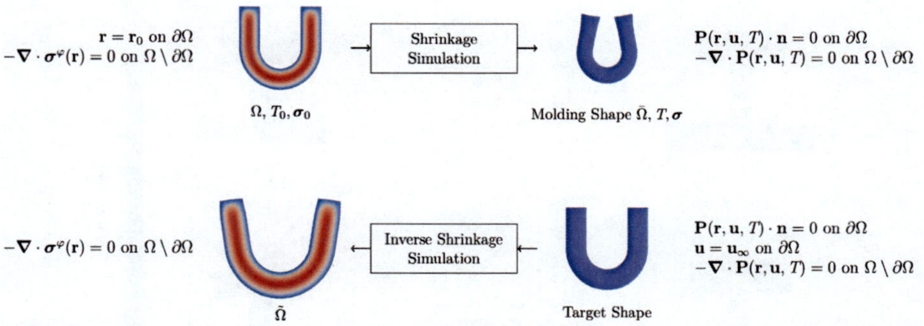

Fig. 8. The equations and boundary conditions used in the forward and inverse simulations are similar, with the exception of the Dirichlet boundary conditions for either the initial or equilibrium displacement field.

4 Numerical Example

In this section, we demonstrate the application of both forward and inverse formulations of the shrinkage and warpage simulation on a simple geometry. For this example, rather than solving the solidification simulation, we solve the simpler heat conduction equation

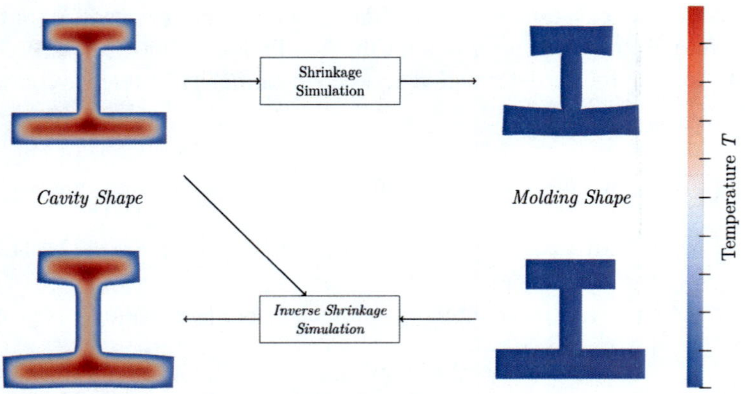

Fig. 9. Initial (left) and equilibrium (right) temperature fields in forward and inverse simulations.

on the geometry to obtain a suitable temperature field. This temperature field is shown in Fig. 9, along with the results of both the forward and inverse simulation runs. The target molding shape has been used as the initial guess for the cavity shape.

The stress fields corresponding to the different configurations and simulation types are shown in Fig. 10. In this example we have not imposed any stresses in the initial configuration. However, residual stresses do result in the molding due to the inhomogeneous initial temperature distribution. The inverse method does not prevent such stresses, since it has been designed with the single goal of achieving the target molding shape.

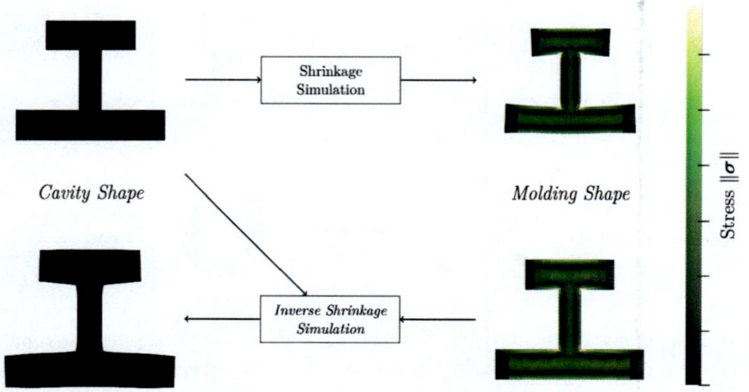

Fig. 10. Initial and equilibrium stress fields (spectral norm) in forward and inverse simulations.

5 Iteration Method

The inverse method described here relies on the approximation of the changes in the temperature and stress fields in a changed boundary. In order to find out how accurate these approximations are, the filling and solidification simulation has to be run a second time on the adjusted cavity shape. If this is followed with a second run of the inverse shrinkage and warpage simulation, the changes in the cavity shape can be assumed to be much smaller. This also means that the effect of the approximation is less significant. This motivates an iterative method, where the forward filling and solidification simulation is combined with the inverse shrinkage and warpage simulation in a loop (see Fig. 11).

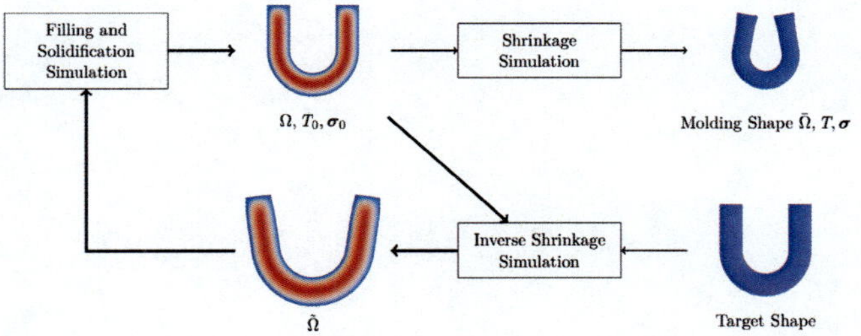

Fig. 11. The forward solidification simulation forms an iteration loop with the inverse shrinkage simulation.

6 Conclusion

In this paper, we have presented a method for the automatized design of an injection molding cavity shape. The purpose of this method was the elimination of geometric faults in the molding shape.

By introducing a few key simplifications, such as purely elastic behavior of the solidified material, and splitting the simulation into one transient and one stationary part, we could create a method with a relatively low computational cost, which requires only the inversion of the second, stationary, part of the simulation.

On a simple example geometry, we have demonstrated both forward and inverse variants of the simulation. While it is not possible to prevent the build-up of residual stresses with this method, we were able to show that the inverse method is capable of reducing geometric defects in the molding shape.

Funding. The presented investigations were carried out at RWTH Aachen University within the framework of the Collaborative Research Centre SFB1120–236616214 "Bauteilpräzision durch Beherrschung von Schmelze und Erstarrung in Produktionsprozessen" and funded by the Deutsche Forschungs-gemeinschaft e.V. (DFG, German Research Foundation). The sponsorship and support is gratefully acknowledged.

References

1. Gedde, U.: Polymer Physics. Springer, Heidelberg (1995)
2. Ward, I.M., Sweeney, J.: Mechanical properties of solid polymers. Wiley, Chichester (2012)
3. Michaeli, W.: Training in injection molding: a text- and workbook. Cincinnati (2001)
4. Zwicke, F., Elgeti, S.: Inverse design based on nonlinear thermoelastic material models applied to injection molding. Finite Elem. Anal. Des. **165**, 65–76 (2019)
5. Johnson, A., Tezduyar, T.: Mesh update strategies in parallel finite element computations of flow problems with moving boundaries and interfaces. Comput. Methods Appl. Mech. Eng. **119**(1), 73–94 (1994)

Cutting

Cutting Whistle – An Original Approach for Nozzle Design in Fiber Laser Cutting of Stainless Steel

M. de Oliveira Lopes[1(✉)], Dirk Petring[2], Dennis Arntz-Schröder[2], Frank Schneider[2], Stoyan Stoyanov[2], and Arnold Gillner[1,2]

[1] Chair for Laser Technology LLT, RWTH Aachen University, Steinbachstr. 15, 52074 Aachen, Germany
marcelo.lopes@llt.rwth-aachen.de
[2] Fraunhofer Institute for Laser Technology ILT, Steinbachstr. 15, 52074 Aachen, Germany

Abstract. Instabilities of the melt flow dynamics at the laser cutting front lead to quality losses, due to the formation of striations at the cut flank, especially when using cost-efficient 1 μm laser radiation. This paper summarizes fundamental studies of the subproject A8 of the SFB 1120 using the concepts of a whistle-like structure as basis for the nozzle design. This approach is an innovative strategy for quality enhancement of cut flanks. The nozzle was designed for the stimulation of resonance between the acoustic waves generated by the gas flow through the nozzle with characteristic melt wave frequencies at the cutting front. The cutting experiments reveal an improvement of cut flank quality, while reducing the assist gas consumption compared to a standard nozzle with diameter of 5 mm.

Keywords: Cutting · 1 micron laser · Acoustic sensing

1 Introduction

Laser cutting with 1 micron laser-systems has been for the past decade the largest market in the laser industry [1]. The advantages for using this type of configuration compared to the old-fashioned CO_2 technology in the cutting process are manifold. However, the cut flank quality is still not matched for sheet thicknesses greater than 3–4 mm. Hence, an efficient strategy for promoting the enhancement of cut flank quality for thick sheets of stainless steels using 1 micron laser-systems is yet to be demonstrated.

In laser fusion cutting, the laser beam and the assist gas are conventionally combined in a coaxial arrangement, where the gas forms a jet which has the main function of expelling the cut kerf material present as steam or melt. The dominant requirement of the gas flow is the coupling and transfer of its momentum to the liquid surface and the debris [2]. Thus, in addition to the properties of the laser beam, the efficiency of the gas flow also exhibits a significant influence on the resulting cut flank quality [3]. The evaluation of cut flank quality is normally done by quantifying its surface roughness with the profile roughness parameters. The surface roughness of a cut flank results from

U. Reisgen et al. (Eds.): EMPOrIA 2020, LNME, pp. 267–276, 2021.
https://doi.org/10.1007/978-3-030-70332-5_24

a striation pattern caused by instabilities of the melt flow dynamics at the laser cutting front.

Further understanding of the mechanisms of striation generation was reached by Arntz et al. [4]. Their investigations about the melt flow dynamics at the cutting front by means of in-situ high-speed video diagnostics have shown that a high melt wave frequency is beneficial for the formation of low surface roughness. In their most recent study [5], it was exposed that the melt waves sliding downwards at the cutting front apex reveal characteristic frequencies in the range of 27–33 kHz in the regions where the lowest roughness values occur on the cut flank. Fig. 1 illustrates the cutting depth-dependent analysis of the melt wave frequencies (f_{mw}) for different assist gas pressures (5 bar, 10 bar, 15 bar and 20 bar). As an explanation for this process response at the cutting front, the kerf was regarded for the very first time as a whistle-like structure. They suggest that these frequencies are produced by longitudinal waves which are excited by the interaction of the gas flow within the kerf. This innovative approach could potentially not only improve the ejection efficiency of the melt, but also reduce the roughness of the cut surfaces.

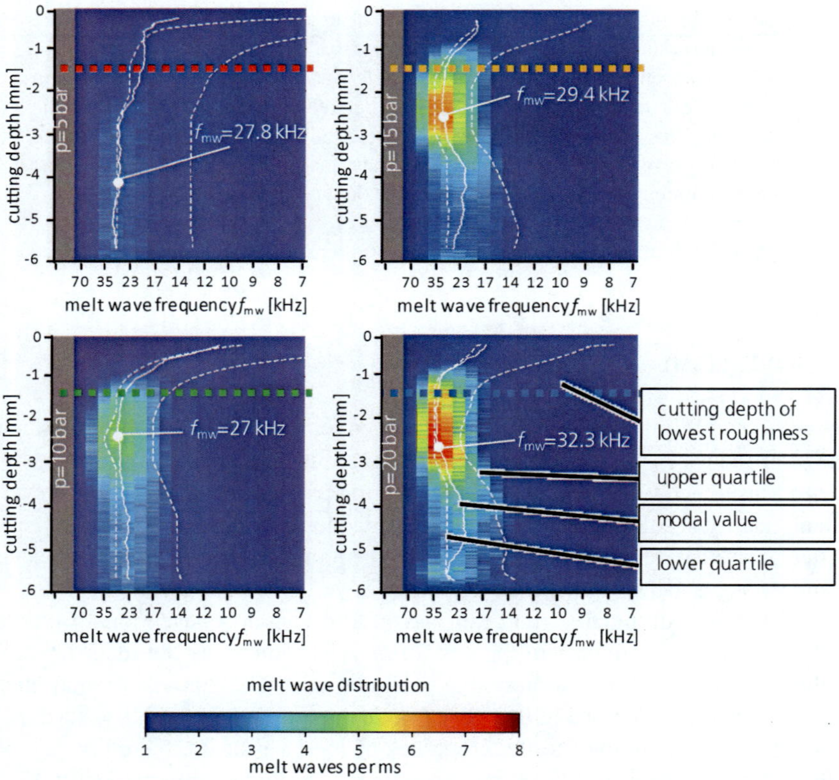

Fig. 1. Cutting depth-dependent analysis of the melt wave frequencies (f_{mw}) for different assist gas pressures. (Source: [5])

2 Fundamentals of Whistling Mechanisms

The knowledge of acoustic wave generation is important if one intends to detect them, and crucial if their fine-tuning is aimed. Whistle-like sounds can be produced by mechanical systems with a number of different geometries. Three prerequisites are described in [6]. There must be an obstruction in the path of a jet, the flow speed must be in an unstable range so that any perturbations in the smooth flow of the jet increase in size, and there must be a feedback path by which specific frequencies of such perturbations are emphasized. Chanaud [7] has classified the various different whistling mechanisms into three categories according to the feedback mechanism:

Class I – Acoustic waves are generated by the principle of vortex shedding behind a long cylinder.
Class II – Acoustic waves are generated by disturbances travelling downstream and propagate back upstream causing further disturbances.
Class III – A resonant structure controls the feedback loop.

Wilson et al. [8] carried out experiments on the fluid mechanics of whistling, where they suggest that the formation of a jet, the subsequent instability producing vortex rings and the interaction of these rings with a rigid boundary are the essential features of the whistling mechanism. They proposed a different classification than Chanaud. Rather than considering the feedback mechanism, they have classified it into two categories according to the nature of the instability mechanism. One class would be characterized by an asymmetric instability of the plane jet and the other, by an instability of a circular jet to the formation of vortex rings, or a plane jet to the formation of symmetric vortex streets. Typical systems of the latter are the hole tone and the Pfeifentöne.

The hole tone system consists of a circular jet impinging on a plate positioned slightly downstream of the jet exit, containing a hole that is both approximately the same size as and axially aligned with the hole from which the jet emanates (see Fig. 2). Important work on this system was published in the 1960s by Chanaud and Powell [9] and more recently in 2013 by Henrywood and Agarwal [10]. It is important to point out that it occurs only at conditions where the jet is unstable to applied disturbances. In this system, the jet is axially symmetric, it is sensitive to pressure disturbances, and the jet disturbances are symmetric. Furthermore, the vortex shedding and scattering of disturbances at the downstream orifice plate radiate acoustic waves (leading to the observed whistling tone), which drives the acoustic waves within the duct, completing the feedback loop.

The Pfeifentöne has been referred by Anderson [11] as acoustic waves produced when a gas moving inside a pipe presents a change in the velocity flow. *"Pfeifentöne to be initiated in, and by the gas discharging continuously from the pipe orifice"*. Moreover, he argued that the Pfeifentöne do not exactly match with the geometrical length of the pipe and a correction in length should be added to the open end of the pipe and even in a case of a cylindrical pipe with a clean-cut, the antinode does not coincide with the end of the pipe. His generalization of the effect of a pipe geometry on Pfeifenton wavelength for a pipe with a diameter equal to the exit orifice is presented in Eq. (1); where n is the number of the harmonic; c, the velocity of sound; l, the length of the pipe; R, the radius

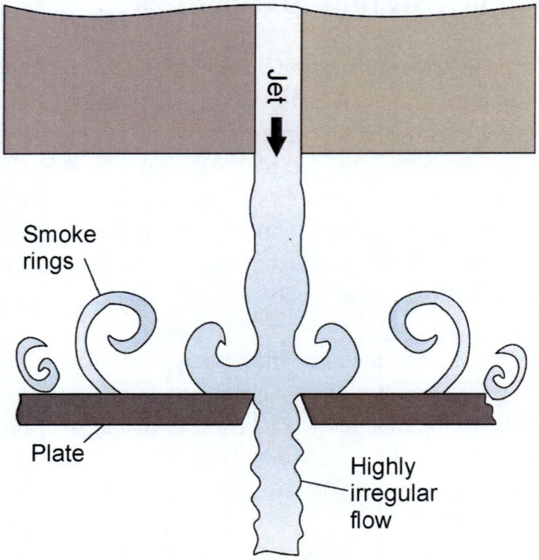

Fig. 2. Sketch of the cross-sectional view of the hole tone system showing the jet disturbances. (Figure adapted from Fig. 1. in Ref. [9])

of the pipe; and f, the frequency of the Pfeifenton.

$$f = \frac{n}{2} \frac{c}{(l + 0.6R)} \tag{1}$$

The correction in the geometrical length, i.e., the diameter component in Eq. 1, was also present in the work of Hutchings et al. [12]. In their development of a modulated ultrasonic whistle, an adjustable spectrum over a wide frequency range was shown to be generated. Their system was supplied by a jet with pressure of 5 bar through a nozzle, similar to the pressure range used in the laser cutting process. The depth of the cavity and also the gap between nozzle and cavity lips were adjustable. The diameter was fixed at a value of 0.5 mm.

Acoustic sensing of a laser cutting process has already been implemented by Sheng and Chryssolouris [13]. They have related the acoustic emission to the geometry of the kerf. Two applications relevant to laser cutting were introduced: regulation of beam break-through and regulation of surface roughness at the kerf walls. It was shown that the resonant frequency is inversely proportional to the kerf depth. However, the physical explanation for this effect was not given. Furthermore, it was discussed that due to the surface roughness on the kerf walls, the impedance of the gas flow is increased and consequentially, the magnitude and frequency of the resonant peak is reduced. De Keuster et al. [14] have also evaluated the applicability of acoustic sensing as an on-line monitoring system for cut quality measurement. They argued that the sound of the cutting process is dominated by the noise of the gas flow, which could make the interpretation of the generated spectrum time consuming.

The aim of the current work is to investigate by means of acoustic sensing whether the laser cutting process can actually be treated as a whistle-like system. Furthermore, it shall be shown that it is possible to adjust the resonant frequency emitted by the nozzle in free space by a suitable nozzle design. The concepts of a whistle-like structure were adopted as basis for a nozzle design which is tailored to improve the quality of the cut flank.

3 Experimental Setup

3.1 Acoustic Sensing Experiments

In order to investigate the acoustic signal emitted by the nozzles as the gas flow exits its outlet orifice and enters the free space, a XARION Eta250 Ultra optical microphone was used. The amplified signal was input into a signal analyzer, which transformed the acoustic emission into a frequency spectrum using a Fast Fourier Transform. A frequency range between 1 kHz and 50 kHz was sampled with a resolution of 95.43 Hz. The sampling time for a single sweep of the frequency range was 84 ms and each measurement lasted a total of 7 s.

For the acoustic sensing experiments, the nozzle was mounted to a Precitec head through an adapter, which provides a continuous edge-free shape from the cutting head to the nozzle, so that the disturbance of the gas flow is minimized. The assist gas pressure was measured by means of a regulating pressure valve and a manometer, which indicates the gas pressure immediately before entering the cutting head. The nitrogen assist gas pressure was varied in the range of 5 bar to 15 bar. The microphone was positioned next to the nozzle, at a distance of 30 mm, in a way that it would not obstruct the ultrasonic jet.

The sketch in Fig. 3(a, b) illustrates the two nozzles compared in the current work, namely a commercially available standard nozzle with an outlet orifice diameter of 5 mm and a newly developed nozzle with outlet orifice diameter of 3 mm. The sketch reveals only the geometrical features used in the calculations presented in this study. The tailored nozzle Ø3 was conceived as an innovative strategy for the quality improvement of cut flanks. This approach aims to stimulate resonance between the acoustic waves produced in the nozzle with characteristic melt wave frequencies in the kerf shown in Fig. 1 and the corresponding acoustic resonance in the kerf as postulated in [5]. To achieve this, the cylindrical component of the nozzle outlet section was considered as a whistle-structure which, as seen in the work of Anderson [11], may develop Pfeifentöne. The cylindrical part of the nozzle was adapted to promote 30 kHz according to Eq. (1), using the sound velocity of nitrogen at room temperature ($c = 333.6$ m/s) and the first harmonic ($n = 1$). In addition to the inner nozzle geometry, the area of the lower outer part of the nozzle was increased in a simplistic attempt to reinforce the resonant effect between the nozzle and the sheet surface.

3.2 Cutting Experiments

The cutting experiments were carried out using a 4 kW fiber laser (YLR-4000 SS Serie CW Ytterbium) with wavelength of 1070 nm at its maximum output power. A Precitec

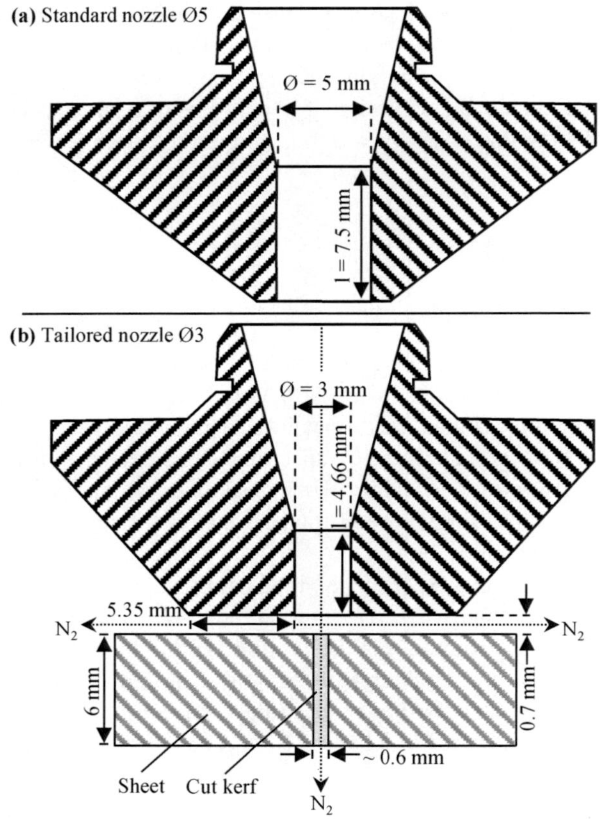

Fig. 3. (a) Sketch of the standard nozzle Ø5 showing the geometrical features used in the calculation of the resonant frequency. (b) Sketch of the tailored nozzle Ø3 in cutting position showing the geometrical features used for its design.

HPSSL cutting head was used and the fiber ($\varnothing = 100\ \mu m$) guided laser beam was collimated and focused with focal lengths of 100 mm and 250 mm, respectively. This configuration provides a focal diameter of 250 μm. The focus was fixed 5 mm below the sheet surface and a stand-off distance of 0.7 mm was used. The cutting speed was set at 2.2 m/min and three values of nitrogen assist gas pressure were applied to cut 6 mm thick stainless-steel samples (5 bar, 10 bar and 15 bar).

The reason for choosing these parameters is that it is previously known that for this optical configuration in combination with a standard Ø5 nozzle they are close to the optimum for achieving a smooth cut flank. The aim of this experiment is to compare this already known standard flank quality with the cut flanks produced with the tailored Ø3 nozzle.

4 Results and Discussion

This section presents the frequency spectra detected by the optical microphone at assist gas pressures of 5 bar to 15 bar (interval of 1 bar between measurements) using the standard nozzle Ø5 and the tailored nozzle Ø3 (Fig. 4(a)).

Although the spectra exhibit a somewhat similar shape regardless of the assist gas pressure, the peak frequency changes with pressure. Between 10 bar and 12 bar, which is a conventional parameter range used in laser fusion cutting of stainless steels (6 mm), the peak frequency only slightly changed in the case of the standard nozzle Ø5. Interestingly, according to the geometric characteristics of this nozzle (Fig. 3(a)), the fundamental frequency calculated from Eq. 1 is 18.5 kHz if the sound velocity of nitrogen 333.6 m/s (room temperature) is used in the calculation. This theoretical frequency corresponds precisely with the peak frequency measured at 10 bar. On the other hand, the tailored nozzle Ø3 has generated a peak rather at a higher frequency of 23.5 kHz at the same assist gas pressure.

An appealing range of assist gas pressure (5 bar – 7 bar) is marked in Fig. 4(a) as the frequency peaks are close to the characteristic melt flow frequencies (27 kHz–33 kHz, see Fig. 1). Figure 4(b) presents a finer interval between measurements, namely 0.4 bar. The measurements with the standard nozzle Ø5 have revealed peak frequencies in the range 25 kHz–28 kHz. In the case of the tailored nozzle Ø3, the peak frequency was in the range 26 kHz–36 kHz. That emphasizes that it is possible to modify the acoustic waves generated by the gas flow exiting the nozzle into free space by suitably designing the internal geometry of the nozzle. Nevertheless, it is obvious that there are other factors, e.g., gas pressure, which must be taken into account to produce a precise and strong spectral peak in the target frequency range.

The average surface roughness for different nozzles and assist gas pressures is plotted in Fig. 5. The measurements were performed at three different cutting depths in 1.5 mm intervals. For each assist gas pressure examined, the surface roughness presents a trend, while it increased from the upper sheet surface towards its lower part.

At 5 bar, the average surface roughness produced with the tailored nozzle at a cutting depth of 1.5 mm shows a reduction of 31% compared to the standard nozzle. However, the difference in the cut surface roughness produced by the different nozzles at the other two, lower cutting depth is less than 10%. At 10 bar, a significant improvement in cutting quality was achieved by using the tailored nozzle. The surface roughness has shown a reduction of 25.6%, 31.1% and 15%, respectively, with increasing cut depth. At 15 bar, the nozzles present a minor influence on the generated roughness of the cut flank. At every cutting depth, the difference was lower than 12%.

Intriguingly, among the three assist gas pressures used for cutting, the largest contrast of peak frequency generated from the distinct nozzle (Fig. 4(a)) is at 10 bar, where the tailored nozzle produces higher frequency than its counterpart. The frequency generated is nearer to the characteristic melt wave frequency, which may have had a positive effect on the smoothering of the cut flanks.

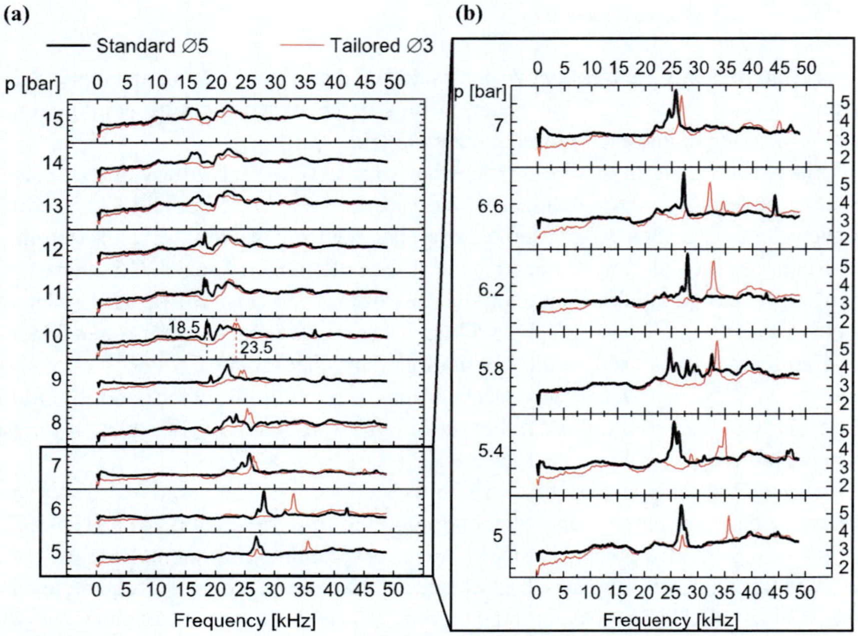

Fig. 4. (a) Frequency spectra for assist gas pressures from 5 bar to 15 bar with interval of 1 bar between the measurements. The y-axis of each layer is adjusted for better visualization of the frequency peaks. (b) Frequency spectra for assist gas pressures from 5 bar to 7 bar with interval of 0.4 bar between the measurements. The y-axes are identically normalized and presented in arbitrary units.

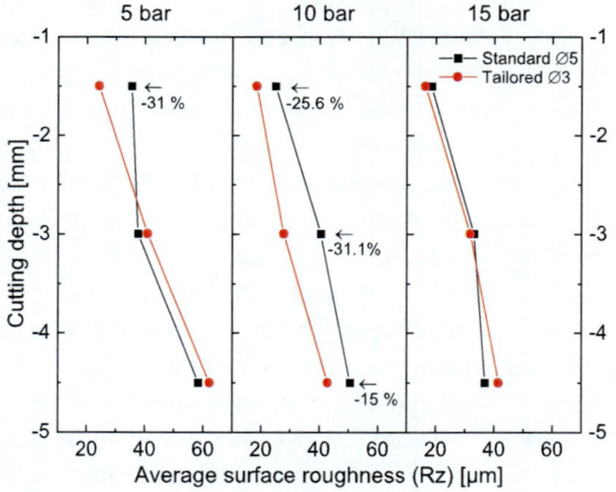

Fig. 5. Average surface roughness measured at three cutting depth in 1.5 mm intervals for different nozzles and assist gas pressures.

5 Conclusion and Further Outlook

The concepts of a whistle-like structure are adopted as fundamental foundation for a new nozzle design. This approach aims the stimulation of resonance between the acoustic waves coming out of the nozzle with characteristic melt wave frequencies at the cutting front and acoustic resonances in the kerf. The current work allows by means of acoustic sensing the following statements:

- It is possible to shift the peak frequency emitted through the nozzle in free space by modifying the nozzle geometry. However, a systematic investigation of determining factors for precisely tuning the target peak frequency other than the simple geometrical features of the nozzle outlet orifice, e.g., gas pressure, shall be carried out.
- At 10 bar, the tailored nozzle Ø3 has generated a peak frequency roughly 20% higher than the standard nozzle and has also shown a significant improvement in cutting quality, where the maximum value of average surface roughness Rz has been reduced by 15% compared to a standard nozzle using otherwise the same process parameters. Furthermore, it presents a reduction in the gas outlet area approximately by a factor of 3, which is an essential aspect for decreasing the gas consumption.

The results presented here provide the basis for further investigations which will lead to a better understanding of the overall resonance system nozzle-kerf and ultimately to the optimization of cut flank quality.

These include:

- On-line acoustic sensing measurements.
- High-speed video analysis of the melt dynamics stimulated by the tailored nozzle.
- A systematic analysis of the influence of other process parameters on the resonance within the kerf (e.g., stand-off distance and area of the lower outer part of the nozzle).

Acknowledgements. The presented investigations were carried out at the Chair of Laser Technology LLT of RWTH Aachen University and Fraunhofer Institute for Laser Technology ILT within the framework of the collaborative Research Centre SFB1120–236616214 "Bauteilpräzision durch Beherrschung von Schmelze und Erstarrung in Produktionsprozessen" and funded by the Deutsche Forschungsgemeinschaft e.V. (DFG, German Research Foundation). The sponsorship and support is gratefully acknowledged.

References

1. Belforte, D.: Annual market review 2020. Industrial Laser Solutions for Manufacturing (2020)
2. Fieret, J., Terry, M. J., Ward, B. A.: Aerodynamic interactions during laser cutting. In: Proceedings of SPIE 0668, Laser Processing: Fundamentals, Applications, and Systems Engineering (1986). https://doi.org/10.1117/12.938884
3. Petring, D., Abels, P., Beyer, E., Herziger, G.: Werkstoffbearbeitung mit Laserstrahlung. Feinmech. Messtech. **96**, 364–372 (1988)

4. Arntz, D., Petring, D., Schneider, F., Poprawe, R.: In situ high speed diagnosis - a quantitative analysis of melt flow dynamics inside cutting kerfs during laser fusion cutting with 1μm wavelength. J. Laser Appl. **31**, 022206 (2019). https://doi.org/10.2351/1.5096091
5. Arntz-Schroeder, D., Petring, D.: Analyzing the dynamics of the laser beam cutting process. PhotonicsViews 43–47 (2020). https://doi.org/10.1002/phvs.202000015
6. Shadle, C.H.: Experiments on the acoustics of whistling. Phys. Teach. **21**, 148 (1983). https://doi.org/10.1119/1.2341241
7. Chanaud, R.C.: Aerodynamic whistles. Sci. Am. **222**(1), 4046 (1970)
8. Wilson, T.A., Beavers, G.S., DeCoster, M.A., Holger, D.K., Regenfuss, M.D.: Experiments on the fluid mechanics of whistling. J. Acoust. Soc. Am. **50**, 366 (1971). https://doi.org/10.1121/1.1912641
9. Chanaud, R.C., Powell, A.: Some experiments concerning the hole and ring tone. J. Acoust. Soc. Am. **37**, 902–911 (1965)
10. Henrywood, R.H., Agarwal, A.: The aeroacoustics of a steam kettle. Phys. Fluids **25**, 107101 (2013). https://doi.org/10.1063/1.4821782
11. Anderson, A.B.C.: Dependence of the primary pfeifenton (pipe tone) frequency on pipe-orifice geometry. J. Acoust. Soc. Am. **25**, 541 (1953). https://doi.org/10.1121/1.1907086
12. Hutchins, D.A., Jones, H.W., Vermeulen, P.J.: The modulated ultrasonic whistle as an acoustic source for modelling. J. Acoust. Soc. Am. **73**, 110 (1983). https://doi.org/10.1121/1.388843
13. Sheng, P., Chryssolouris, G.: Investigation of acoustic sensing for laser machining processes. Part 2: Laser grooving and cutting. J. Mater. Process. Technol. **43**, 45–163 (1994)
14. De Keuster, J., Duflou, J.R., Kruth, J.-P.: Monitoring of high-power CO_2 laser cutting by means of an acoustic microphone and photodiodes. Int. J. Adv. Manuf. Technol. **35**, 115–126 (2007). https://doi.org/10.1007/s00170-006-0695-z

Optimization of Beam Shapes for Laser Fusion Cutting by 3D Simulation of Melt Flow

Ulrich Halm[1(✉)] and Wolfgang Schulz[1,2]

[1] Nonlinear Dynamics of Laser Processing, RWTH Aachen University, Steinbachstr. 15, 52074 Aachen, Germany
ulrich.halm@nld.rwth-aachen.de

[2] Fraunhofer Institute for Laser Technology, Steinbachstr. 15, 52074 Aachen, Germany

Abstract. The mean profile height and the angular tolerance of the cut surface are two important quality criteria in laser fusion cutting of sheet metals. Especially for laser sources with 1 μm wave length, the concurrent optimization for both quality features cannot be performed by plain Gaussian or tophat beam profiles. A dynamical 3D simulation reflects this behavior and enables testing of different beam configurations. It is shown that pure Gaussian beam profiles create the lowest mean profile height, while tophat shapes create the lowest angular tolerance. Comparing 10 μm laser sources it is shown that Gaussian beam profiles can reduce angular tolerance and mean profile height at the same time due to differing absorption characteristics. To optimize the cutting result for 1 μm laser sources, the effect of mixed beam profiles should be analyzed in prospective research.

Keywords: Laser fusion cutting · Boundary layer approximation · Beam shaping · Melt flow dynamics

1 Introduction

Laser fusion cutting is one of the most important manufacturing applications for industrial lasers today [1] Quality reducing features in laser cutting are, inter alia, inclined cut surfaces, pronounced striation patterns, like seen in Fig. 1, and adherent dross. The first two features are subjected in this paper. EN ISO9013 [2] defines the angular tolerance u and the mean profile height R_{z5} as their measures. The optimization of process parameters for technical applications focusses on the reduction of these two values. Besides adjustable process parameters like feed velocity, laser power or focal position, the wavelength λ is defined by the selected laser system. While CO_2 lasers offer excellent beam quality at high powers [3], they consume roughly three times more power than fiber lasers [4]. High-power multi-mode fiber lasers on the other hand provide reduced beam quality [5], while power and cost efficiency is significantly better. Besides different beam profiles, the wavelength of the emitted laser light varies by one order of magnitude. This leads to drastically changed absorption characteristics depending on the cosine μ of the angle between the surface normal $\vec{n}$ and the incidence direction of the laser light $\vec{S}$. As shown in Fig. 2, the CO_2 laser light shows a pronounced absorption maximum for

U. Reisgen et al. (Eds.): EMPOrIA 2020, LNME, pp. 277–285, 2021.
https://doi.org/10.1007/978-3-030-70332-5_25

grazing incidence, while the absorption for fiber laser radiation starts at angles beyond the Brewster's maximum of the CO_2 laser and remains nearly constant for the rest of the angle-spectrum. This leads to a de-stabilization of the melt film by the excitement of melt waves [6] The fast decay of the absorption beyond the Brewster's angle for CO_2 lasers reduces this de-stabilizing effect drastically in contrast to the fiber laser. The differing positions of the Brewster's maximum also affect the optimal beam shapes for cuts with low angular tolerance. While CO_2 lasers allow the use of high-quality nearly Gaussian beam shapes [3] high power fiber lasers operate with a multi-mode tophat-like beam shape [5].

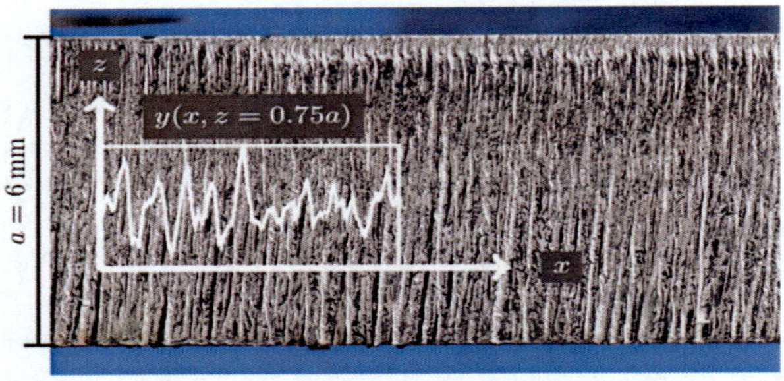

Fig. 1. Cut surface in laser cutting of stainless steel with thickness $a = 6\,\text{mm}$. The surface profile $w(x, z)$ shows a certain roughness due to instabilities in the melt.

The purpose of this paper is to show that the optimization of the beam shape for angular tolerance depending on the wavelength leads to a contrary effect for the mean profile height. The larger slope of the tophat beam profile near the cut surface is identified as one root cause for the increased mean profile height.

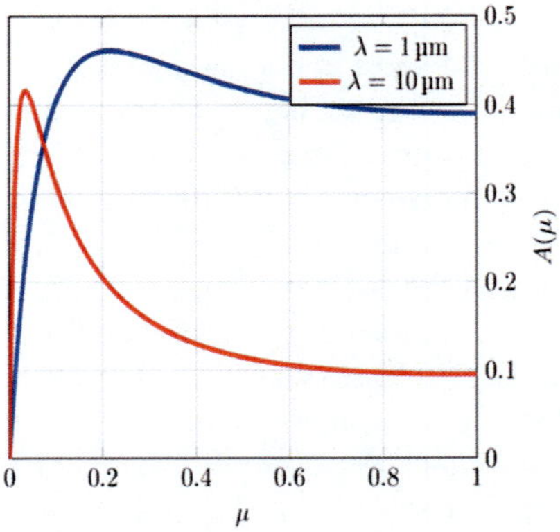

Fig. 2. Absorption characteristics $A(\mu)$ for fiber ($\lambda = 1\,\mu\text{m}$) and CO_2 laser ($\lambda = 10\mu\text{m}$) radiation for mixed polarization $s = p = 0.5$. The cosine of the angle of incidence is $\mu = -\vec{n} \cdot \vec{S}/|\vec{S}|$, with surface normal $\vec{n}$ and Poynting vector $\vec{S}$.

2 Related Work

Stelzer et al. [7] compared mean profile heights for CO_2 and fiber lasers using comparable beam settings. For stainless steel with a sheet thickness $a = 10\,\text{mm}$ the obtained mean profile heights were $R_{z5} = 30\,\mu\text{m}$ for the CO_2 laser with a beam quality of $M^2 = 1.6$ and $R_{z5} = 80\,\mu\text{m}$ for the fiber laser with a beam quality of $M^2 = 6.2$.

Hirano and Fabbro [8] observed a stabilization of the melt film by increasing the cutting speed and thus flattening the melt front by using disc lasers that show equal optical properties to fiber lasers. The parameters for this study are given in Table 1 and used for the findings in this paper. Amara et al. [9] simulated these parameters using the *FLUENT* solver and were able to reproduce the findings of Hirano and Fabbro. By changing the optical properties to those of CO_2 lasers, a decreased inclination of the cut surface could be observed.

3D modelling of melt film dynamics is a demanding multi-scale task that couples two freely moving interfaces: solid-liquid and liquid-gaseous. Typical velocities in the melt film exceed $v_0 \geq 10\,\text{ms}^{-1}$ while the melt film thickness can drop below $t_m \leq 10\,\mu\text{m}$. Using proper mathematical, physical and phenomenological reduction techniques, we developed a reduced 3D simulation of the melt film that could be solved with a reasonable computational effort [10].

Table 1. Seed point parameters used for simulations.

Parameter	Symbol/Unit	Seed-point value
Material		1.4301
Sheet thickness	a/mm	3
Laser power	P_L/kW	8
Cutting speed	v_0/m min^{-1}	1
Focal point	z_0/mm	0
Rayleigh length	z_R/mm	30
Beam radius	w_0/μm	850
Wave length	λ/μm	1

3 Methodology

The mathematical task for laser fusion cutting given by Schulz et al. [1] is subjected to proper reduction techniques to develop a numerical simulation that can reflect the melt film dynamics for the parameters given in Table 1 and a cut length of $L = 7.5\,\text{mm}$. The applied reduction techniques are given by Jansen et al. [10]. The basic assumptions for the model include:

- Constant parametrized assist gas driving forces

- Beam propagation via ray tracing
- Surface tension neglected
- Evaporation neglected
- Constant material parameters
- Quadratic velocity profile for melt flow in radial direction
- Essential characteristics of melt flow given by leading order of perturbation series

The assumption of a velocity profile enables the integration of the Navier-Stokes equations in the melt film along the radial direction. This so-called boundary layer approximation reduces the dimensionality of the mathematical task from 3D to 2D and drastically speeds up the computation.

3.1 Simulation of Laser Fusion Cutting

The parameters given in Table 1 are used as seed-point for the computational analysis. The resolution near the melt front is $\Delta\alpha r \approx 25\,\mu\text{m}$ in azimuthal direction, $\Delta z = 50\,\mu\text{m}$ in axial direction and $\Delta r = 20\,\mu\text{m}$ in radial direction of the solid phase. Inside the solid phase the resolution is exponentially coarsened by a factor of 1.1. The 2D region of the integrated melt film equation is approximated by 3660 3D quad elements, the solid phase by 346860 hexahedron elements to simulate the heat conduction. The beam propagation inside the cutting kerf is modelled by 8×8 incident rays for each melt surface element. The simulations are run on Dual Xeon-E5649 systems and the computation duration is around 60 h for the given parameters and a cut length of $L = 7.5\,\text{mm}$. The motion of the melt film and the melt front is recorded at a frame rate of $f = 25\,\text{kHz}$, so that the structure of the cut surface can be resolved with a resolution below $\Delta x \leq 1\,\mu\text{m}$ in cutting direction.

3.2 Extraction of Quality Features

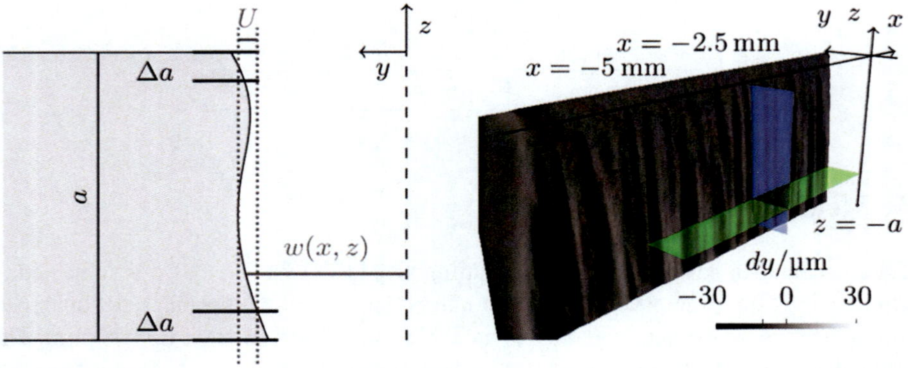

Fig. 3. Quality features in laser fusion cutting. **Left**: Angular tolerance by vertical cut. **Right:** cut surface with vertical cut (blue) for angular tolerance and horizontal cut (green) for mean profile height.

The developed laser cutting simulation allows the extraction of two basic quality features as defined in EN ISO9013 [2], the mean profile height R_{z5} and angular tolerance u. The mean profile height is calculated by a surface profile at a certain height of the cut $z = -0.75a$. The green plane in FIg. 3 indicates this cut, used for calculation of R_{z5}. It is the mean of the difference between five bordering minimum w_n^- and maximum values w_n^+ of the profile $w(x, z = \text{const})$

$$R_{z5} = \frac{1}{5}\sum_{n=1}^{5} w_n^+ - w_n^-.$$

The angular tolerance is the difference between minimum and maximum values of the surface profile at a constant $x-$ position. To prevent errors due to rounding at top or dross formation at bottom, the lower- and topmost 10% of the cut surface are neglected.

$$u = \max(w(x, z)) - \min(w(x, z)); x = \text{const}, z \in [-(a - \Delta a), -\Delta a]$$

A mean angular tolerance with standard error can be determined by evaluating u at several $x-$ positions.

3.3 Beam Shape

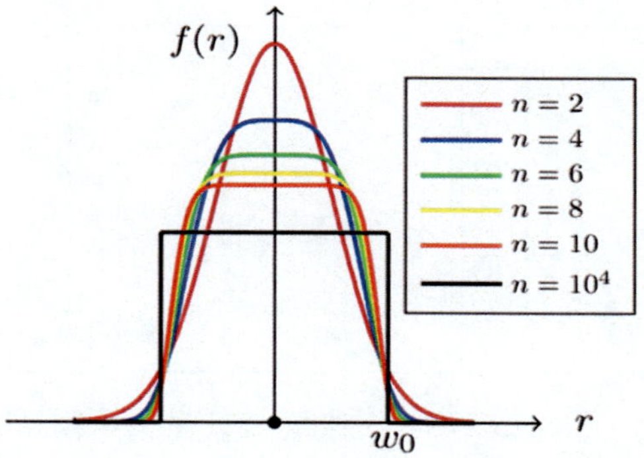

Fig. 4. Beam shapes used in simulation. The parameter n is the exponent of the super-Gaussian beam distribution.

A radially symmetrical super-Gaussian intensity distribution function at an axial plane $\mathrm{I(r, z = const)}$ is given by

$$\mathrm{I(r, z)} = \frac{\mathrm{P_L}\sqrt[n]{4}}{\pi \mathrm{w^2(z)}\Gamma\left(\frac{2+n}{n}\right)}\exp\left\{-2\left(\frac{\mathrm{r}}{\mathrm{w(z)}}\right)^{\mathrm{n}}\right\}$$

with the beam waist function

$$\mathrm{w(z)} = \mathrm{w_0}\sqrt{1 + \left(\frac{\mathrm{z - z_0}}{\mathrm{z_R}}\right)^2}$$

for beam radius at focal point w_0, focal position along beam axis z_0, Rayleigh length $z_R = \pi w_0^2/(\lambda M^2)$, laser power P_L and gamma function Γ. The intensity distributions at focal point for different super-Gaussian exponents n are shown in Fig. 4.

4 Findings

The slope of the beam distribution is varied by the exponent n of the super-Gaussian beam distribution between $n = 2$ and $n = 10$. Additionally a tophat distribution is modelled by $n = 10^4$. The process parameters given in Table 1 are used and the simulation is run and evaluated for angular tolerance and mean profile height as described in the previous section. As fiber and disc lasers tend to produce a greater mean profile height, a wavelength of $\lambda = 1\,\mu$m is chosen. To analyze the difference between CO_2 ($\lambda = 10\,\mu$m) and fiber lasers ($\lambda = 1\,\mu$m), both wavelengths are analyzed for $n = 2$ (Gaussian profile) and $n = 10$ (nearly tophat).

4.1 Variation of Beam Shape for Fiber Lasers

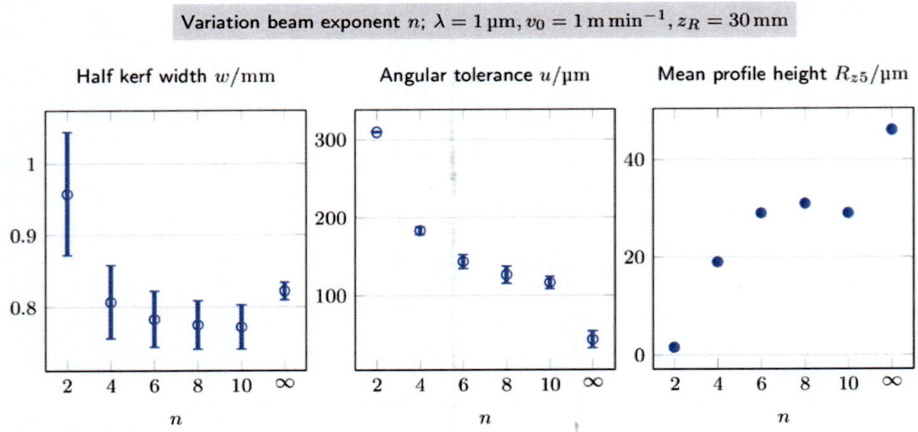

Fig. 5. Change of quality features with change of beam exponent for fiber lasers ($\lambda = 1\,\mu$m).

Figure 5 shows the dependency of the angular tolerance u and the mean profile height R_{z5} against the exponent of the super-Gaussian beam distribution n. Additionally the half kerf width $\bar{w}$ as mean of the profile $w(x, z)$ and its standard deviation is drawn.

The mean profile height R_{z5} shows a clear tendency to increase with the order of the exponent of the super-Gaussian beam distribution. This means, the greater the slope of the beam distribution, the greater the mean profile height. To reduce the mean profile height, the slope of the beam distribution should be reduced. A contrary effect is seen for the angular tolerance. The angular tolerance drops with increasing exponent n and a tophat distribution leads to smallest angular tolerance. For both criteria u and R_{z5}, usually smaller means better and parameter optimization focusses on smallest possible values for both quantities. Figure 1 clearly indicates that a simultaneous optimization of u and R_{z5} by changing the beam shape from Gaussian to tophat is not possible for fiber and disc lasers. New approaches for these laser systems aim on special transport fibers, that create mixed beam distributions that combine Gaussian and tophat beam profiles.

4.2 Comparison of CO_2 and Fiber Lasers for Gaussian and Top-Hat Beams

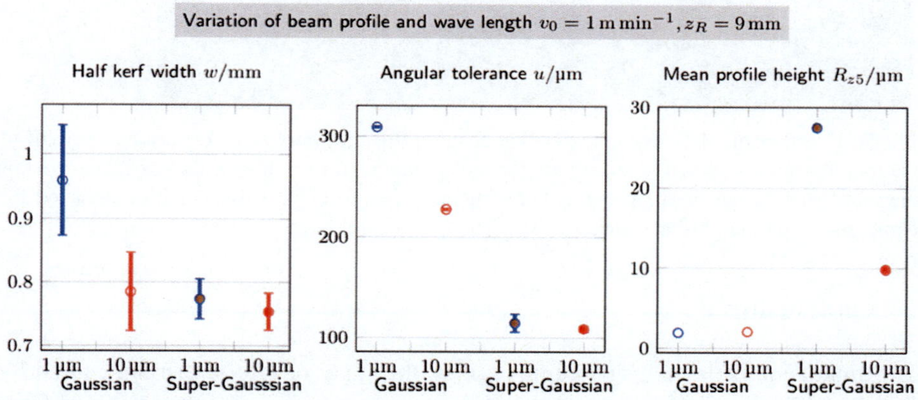

Fig. 6. Comparison of angular tolerance and mean profile height for Gaussian and tophat profiles for CO_2 ($\lambda = 10\,\mu$m) and fiber lasers ($\lambda = 1\,\mu$m).

To show the effect of the different absorption characteristics as shown in Fig. 2, the parameter for $A(\mu)$ is changed in the simulation for Gaussian ($n = 2$) and a tophat ($n = 10$) beam distribution. The tendency for angular tolerance (greater n is better) and mean profile height (smaller n is better) is reproduced for both laser systems. The absolute values for angular tolerance and mean profile height are smaller for the CO_2 laser, as expected from literature (See Fig. 6).

To depict the qualitative difference between both laser systems, simulation snapshots of the melt front and the malt film thickness are show in Fig. 7. For the Gaussian beam profiles ($n = 2$) on the left-hand side, pronounced melt waves begin to form up in the lower part of the melt front. The inclination of the cut-surface and the overall kerf width is significantly greater for the fiber laser ($\lambda = 1$ µm). The inclination for the CO_2 laser is comparable to tophat beam profiles on the right-hand side.

The tophat beam profiles cause pronounced melt waves on the whole melt front. Meanwhile the quality of the waves seems to be different. While CO_2 laser form hoof-shaped melt waves that are connected over the whole melting front, the fiber laser produces a more step-like expulsion of the melt. This significant difference in the shaping of the melt waves is caused by the different absorption characteristics as it is the only parameter changed in the simulation and might be one root-cause for irregular melt expulsion in cutting with fiber lasers that cause a larger mean surface roughness compared to CO_2 lasers. Measures to reduce the mean profile height aim on a regularization of the melt expulsion.

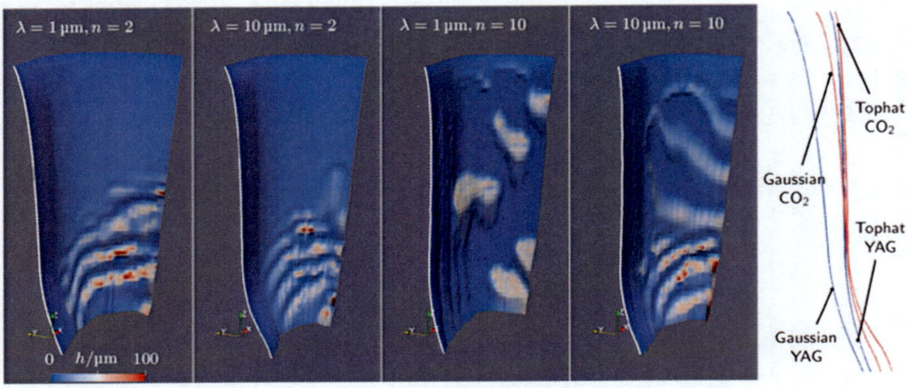

Fig. 7. Comparison of Gaussian and tophat beam profiles for fiber and CO_2 lasers. One half of the melting front is shown together with the melt film thickness in false-colors. The rear part of the melt front (white line) represents the vertical shape of the cut surface and is extracted for comparison on the right-hand side.

5 Conclusion

The simulation of laser fusion cutting allows the study of effects that are caused by the changing of individual parameters. In this work, we analyzed the transition from Gaussian beam shapes (exponent $n = 2$) to tophat profiles ($n = 10$). Additionally the effect of differing absorption characteristics for CO_2 and fiber lasers can be investigated separately without changing any other parameter. The main findings of this paper match the current state of the literature and show up an optimization conflict for fiber lasers. Better angular tolerance is reached by tophat beam profiles, while better mean profile height is achieved with smooth beam profiles that show a smaller slope. Another finding of the presented study is that the melt expulsion for fiber lasers and tophat beam profiles exhibits a significant qualitative difference compared to CO_2 lasers and to Gaussian beam profiles. The melt waves are separated and more pronounced in this case. This finding seems to drive an irregular melt expulsion and causes a larger mean profile height.

5.1 Limitations

The parameter set given in Table 1 that is analyzed by Hirano and Fabbro [8] is optimized for studying purposes and has no real technical relevance. However, this parameter set is well suited for simulation as the scales of these experiments are easier to manage for a numerical simulation. The comparable large beam diameter and low assist gas flow support the numerical stability of the simulation. For technically more relevant parameter sets, the simulation has to be enhanced in terms of stability and effects like surface tension and evaporation. The quality of the main findings does not suffer from these limitations.

Acknowledgments. All presented investigations are conducted in the context of the Collaborative Research Centre SFB1120 "Precision Melt Engineering" at RWTH Aachen University and funded by the German Research Foundation (DFG). For the sponsorship and support we wish to express our sincere gratitude.

Conflict of Interests Statement. The authors declare that there is no conflict of interest.

References

1. Belforte, D.: 2017 was a great year -- for industrial lasers. Industrial Laser Solutions 33 (2018)
2. International Organization for Standardization: Thermal cutting -- Classification of thermal cuts -- Geometrical product specification and quality tolerances. Geneva (2017)
3. Green, L.I., Herrit, G.L., Reedy, H.E.: Novel method for imaging high-power CO 2 laser beams. In: Kudryashov, A.V., Paxton, A.H. (eds.) Laser Resonators and Beam Control VI, p. 97. SPIE (2003)
4. Kellens, K., Rodrigues, G.C., Dewulf, W., Duflou, J.R.: Energy and resource efficiency of laser cutting processes. Phys. Proc. **56**, 854–864 (2014)
5. Cho, W.-I., Na, S.-J., Thomy, C., Vollertsen, F.: Numerical simulation of molten pool dynamics in high power disk laser welding. J. Mater. Process. Technol. **212**, 262–275 (2012)
6. Vossen, G., Schüttler, J.: Mathematical modelling and stability analysis for laser cutting. Math. Comput. Modell. Dyn. Syst. **18**, 439–463 (2012)
7. Stelzer, S., Mahrle, A., Wetzig, A., Beyer, E.: Experimental investigations on fusion cutting stainless steel with fiber and CO2 laser beams. Phys. Proc. **41**, 399–404 (2013)
8. Hirano, K., Fabbro, R.: Experimental investigation of hydrodynamics of melt layer during laser cutting of steel. J. Phys. D: Appl. Phys. **44**, 105502 (2011)
9. Amara, E.H., Kheloufi, K., Tamsaout, T., Fabbro, R., Hirano, K.: Numerical investigations on high-power laser cutting of metals. Appl. Phys. A **119**, 1245–1260 (2015)
10. Jansen, U., Niessen, M., Hermanns, T., Arntz, D., Poprawe, R., Schulz, W.: Boundary layer approximation for melt film dynamics in laser fusion cutting. In: Syme, G., Hatton MacDonald, D., Fulton, B., Piantadosi, J. (eds.) MODSIM2017, 22nd International Congress on Modelling and Simulation. Modelling and Simulation Society of Australia and New Zealand (MSSANZ), Inc. (2019)
11. Schulz, W., Nießen, M., Eppelt, U., Kowalick, K.: Simulation of laser cutting. In: Hull, R., Osgood, R.M., Parisi, J., Warlimont, H., Dowden, J. (eds.) The Theory of Laser Materials Processing, vol. 119, pp. 21–69. Springer, Dordrecht (2009)

Author Index

U. Reisgen et al. (Eds.): EMPOrIA 2020, LNME, pp. 287–288, 2021.
https://doi.org/10.1007/978-3-030-70332-5